Women in Wildlife Science

Women

in

Wildlife Science

Building Equity, Diversity, and Inclusion

EDITED BY Carol L. Chambers
and Kerry L. Nicholson

JOHNS HOPKINS UNIVERSITY PRESS
Baltimore

Johns Hopkins University Press
2715 North Charles Street
Baltimore, Maryland 21218
www.press.jhu.edu

Library of Congress Cataloging-in-Publication Data

Names: Chambers, Carol L., editor. | Nicholson, Kerry L., 1978– editor.
Title: Women in wildlife science : building diversity, equity, and
    inclusion / edited by Carol L. Chambers and Kerry L. Nicholson.
Description: Baltimore : Johns Hopkins University Press, 2022. |
    Includes bibliographical references and index.
Identifiers: LCCN 2021062993 | ISBN 9781421445021 (hardcover) |
    ISBN 9781421445038 (ebook)
Subjects: LCSH: Natural resources—Management. | Women in natural
    resources management. | Women and the environment. | Wildlife
    conservation.
Classification: LCC HC85 .W65 2022 | DDC 333.7082—dc23/
    eng/20220714
LC record available at https://lccn.loc.gov/2021062993

A catalog record for this book is available from the British Library.

*Special discounts are available for bulk purchases of this book. For more
information, please contact Special Sales at specialsales@jh.edu.*

CONTENTS

ACKNOWLEDGMENTS

Many of us worked together to write this book. We represent people from different cultures and with different perspectives and experiences. We learned from each other as we wrote these chapters and have many to thank for supporting us through the process.

We have deep gratitude for the women who trusted us with their stories and shared them with honesty and candor. We appreciate the intimate sharing of parenting adventures from the wildlife moms. We thank all of those in the LGBTQ+ family who are out and proud, getting along, or simply managing to survive as they continue on their journey. We thank all those authors from marginalized groups, including Black, Indigenous, and women of color in wildlife, without whom producing this book would have been impossible. We thank those allies who stepped up when they saw a disparity and those future allies who are willing to take the extra moment to consider others. We acknowledge the Fellows of the National Conservation Leadership Institute's Cohort 13 for their passion and dedication keeping the fire lit. We thank the participants who took the time to compile their university demographics and histories for the survey. Those mentioned here are just some of the people who inspired the authors who contributed to this book. While some of our stories may be uncomfortable, they need to be heard.

We are indebted to the spouses, partners, families, friends, personal mentors, and leaders who have supported our unique journeys. We thank our colleagues and peers for their insightful comments and discussions that sharpened our thinking and presentation of information. Specifically, this book is successful because of the magnificent and most fabulous Mary DeJong, Northern Arizona University librarian for the College of the Environment, Forestry, and Natural Sciences. We appreciate the incredible efforts from Jill Baron, Nina Burkardt, Geneva Chong, Galen Hench, Anne Kretschmann, Diana Lafferty, Tom O'Connell, Allan O'Connell, Robin White, and Todd Wojtowicz. Authors Jennifer Malpass, Katherine O'Donnell, and Stephanie Romañach not only wrote but also provided support for other chapters. Brenda McComb generously provided her knowledge and support, both writing and reviewing.

The authors of Chapters 5 and 6 provided acknowledgments that speak to and for many of us: The authors of Chapter 5 want to thank all of the women who have experienced challenges in their wildlife careers firsthand and helped create the navigational map for those who followed—and continue to follow. We acknowledge all Indigenous peoples, who were the first stewards of the land and have been removed or displaced—on all continents—but continue to maintain relationships with the lands upon which our institutions, agencies, and homes now exist. The authors of Chapter 6 thank the creator for all our relations; Mother Earth, who nourishes us, and her soil; the plants, who teach us; and the animals, who protect us. They acknowledge all their ancestors for thriving, surviving, and passing on knowledge vital to the stewardship of our homelands. They pray for future generations that we are able to provide a healthy planet for them.

The Alaska Department of Fish and Game provided funding and support to Kerry Nicholson for her work writing and editing. We would like to acknowledge Paul Krausman and The Wildlife Society, who helped connect our voices with our publisher, Johns Hopkins University Press (JHUP). We thank our JHUP team for all they contributed to this project. Tiffany Gasbarrini, senior acquisitions editor extraordinaire, supported us from start to finish, which was a huge job considering this was our first book (no, really, you have no idea if you haven't done this before). Thank you, Tiffany; you helped make this fun. We appreciate the help of Ezra Rodriguez, assistant editor, who, in record time, wrangled chapters, figures, tables, and forms to generate a book. We thank Juliana McCarthy, managing editor, and Carrie Watterson, copy editor, who with grace and a lot of patience smoothed this book into its present form. Bea Jackson skillfully incorporated pictures of our authors to create the cover.

Any use of trade, firm, or product names is for descriptive purposes only and does not imply endorsement by the US government. The findings and conclusions in this book are those of the author(s) and do not necessarily represent the views of the US Fish and Wildlife Service.

CONTRIBUTORS

Gordon R. Batcheller is a Certified Wildlife Biologist, TWS Fellow, past president of The Wildlife Society, and retired chief wildlife biologist for New York State.

Jamila Blake is a TWS Associate Wildlife Biologist and the professional development manager at The Wildlife Society.

Travis L. Booms is a research wildlife biologist at the Alaska Department of Fish and Game.

Rena Borkhataria, PhD, is the director of the Doris Duke Conservation Scholars Program Collaborative and courtesy faculty in the Department of Wildlife Ecology and Conservation at the University of Florida.

Jenny ("Boomer") Bryant, Koyukon Athabascan, is a Louden Tribal citizen and US Fish and Wildlife Service biologist for Kukukuk National Wildlife Refuge.

Joanne Bryant Ditsuh Taii, Gwich'in Athabascan, is a Native Village of Venetie Tribal Government member and US Fish and Wildlife Service tribal communication and outreach specialist.

Lubia N. Cajas-Cano is a PhD in environmental science and affiliated faculty at the University of San Carlos in Guatemala. ORCID ID: 0000-0001-6395-3990.

Dale Caveny is the retired chief of law enforcement with the North Carolina Wildlife Resources Commission.

Modeline Celestin is a PhD student in the ecology and evolutionary biology program at Rice University. ORCID ID: 0000-0003-3504-5324.

Carol L. Chambers, PhD, is a professor of wildlife ecology at Northern Arizona University School of Forestry, TWS fellow, and past president of The Wildlife Society. ORCID ID: 0000-0002-2524-4672.

Erin Considine is a wildlife biologist with the US Forest Service on the Nebraska National Forests and Grasslands.

Claire Crow has worked on the management and conservation of multiple wildlife taxa for federal agencies.

Diana Doan-Crider, PhD (O'dam), is the director of Animo Partnership in Natural Resources, coordinator for the Native American Rangeland Partnership, and research personnel at Salish Kootenai College. She holds dual citizenship in the United States and México.

Elizabeth A. Flaherty is a Certified Wildlife Biologist and associate professor of wildlife ecology and habitat management at Purdue University. ORCID ID: 0000-0001-6872-7984.

Selma N. Glasscock, PhD, is the assistant director of the Rob and Bessie Welder Wildlife Foundation, a Certified Wildlife Biologist, TWS fellow, and past president of the TWS Southwest Section and the Texas Chapter of TWS. ORCID ID: 0000-0002-5831-960X.

Kathy A. Granillo, who earned an MS in wildlife resources, is a US Fish and Wildlife biologist (retired), TWS fellow, and 2021–2024 Southwest Section representative to council.

Tabitha A. Graves, who earned a PhD in forest science, is a member of The Wildlife Society's Inclusion, Diversity, Equity and Awareness Working Group. ORCID ID: 0000-0001-5145-2400.

Princess Hester serves as director of administration at the Riverside County Habitat Conservation Agency (RCHCA) in Riverside, California.

Serra J. Hoagland, PhD, Certified Wildlife Biologist, is a Laguna Pueblo Tribal member and US Forest Service liaison officer for the Fire, Fuels, and Smoke Science Program at the Rocky Mountain Research Station.

Jessica A. Homyack is the director of environmental research and operational support for Weyerhaeuser and a Certified Wildlife Biologist. ORCID: 0000-0002-6889-3488.

Sheila Minor Huff is a retired environmental protection specialist and became a TWS Certified Wildlife Biologist in 1979.

Maria Johnson received a BS in natural resources from the University of Arizona in 2018. There, she is the program coordinator for resilience internships in the Arizona Institute for Resilient Environments and Societies (AIRES).

Patricia L. (Pat) Kennedy is professor emerita in the Department of Fisheries and Wildlife and Eastern Oregon Agricultural Research Center at Oregon State University. ORCID ID: 0000-0002-2090-1821.

Winifred B. Kessler, PhD, is a Certified Wildlife Biologist, TWS fellow, and past president of The Wildlife Society.

John L. Koprowski is the dean and a professor of the Haub School of Environment & Natural Resources, a Certified Wildlife Biologist, and a TWS, AAAS, and Linnean Society of London fellow. ORCID ID: 0000-0003-1406-9853.

Crystal Ciisquq Leonetti, Yup'ik, is a Curyung Tribal citizen and Alaska Native affairs specialist for the US Fish and Wildlife Service.

Jennifer Malpass, PhD, is an Associate Wildlife Biologist and 2020–2022 chair of The Wildlife Society's Inclusion, Diversity, Equity and Awareness Working Group.

Brenda McComb, PhD, is a professor and dean emerita at Oregon State University. ORCID ID: 0000-0001-7013-3640.

Evelyn H. Merrill is a professor of wildlife ecology in the Department of Biological Sciences at the University of Alberta. ORCID ID: 0000-0001-7737-958X.

Samantha Miller currently serves as the communications and partnerships manager at the National Wildlife Federation.

Charisa Morris is the national science advisor at the US Fish and Wildlife Service.

Kerry L. Nicholson, PhD, is a Certified Wildlife Biologist working as a furbearer and carnivore research biologist for the Alaska Department of Fish and Game and a TWS fellow. ORCID ID:0000-0001-9951-9897.

Alexander Novarro, PhD, is a program director for The Nature Conservancy in Connecticut. ORCID ID: 0000-0003-1806-7273.

Katherine M. O'Donnell, PhD, is an ecologist and decision analyst with Compass Resource Management in Vancouver, BC, and a cofounder of the TWS Out in the Field initiative. ORCID ID: 0000-0001-9023-174X.

Mamie Parker is a commissioner of the Virginia Department of Wildlife Resources and a former Fish and Wildlife Service executive.

Audrey Peterman is a longtime advocate for the integration of our public lands and author of three books on the subject. She is the recipient of the National Parks Conservation Association's Centennial Leadership Award 2022.

Ruth Plenty Sweetgrass-She Kills, Hątáde Wíyą (Night Wind Woman), PhD (Hidatsa, Mandan, Dakota, Nakota), is the food sovereignty director at Nueta Hidatsa Sahnish College and a senior social science researcher at the University of Montana. ORCID ID: 0000-0002-5738-6480.

Camya Robinson, a University of Florida graduate with a BSc degree in wildlife, ecology, and conservation, is a Doris Duke Conservation Scholar Program alumna.

Stephanie S. Romañach is a research ecologist working in Florida's Everglades and Africa's savannas. ORCID ID: 0000-0003-0271-7825.

Paige Schmidt is a zone biologist with the US Fish and Wildlife Service.

Susan K. Skagen is a research wildlife biologist emerita with the US Geological Survey, Fort Collins Science Center, Fort Collins, Colorado. ORCID ID: 0000-0002-6744-1244.

Lisette P. Waits is a distinguished professor in the Department of Fish and Wildlife Sciences at the University of Idaho. ORCID ID: 0000-0002-1323-0812.

PART I

Breaking Ground and Presenting Facts

Acknowledging the Past and Defining Present Challenges

CHAPTER

1

Carol L. Chambers and Kerry L. Nicholson

The Importance of Diversity

Life is not easy for any of us. But what of that? We must have perseverance and above all confidence in ourselves. We must believe that we are gifted for something and that this thing must be attained. —*Marie Curie*

The True Meaning of Diversity

We are taught that biodiversity is good. Intact ecological systems consisting of populations, communities, and ecosystems are adaptable, resistant, and resilient. We learn that diversity is not just about richness, the number of species present, but includes functional diversity, which is key to a resilient ecosystem. We read examples from terrestrial and aquatic ecosystems demonstrating that productivity, as well as resistance to and recovery from disturbance, are greater in diverse communities than in depauperate ones. In a resilient ecosystem, the diversity of community members, structural processes, and biological legacies are critical to continuity and can create pathways of reorganization and self-renewal, as well as new evolutionary directions. Functional diversity helps maintain ecosystems and their services. The reintroduction of gray wolves (*Canis lupus*) in Yellowstone National Park, for example, restored a trophic cascade by altering elk (*Cervus elaphus*) density, leading to an increase in woody plant species and herbaceous plants, which, in turn, increased forage for beaver (*Castor canadensis*) and bison (*Bison bison*) (Ripple and Beschta 2012).

Beaver serve as keystone species affecting wetlands and wetland-dependent species (e.g., amphibians), so the effect of wolves on diversity rippled through the ecosystem (Hossack et al. 2015). Examples such as these show the importance of biological diversity and intact ecosystems. Each mammal, bird, plant, fungus, insect, amphibian, reptile, and soil microbe has a unique part to play. These are basic and fundamental concepts for biologists.

Aldo Leopold, considered the father of the scientific field of wildlife management in the United States, encouraged us to understand the "land-community" and the role that people played: "*A land ethic changes the role of* Homo sapiens *from conqueror of the land-community to plain member and citizen of it. It implies respect for his fellow members, and also respect for the community as such*" (Leopold 1949). Wildlife management depends on a triad of wildlife, habitat, and humans. When making conservation and management decisions, wildlife biologists work not just with wildlife and habitat but also with a wide range of public stakeholders. Yet members of the wildlife profession are not representative of our diverse public. In the United States, over 75% of wildlife biologists are white and only 37% are women (Zippia n.d.). Currently, most degrees (>80%) are awarded to white students; more women (58%) are earning bachelor's degrees than men at many institutions (DataUSA n.d.). We all have a stake in natural resources. By creating a diverse profession, our wildlife goals become more relatable to the diverse members of the public.

The Art of Thinking Independently, Together

"*We all should know that diversity makes for a rich tapestry, and we must understand that all the threads of the tapestry are equal in value no matter their color*" (Angelou 1994). The same is true of wildlife conservation and management. But what does peer-reviewed science say? For starters, the collective intelligence of groups is greater than that of an individual (Woolley et al. 2010). Collective intelligence, or "group think," is not dependent on the maximum or even average intelligence of individuals in the group but instead is related to the average social sensitivity, parity (equality of turn taking), and proportion of women among its members. Second, groups that are cognitively diverse are better at solving complex problems (Valantine and Collins 2015). Third, gender-diverse groups are better problem solvers. Nielsen et al. (2017) found that women were better at recognizing the expertise of team members than men. When evaluating group members, women focused on research expertise, while men focused on irrelevant issues such as gender. Gender-diverse teams are better able to dispel bias and misconceptions about others. Nielsen et al. (2017) found that the questions asked by female-dominated groups differed from those generated by male-dominated groups. Groups of diverse team members will develop more creative questions and better approaches to problem solving. And for those who publish in the scientific field, gender-heterogeneous authorship teams receive 34% more citations than publications by gender-uniform teams (Campbell et al. 2013).

Likewise, people of different cultures also bring in complementary and contrasting ideas. *Cultural capital* is the knowledge, skills, and competence that children gain from their parents, peers, and schools (Throsby 1999). Cultural practices influence how we construct knowledge and conduct science. Cultural practices can affect the questions we ask, populations we study, and procedures we use. As a graduate student, Flo Gardipee was interested in the genetics of bison, but as a Native American, she believed that the methods for gathering DNA samples were intrusive and disrespectful to the animals. Instead of using the standard approach of capturing, tranquilizing, or killing individual animals, she led the development of a noninvasive approach to collect DNA from feces (University of Montana 2007; Medin and Bang 2014). Her work gained media attention as "pooper-scooper" research and led to knowledge of the population structure of bison in Yellowstone and Grand Teton National Parks (Gardipee 2007; University of Montana 2007). Today the extensive use of noninvasive DNA collection from feces, saliva, and the environment is testament to the functionality and practicality of this method. Diverse cultural perspectives can influence the construction of knowledge and lead to more creative, novel approaches to problem solving. If we think of team members as tools, problem solving with a tool kit composed of hammers is much more limiting than a tool kit with hammers, screwdrivers, wrenches, drills, saws, and grinders (Valantine and Collins 2015).

From a practical standpoint, diversity is the 21st-century reality. Generation Z (or iGen) is the group of people born after 1996. Although millennials were once considered a diverse generation, Generation Z is the most racially and ethnically diverse to date; just 52% are non-Hispanic white. Its members are also enrolling in college at a higher rate than those of other generations at the same age (Fry and Parker 2018). It seems certain that both millennials and Generation Z will have high expectations of diversity in the workplace.

Who's Dancing at the Party?

You cannot truly have diversity without equity and inclusion. Diversity is the presence of a range of human qualities and attributes—ages, genders, races, ethnicities, cultures, skills, and more—both visible and invisible, within a group, organization, or society. For example, some of the race-ethnic groups described in this book include African American, Asian, Black, Caucasian, First Nations, Hispanic, international, Indigenous, Latino, Latinx, Native, and white. Information on some aspects of diversity is easier to find than information on others. Our knowledge of gender representation is less complete because gender statistics are generally limited to a binary system of male and female.

Understanding and acknowledging diversity alone is not sufficient. Inclusion means creating an environment in which diversity is respected, accepted, and appreciated and in which all are able to participate equally. It should be distinguished from tokenism, the expectation that one or a few people represent and speak for an entire community.

There is another important component of diversity: equity. Biases and obstacles exist for some but not for others. We cannot assume an even playing field with the same equipment for all (Figure 1.1). In a work environment, some arrive with background and coursework in conservation ecology, mammalogy, or wildlife habitat management, and experience through unpaid internships that give them familiarity in a discipline. Others cannot afford to work for no pay and, with fewer opportunities, possess a shorter resume and will rank low when applying for wildlife jobs. We must recognize that not everyone has the same access to education, job training, and work experience. Achieving equity requires actively correcting for the disparity of advantages enjoyed by some and not others. If we recognize that not everyone has equal access to the same resources and compensate for this fact, then we are providing equity. As Vernā Myers (Myers n.d.) put it, "*Diversity is being asked to the party. Inclusion is being asked to dance.*" We can take it a step further and level the dance floor to promote equity.

In Retrospect

The disparity in gender representation in scientific fields is centuries old (see Chapter 2). In 1790, Judith Sargent Murray questioned whether nature was partial. Did nature distribute mental superiority to one gender over another? Were the minds of women deficient? Why were women denied the same educational opportunities as men? She argued that women were just as capable (Harris 1995). Yet women were denied similar education, and those who violated behavioral expectations often lost societal respect. In an 1871 *Atlantic Monthly* essay, Wilson Flagg (Flagg 1871) wrote, "*Women cannot conveniently become hunters or anglers, nor can they without some eccentricity of conduct fol-*

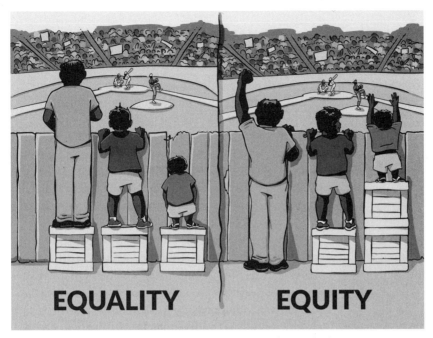

FIGURE 1.1. Equality isn't enough. Equal treatment doesn't provide the same opportunities to each person. We continue to disadvantage women and underrepresented groups when we don't recognize that equity is an important part of the picture. Opportunity gaps caused by systemic racism, oppression, and other aspects of our history have contributed to disparities that we now need to correct (http://www.socialventurepartners.org/wp-content/uploads/2018/01/Problem-with-Equity-vs-Equality-Graphic.pdf). Reproduced with thanks from Interaction Institute for Social Change. Artist: Angus Maguire.

low birds and quadrupeds to the woods. . . . [T]he only part of natural history which they can pursue out of doors is the study of plants."

Fast-forward a few centuries, and opinions about the role of women and minorities in the sciences and wildlife fields had not improved greatly, despite the women's movement and growing acceptance of gender equality (Box 1.1). Women are still considered too delicate to handle the rough work required for a job or too distracting for male colleagues. For respect and acceptance, women and minorities have had to overcome obstacles such as lack of access to education, tools, and jobs; lack of interest in or credibility conferred on their findings; harassment; and bullying.

Have Things Changed?

We still encounter many challenges to inclusion. We lack role models and mentors to inspire and support members of underrepresented groups in the wildlife

BOX 1.1 Change: How Long Must We Wait?

Women's rights are still not legally protected by the US Constitution. In 1848, the first visible public demand for women's suffrage was made by Elizabeth Cady Stanton and Lucretia Mott at the first Woman's Rights Convention in Seneca Falls, New York. In 1923, Alice Paul from the National Woman's Party first proposed the Equal Rights Amendment (ERA) to prohibit discrimination on the basis of sex, providing for the legal equality of the sexes (Figure 1.2). It took more than 45 years before it gained traction and was brought forward in the late 1960s to Congress. In 1971 the ERA won the needed two-thirds vote to pass the House of Representatives and on March 22, 1972, it passed in the Senate. This amendment would end the legal distinctions between men and women in divorce, property ownership, employment, and other matters. Passage in Congress required 38 states had to ratify the amendment within seven years. Hawaii was the first, and within the first year of the proposal, 22 states ratified it. However, the process slowed, and states started to rescind their prior support. In 1978, the seven-year deadline was extended to 1982, but even then, only 35 states had signed on. Two additional states ratified the amendment: Nevada in 2017 and Illinois in 2018. Sitting one state shy of ratification, in 2019, Senators Ben Cardin (D-MD) and Lisa Murkowski (R-AK) proposed to immediately reopen the consideration of the ERA and eliminate the arbitrary deadline (Figure 1.3).

FIGURE 1.2. Alice Stokes Paul (January 11, 1885–July 9, 1977) was an American suffragist, feminist, women's rights activist, and one of the main leaders and strategists of the campaign for the Nineteenth Amendment to the US Constitution, which prohibits sex discrimination in the right to vote.
Photo by Harris & Ewing (1915); restoration courtesy of Adam Cuerden, commons.wikimedia.org /wiki/File:Alice_Paul _(1915)_by_Harris_%26 _Ewing.jpg.

Some people would say we are 173 years into the fight to afford women the same legal rights men. However, women like Phyllis Schlafly see it differently. She took up the fight against the ERA in the 1970s. She suggested that if the amendment were passed, women would be forced to go to war and would lose their right to child support and alimony, the heterosexual world would collapse, morality would fall by the wayside, and women would lose their femininity and the opportunities that marriage could provide. Today, opposition to ratification warns it will, for example, strengthen gay or transgender rights, undermine the rights of churches, and prevent states from restricting abortion.

While the United States continues its debate, Canada resolved this issue in the 1980s. However, although the Canadian Bill of Rights of 1960 guaranteed the right of equality, it has no enforcement in the provinces and can easily be amended by Parliament. It is Section 15 of the Canadian Charter of Rights and Freedoms that has the specific wording affording equal rights to every individual. It is part of the Constitution of Canada and was signed into law in 1982, but Section 15 did not take effect until 1985, to allow the provinces to review and remove unconstitutional contradictions between provincial and federal law.

In 2020, Virginia became the 38th state to ratify the Equal Rights Amendment, thus reaching the three-quarters threshold to add it to the US Constitution. As of 2021, this has not happened. Several states have acted to rescind their prior approval, and Alabama, Louisiana, and South Dakota are suing the federal government to block the ERA. With today's discussions around diversity, equity, and inclusion, can we resolve this issue? As we change the culture and the dialogue, what was once unacceptable can become acceptable. Alice Paul's poignant statement is still relevant today: *"If we keep on this way they will be celebrating the 150th anniversary of the 1848 Convention without being much further advanced in equal rights than we are. . . . We shall not be safe until the principle of equal rights is written into the framework of our government."* Enshrining women's rights in the US Constitution would send a clear message and have a major cultural impact.

FIGURE 1.3. Kate Campbell Stevenson brought to life Abigail Adams, Harriet Beecher Stowe, Alice Paul, and other women suffragists through costumes and music for the Norfolk District's Women's Equality Day celebration on August 22, 2012.
US Army Photo / Patrick Bloodgood.

profession. These unintentional or intentional roadblocks to increasing diversity vary, but they create an atmosphere of exclusion. From our natural desire to "silo" to our unawareness of implicit biases and privilege, each factor impedes greater human diversity in our organizations, schools, and workplaces.

Humans are hardwired to be attracted to people like ourselves (Rivera 2020). This "silo effect" keeps us fragmented in homogenous social groups or specialist teams. We separate ourselves without an effective way to communicate among groups. This disconnect causes collaboration and productivity to suffer (Tett 2015; Hodgson et al. 2019).

Implicit bias refers to attitudes or stereotypes we hold unconsciously that can affect our actions and decisions (FitzGerald and Hurst 2017). Privilege comes from systemic dominance that is maintained but often not recognized by those with it. Role models and mentors can inspire and help nurture newcomers. But roadblocks can hinder participation.

Without an inclusive environment, women and those from other underrepresented groups may get the message "you don't belong here." We conducted short interviews with women working in the wildlife profession ranging in age from 24 to 65 (average age 47). The most notable change in women's roles that they described was the increase in the number of women, particularly in leadership, in the profession. However, when asked about the biggest challenges for women or other members of marginalized groups, women spoke of not being taken seriously and of isolation and tokenism. They talked about having to be twice as good to get half the recognition of their male counterparts. They heard, "You do a pretty good job for a girl," and were called "the Token." They tried to conform to expectations such as dressing "correctly" and being "nice," but experienced the dejection of being called "a bitch" if their behaviors were perceived as "too aggressive." Their success was laced with exhaustion.

Women also described shocking experiences of discrimination and harassment. A coworker tried to sexually assault one interviewee. When she reported the assault, she was stigmatized and received no support from her supervisor or the human resources department. Women described being sexually harassed both on the job and in graduate school by coworkers, supervisors, and graduate advisors. One woman described feeling trapped when her supervisor repeatedly massaged her neck, arms, and shoulders over the course of a multi-week field project. She needed a letter of support from him to succeed, so, as she explained it, "I just sat there and took it." Women also experienced being passed over for promotions and being excluded from decision making. They were passed over for jobs because of the expectation that they couldn't handle working in the field. They were paid less than men in similar positions. Some were told, because of their gender, that they would fail in their job. Having children, they heard, would give the message they were not serious about their career; it would reduce work productivity. One mentioned losing a promotion because a male colleague had a family to support, despite her advanced expertise and experience. This range of experiences indicates that we have changes to make in all environments where women work in wildlife.

Although the nature of discrimination and harassment had not changed for many we interviewed, its subtlety had increased. Overt discrimination changed to microaggressions, indirect or unintentional slights against members of marginalized groups that convey a sense of exclusion from the dominant group. A Black woman spoke of frequent comments made about her income. People assumed she grew up in a low-income environment; she hadn't. Another was constantly called "kid" instead of her name. Other subtle discrimination ranged from expectations that women shouldn't work in the field without a man present to the assumption that a woman couldn't lead. One woman, who worked with game animals, was asked by a supervisor whether she hunted. He said he didn't expect her to because his wife and daughters didn't do "outdoorsy stuff." Another woman was asked, while conducting remote fieldwork, "Your husband lets you do this work?" Some women received unwanted comments about their appearance while on the job or in professional meetings. One of the women we interviewed spoke of being asked, "Do you know how beautiful you are?" by a conference attendee instead of being able to hold a conversation about the wildlife work they shared.

Despite the continued challenges, there is recognition that things are better. More women are moving into leadership roles. Women are being treated more equally and being viewed as equally capable. But universally, our interviewees recognized that diverse representation and true inclusion are works in progress. The wildlife profession needs to appeal to diverse audiences. To do so, employers must hire people with diverse backgrounds. The culture must shift to be more inclusive, empowering minority women. As one woman stated, "The wildlife field is still very, very white."

Roadblocks

In science and engineering occupations, women and most minority groups are underrepresented in proportion to the US population (NCSES 2019). Women and minority students continue to experience roadblocks to obtaining a science degree. Mervis (2011) reported that, although universities want to broaden participation in science, technology, engineering, and mathematics (STEM) degrees, introductory courses continue to "weed out" potential students from majoring in these fields, thus undermining diversity. Students who are well prepared may nevertheless leave programs because of isolation, lack of support, low expectations, and discrimination (Summers and Hrabowski 2006; Puritty et al. 2017). Anderson (2020) noted that in the wildlife profession, men still dominate, and there are few women of color or from other cultural groups. Women are still underrepresented as the first or sole authors of scientific papers and in higher-level positions. Of 437 wildlife and fisheries faculty from universities in the United States, just 21% compared to 79% of tenure-track faculty were female and male professors, respectively. In addition, women faculty were more likely to hold the rank of assistant than full professor compared to men (Swihart et al. 2016). Although women are now half of the college-educated workforce, less than a third work in science and engineering occupations. Women also receive less support for leadership advancement throughout their careers. The COVID-19 pandemic has thrown up new roadblocks for women. Myers et al. (2020) described the loss of research time that women with young children experienced compared to men in their fields.

Overlay these conditions with cultural, racial, or other biases, and the situation becomes more isolating in natural resource management. One person described a subtle (and sometimes not-so-subtle) cultural bias to having been born in West Virginia. In both work and personal environments, he encountered assumptions about his education, and he was "jokingly" called a hillbilly and asked whether one leg was shorter than the other because of the state's mountainous terrain (Patton 2020). A Black colleague described avoiding eating watermelon in public settings because of the racist US stereotype that associates eating watermelon with being lazy, childlike, or foolish (Black 2014, 2018). Yet this trope persists, as do others, and may be intentionally or inadvertently expressed, as was the case of a cartoonist depicting a White House intruder recommending watermelon-flavored toothpaste to Barack Obama (Killough 2014). These "jokes" wear thin and constitute a form of subtle discrimination.

How Lucky Are You?

Well-meaning people don't intend to discriminate and may be unaware of the privilege or biases they hold. Societies confer greater advantages to some groups than others. Despite having the skills for a job, when an employer doesn't share similar interests, an applicant is assumed to be a poor fit (Rivera 2020).

Those born into the dominant social group benefit simply because of birthright rather than characteristics such as intelligence or abilities (Black and Stone 2011). Race, gender, and ethnicity affect one's position in society, but able-ness, age, education level, wealth, and religious affiliation can also influence who carries power. Think about how your life might differ if your race, gender, or religion differed from that of the dominant group. Take a privilege test online. You can find many, for instance, from the Social Justice Toolbox (www.socialjusticetoolbox.com). Below are five examples of items on a privilege test:

1. I can go shopping alone most of the time, pretty well assured that I will not be followed or harassed.
2. I can do well in a challenging situation without being called a credit to my race.
3. I am never asked to speak for all the people of my racial group.
4. I feel unthreatened and safe in my interactions with authority figures and police officers.
5. I can go home from most meetings of organizations I belong to feeling somewhat tied in, rather than isolated, out-of-place, outnumbered, unheard, held at a distance, or feared.

These are just a few concrete examples to help you identify the privilege you experience. The more often you reply yes to questions like these, the more likely you are to experience being privileged. By understanding the privilege that we carry, we can understand that personal perspectives influence the types of choices we have.

Cultures tell stories to reflect our traditions and values. Stories connect us to the past and provide social norms. In many stories, listeners are introduced to the "Other," people who are different in customs, beliefs, behavior, or other characteristics. Our inclination to focus on differences rather than similarities continues the construct of privilege; it can root us in ethnocentrism (Medeiros and Cowall 2020).

This Book's Purpose and How to Use It

In this book, we examine aspects of "intersectionality," or how power relations influence social relations as they apply to individuals or groups of women, across categories of race, gender, sexuality, and ethnicity (Collins and Bilge 2020). We want this book to serve as a pragmatic guide to ensuring a more diverse future in the wildlife profession and for the natural world itself. We were moved to write this book because of our own experiences and the stories we heard, some expressed above and others in ensuing chapters. We offer ideas and approaches for building representation. To do this, we brought together a diverse group of contributors from academic, public, and private professional spheres, with the aim of addressing challenges and opportunities for women and minorities in the wildlife profession.

As you read, consider the similarities and differences between your perspectives with those of chapter authors. You may share many experiences they describe. Look for the variation in voices that reflect cultural approaches of communicating. Some authors speak metaphorically, others more literally. Also notice differences in chapter structure. Several follow a traditional academic format and present tables, facts, and figures, while others use a storytelling approach with personal anecdotes. Because privileges and experiences vary, sometimes dramatically, among authors, do not expect to hear the same voice throughout this book. The perspectives expressed may not be yours or represent your cultural background, but they are real for others and important for recognizing diverse perspectives. Listen patiently.

Some authors have placed themselves at risk simply by providing their personal perspectives about gender in the wildlife environment. We found it very difficult to recruit authors for our chapter on lesbian, gay, bisexual, transgender, and queer or questioning (LGBTQ+) perspectives because few feel safe from discrimination in the current environment. Despite their invisibility in the wildlife field, however, according to Booms (2019), *"LGBTQ biologists work in every agency, every region, and every state. . . . And we are also sitting in every undergraduate and graduate wildlife biology class."* So as you read this book, read with empathy.

This book is meant to start a conversation rooted in fact, research, and real experiences. The data presented focus primarily on gender diversity and racial and ethnic diversity in the wildlife profession. The reason for that focus is two-fold: (1) systemic racism, sexism, bias, discrimination, and harassment continue to be issues despite decades of effort to prevent them; and (2) generally more data are available on gender and race than on other demographic groups. However, diversity extends far beyond these categories to aspects such as age, geography, nationality, immigration status, language, religion, socioeconomic status, sexual orientation, (dis)ability, and life experience. Creating inclusive environments and equitable outcomes will help not just individual demographic groups but all people. The goal is to create opportunities for everyone to pursue, persist, and thrive in the wildlife profession. Moving forward, we need to identify solution sets and best practices and continue to collect and critically examine evidence that emerges from our efforts. We seek to expand our understanding through analysis, storytelling, and sharing the experiences of others as we work to be allies in equity and justice work.

It's about All of Us

One of the most important points to take from this book is that, to increase diversity, each of us needs to act. We can't remain passive and believe that others will do the diversity, equity, and inclusion (DEI) work. Simply paying lip service to a more diverse future is no longer enough. Each of us can and should contribute actively to DEI efforts. In subsequent chapters, we recommend helpful approaches and actions you can take for yourself and others in a "How to Be an Ally" section.

One idea is to adopt a SMART goal approach (Doran 1981), with objectives that are specific, measurable, attainable, relevant, and time-bound (Yemm 2012). SMART goals should motivate, be strategic, realistic, and be conducted within a specific time frame. They should be created with and agreed upon by all stakeholders. Although initially developed for use in business management, SMART goals are employed in a variety of fields, ranging from rehabilitation to student learning to physical fitness (e.g., Siegert and Taylor 2004; O'Neill and Conzemius 2006; Swann et al. 2020). Setting action-oriented goals provides individuals and groups with concrete approaches to accomplishing desired outcomes, so why not apply them here?

No discipline remains stagnant. In wildlife conservation and management, we see change in how the public interacts with and values wildlife. The North American human population is becoming more urban centered. We see a reduction in consumptive use (trapping, hunting, and fishing) of wildlife (Robison and Ridenour 2012; Manfredo et al. 2020). People are less likely to interact directly with wildlife and instead learn about animals through a lens or computer screen, on social media or television programs (Manfredo et al. 2020). The wildlife profession is still dominated by white men (91% of agency personnel identified as white and 72% as men). Current minority populations will outnumber white, non-Hispanic populations by 2050. Diversification of the wildlife profession is clearly needed. Asking questions drives science forward and different perspectives lead to asking different questions and to new insights (Cochran 2017).

What You'll Find in the Chapters

This book is structured in two sections, in which we describe women's roles and occupations in the wildlife profession in the past and present (Part I), and future (Part II). In Part I, "Breaking Ground and Presenting Facts," we describe why it is important to diversify (Chapter 1). Chapter 2 reviews the history and status of women in natural resources, chronicling events that led to women entering the profession. We examine why so few women are mentioned among the famous scientists of the past and how women in previous centuries found opportunities to study and work given the limitations placed on their fitness for intellectual pursuits and the expectations of caring for home and family as their highest purpose. This chapter describes the golden age of natural history study and how it helped women become naturalists and botanists and transition into the professional world of the 20th century. Stories of many women pioneers are included in this chapter.

We next highlight where we are today in professional and university settings. The lack of diversity among faculty and students in wildlife programs leads to a workforce that is likewise unrepresentative of the US population (NCSES 2019). In Chapter 3, current diversity statistics and trends among academia and students are examined in the recent past and present to identify programs fostering diversity. Opportunities for university programs and individuals to increase diversity and encourage positive change for women and people of color are described. Chapter 4 examines institutional barriers to employment at universities and within the US federal government that women and minorities face and provides suggestions for generating a more inclusive environment. Chapter 5 describes social barriers that are currently experienced by underrepresented groups in their historical context.

In Part II, "Diverse Perspectives and Practical Acts," we talk about the challenges women face entering and excelling in the field of wildlife conservation across spectrums of race and ethnicity (Chapters 6, 7), sexuality and gender identification (Chapter 8), and age (Chapter 9). We discuss proven models and proposals for new methods to increase engagement of women in the workplace (Chapter 10). We raise awareness of gendered interactions and approaches to reduce inequities (Chapter 11). We describe additional, often subtle, barriers that women still face (Chapter 12) and conclude with a question for you: How will you take action? (Chapter 13). We seek to educate about ethnic, cultural, and gender differences; improve working conditions; and strengthen diversity in the wildlife profession. And although this book is an umbrella for women's issues in the wildlife profession, we touch on other unrepresented groups (gay and transgender men) as appropriate. Our book's intent is to describe allyship and actions we can all take to diversify the wildlife profession.

We have written this book for wildlife professionals, professors, and students. We hope to reach people in the public and private sectors. Use this book in your classroom, workplace, organization, or for public discussion. Each chapter includes questions and interactive learning activities to stimulate thinking, discussion, and action. We hope this book will encourage you to better understand and encourage people to join the profession of wildlife work and natural resource management.

Look at what I bring to the table and what I produce. Not what I'm wearing, my hairstyle, or my skin color. —Sheila Minor Huff

Discussion Questions

1. What does diversity, equity, and inclusion mean to you, and what do you consider the most challenging aspect of working in a diverse environment?
2. How would you advocate for diversity, equity, and inclusion with those who don't understand its importance?
3. In the United States, should the Equal Rights Amendment be ratified? Why or why not?
4. What do you do to make sure everyone feels included?

Activity: SMART Goals

Learning Objectives: Participants will develop SMART goals that focus on actions and identify specific steps that lead to increasing diversity in the wildlife field.

Time: 40 min.

Description of Activity

Materials needed: Copies of the worksheet for anyone without a book and writing utensils.

Process:

Part 1: Personal Reflection and SMART Goal Development (10 min.)

Participants will reflect on the information in Chapter 1 and develop SMART goals for themselves, a group they lead, or their organization. This goal should focus on increasing diversity in the wildlife field.

Part 2: Think-Pair-Share (15 min.)

Participants next will partner with someone sitting near them to (1) review their barriers and SMART goal and (2) identify any potential challenges related to meeting it. Identifying challenges should be viewed as an anticipation activity to encourage participants to think about solutions.

Part 3: Debriefing and Discussion Questions (15–20 min.)

1. What did you learn?
2. What was most surprising about this exercise? How did your SMART process or your goal change after discussing this with your partner?
3. If you anticipate any challenges to meeting your goal, what are possible solutions to overcoming them?
4. What are the benefits of using the SMART process of setting goals to address systemic challenges like diversity?
5. How will you use the information you learned or your experience from this activity in your life as a wildlife professional, student, or citizen?

Summarize your key takeaways from Chapter 1. Then complete the following table to develop a personal SMART goal.

Identify one potential barrier in the wildlife field that may limit diversity, equity, and/or inclusion:

Next, develop a personal goal for reducing or eliminating this barrier by working through the SMART goal process using the table below.

S	Specific	What will you do? Who will be involved or will help you? Why are you doing this?	
M	Measurable	How will you measure progress and success?	
A	Achievable	Do you have the resources and skills to achieve your goal? If not, how will you acquire them?	
R	Relevant	Why are you doing this? Why is this significant?	
T	Timely	What is your timeline and how can it be met?	
GOAL:			

Literature Cited

Anderson, W. S. 2020. "The Changing Face of the Wildlife Profession: Tools for Creating Women Leaders." *Human-Wildlife Interactions* 14(1), article 15. doi: 10.26077/e3e1-nf19.

Angelou, M. 1994. *Wouldn't Take Nothing for My Journey Now*. New York: Bantam.

Black, L. L., and D. Stone. 2011. "Expanding the Definition of Privilege: The Concept of Social Privilege." *Journal of Multicultural Counseling and Development* 33(4): 243–255. doi: 10.1002/j.2161-1912.2005.tb00020.x.

Black, W. R. 2014. "How Watermelons Became a Racist Trope." *Atlantic*. December 8, 2014. www.theatlantic.com/national/archive/2014/12/how-watermelons-became-a-racist-trope/383529/.

Black, W. R. 2018. "How Watermelons Became Black: Emancipation and the Origins of a Racist Trope." *Journal of the Civil War Era* 8(1): 64–86.

Booms, T. 2019. "I Am One of You." *Wildlife Professional*. March/April.

Campbell, L. G., S. Mehtani, M. E. Dozier, and J. Rinehart. 2013. "Gender-Heterogeneous Working Groups Produce Higher Quality Science." *PLoS ONE* 8(10): e79147. doi: 10.1371/journal.pone.0079147.

Cochran, G. L. 2017. "Understanding and Promoting Diversity and Inclusion in Physics." *SPS Observer* (Winter). www.spsnational.org/the-sps-observer/winter/2017/understanding-and-promoting-diversity-and-inclusion-physics.

Collins, P. H., and S. Bilge. 2020. *Intersectionality*. 2nd ed. Key Concept Series. John Wiley & Sons.

DataUSA. N.d. "Wildlife Biology STEM Major." Accessed June 28, 2021. https://datausa.io/profile/cip/wildlife-biology#demographics.

Doran, G. T. 1981. "There's a S.M.A.R.T. Way to Write Management's Goals and Objectives." *Management Review* 70(11): 35–36.

FitzGerald, C., and S. Hurst. 2017. "Implicit Bias in Healthcare Professionals: A Systematic Review." *BMC Medical Ethics* 18, article 19. doi: 10.1186/s12910-017-0179-8.

Flagg, W. 1871. "Botanizing." *Atlantic Monthly* 27(164): 657–64.

Fry, R., and K. Parker. 2018. "Early Benchmarks Show 'Post-millennials' on Track to Be Most Diverse, Best-Educated Generation Yet." Pew Research Center, November. www.pewsocialtrends.org/2018/11/15/early-benchmarks-show-post-millennials-on-track-to-be-most-diverse-best-educated-generation-yet/.

Gardipee, F. M. 2007. "Development of Fecal DNA Sampling Methods to Assess Genetic Population Structure of Greater Yellowstone Bison." Master's thesis, University of Montana, Missoula.

Harris, S. M., ed. 1995. *Selected Writings of Judith Sargent Murray*. Oxford University Press. http://nationalhumanitiescenter.org/pds/livingrev/equality/text5/sargent.pdf.

Hodgson, E. E., B. S. Halpern, and T. E. Essington. 2019. Moving Beyond Silos in Cumulative Effects Assessment." *Frontiers in Ecology and Evolution* 7: 211. doi: 10.3389/fevo.2019.00211.

Hossack, B. R., W. R. Gould, D. A. Patla, E. Muths, R. Daley, K. Legg, and P. S. Corn. 2015. "Trends in Rocky Mountain Amphibians and the Role of Beaver as Keystone Species." *Biological Conservation* 187:260–69. doi: 10.1016/j.biocon.2015.05.005.

Killough, A. 2014. "Boston Herald Apologizes for Obama Cartoon after Backlash." CNN, October 1. www.cnn.com/2014/10/01/politics/boston-herald-cartoon/index.html.

Leopold, A. 1949. *A Sand County Almanac: And Sketches Here and There*. New York: Oxford University Press.

Manfredo, M. J., T. L. Teel, A. W. Don Carlos, L. Sullivan, A. D. Bright, A. M. Dietsch, J. Bruskotter, and D. Fulton. 2020. "The Changing Sociocultural Context of Wildlife Conservation." *Conservation Biology* 34(6): 1549–59. doi: 10.1111/cobi.13493.

Medeiros, P., and E. Cowall. 2020. "The Culture Concept." In *Perspectives: An Open Intro-duction to Anthropology*, edited by N. Brown, T. McIlwraith, and L. Tubelle de González, 2nd ed., 29–44. Arlington, VA: American Anthropological Association.

Medin, D. L., and M. Bang. 2014. *Who's Asking? Native Science, Western Science, and Science Education*. Cambridge, MA: MIT Press.

Mervis, J. 2011. "Weed-Out Courses Hamper Diversity." *Science* 334:1333.

Myers, K. R., W. Y. Tham, Y. Yin, N. Cohodes, J. G. Thursby, M. C. Thursby, P. Schiffer, J. T. Walsh, K. R. Lakhani, and D. Wang. 2020. "Unequal Effects of the COVID-19 Pandemic on Scientists." *Nature Human Behaviour* 4:880–83. doi: 10.1038/s41562-020-0921-y.

Myers, V. N.d. The Vernā Myers Company. Accessed January 3, 2021. www.vernamyers.com/.

National Center for Science and Engineering Statistics (NCSES). 2019. "Women, Minori-ties, and Persons with Disabilities in Science and Engineering." Naional Science Foundation. https://ncses.nsf.gov/pubs/nsf19304/digest.

Nielsen, M. W., S. Alegria, L. Börjeson, H. Etzkowitz, H. J. Falk-Krzesinski, A. Joshi, E. Leahey, L. Smith-Doerr, A. W. Woolley, and L. Schiebinger. 2017. "Opinion: Gender Diversity Leads to Better Science." *PNAS* 114(8): 1740–42. doi: 10.1073/pnas.1700616114.

O'Neill, J., and A. Conzemius. 2006. *The Power of SMART Goals: Using Goals to Improve Student Learning*. Bloomington, IN: Solution Tree Press.

Patton, D. R. 2020. "Cultural Bias in the Workplace." *Forestry Source*, September.

Puritty C., L. R. Strickland, E. Alia, B. Blonder, E. Klein, M. T. Kohl, E. McGee, M. Quintana, R. E. Ridley, B. Tellman, and L. R. Gerber. 2017. "Without Inclusion, Diversity Initiatives May Not Be Enough." *Science* 357(6356): 1101–2. doi: 10.1126/science.aai9054.

Ripple, W. J., and R. L. Beschta. 2012. "Trophic Cascades in Yellowstone: The First 15 Years after Wolf Reintroduction." *Biological Conservation* 145(1): 205–13.

Rivera, L. 2020. "Stop Hiring for 'Cultural Fit.'" Kellogg School of Management at Northwestern University. https://insight.kellogg.northwestern.edu/article/cultural-fit-discrimination.

Robison, K. K., and D. Ridenour. 2012. "Whither the Love of Hunting? Explaining the Decline of a Major Form of Rural Recreation as a Consequence of the Rise of Virtual Entertainment and Urbanism." *Human Dimensions of Wildlife* 17(6): 418–36. doi: 10.1080/10871209.2012.680174.

Siegert, R. J., and W. J. Taylor. 2004. "Theoretical Aspects of Goal-Setting and Motivation in Rehabilitation." *Disability and Rehabilitation* 26:1–8. doi: 10.1080/09638280410001644932.

Summers, M. F., and F. A. Hrabowski III. 2006. "Preparing Minority Scientists and Engineers." *Science* 311:1870–71.

Swann, C., A. Hooper, M. J. Schweickle, G. Peoples, J. Mullan, D. Hutto, M. S. Allen, and S. A. Vella. 2020. "Comparing the Effects of Goal Types in a Walking Session with Healthy Adults: Preliminary Evidence for Open Goals in Physical Activity." *Psychology of Sport and Exercise* 47(March): 101475. doi: 10.1016/j.psychsport.2019.01.003.

Swihart, R. K., M. Sundaram, T. O. Hook, and J. A. Dewoody. 2016. "Factors Affecting Scholarly Performance by Wildlife and Fisheries Faculty." *Journal of Wildlife Manage-ment* 80:563–72.

Tett, G. 2015. *The Silo Effect: The Peril of Expertise and the Promise of Breaking Down Barriers*. New York: Simon and Schuster.

Throsby, D. 1999. "Cultural Capital." *Journal of Cultural Economics* 23:3–12.

University of Montana. 2007. "Distinct Bison Herds Roam Yellowstone." *ScienceDaily*, January 29. www.sciencedaily.com/releases/2007/01/070128104947.htm.

Valantine, H. A., and F. S. Collins. 2015. "National Institutes of Health Addresses the Science of Diversity." *PNAS* 112(40): 12240–42.

Woolley, A. W., C. F. Chabris, A. Pentland, N. Hasmi, and T. W. Malone. 2010. "Evidence for a Collective Intelligence Factor in the Performance of Human Groups." *Science* 330(6004): 686–88.

Yemm, G. 2012. *Essential Guide to Leading Your Team: How to Set Goals, Measure Performance and Reward Talent*. The FT Guides. London: Pearson UK.

Zippia. N.d. "Wildlife Biologist Demographics in the US." Accessed June 27, 2021. www.zippia.com/wildlife-biologist-jobs/demographics/.

Winifred B. Kessler and Selma N. Glasscock

2

Trailblazers

Women Who Forged a Path in Wildlife History

*Appropriate education of the two sexes, carried as far as possible, is a consumma-
tion most devoutly to be desired; identical education of the two sexes is a crime
before God and humanity, that physiology protests against, and that experience
weeps over. . . . It cultivates mediocrity, and cheats the future of its rightful legacy
of lofty manhood and womanhood. It emasculates boys, stunts girls; makes
semi-eunuchs of one sex, and agenes of the other.*
　　　　—E. H. Clark, "Sex in Education or, a Fair Chance for Girls," 1873

These days the opening quote is laughable; we know there are no biological reasons why women and men should not have the same educational
opportunities. But the view presented is key in the history leading up to women's
entry into the wildlife profession. In this view, women's purpose and highest
calling was home and family. Women had no purpose in seeking higher education; in fact, it threatened their well-being and society's too.

These widespread beliefs gave rise to two parallel streams in this history.
One stream, the progression of individuals who made their mark in history as
professional scientists, is notable for its lack of women. We'll see that, over the
centuries, various social, cultural, and economic factors conspired to exclude
all but a few women from achieving professional acclaim through scientific
study, discovery, and academic pursuits. In contrast, women are prominent in
the second stream, individuals who advanced society's understanding of nature through personal study, collecting, observation, drawing, and writing. The
terms "amateur," "naturalist," or "hobbyist" are often applied to such people.
But as we'll see, the women so engaged made enormous contributions to the
knowledge base in the fields of botany, zoology, ornithology, and more.

We begin our historical tour by exploring why so few women stand among the famous men of science and how special their accomplishments were considering the circumstances faced. Next, we examine the "golden age" of natural history study that prevailed for three centuries in Britain and spread across the ocean to produce some of America's finest women naturalists and biologists. Arriving at the 20th century, we consider the male-dominated origins of the wildlife profession and how women have come into their own.

Famous Women Scientists, a Sparse History

The history of science includes substantial contributions by women, but they are less well known and celebrated than contributions by men (Abir-Am and Outram 1989). One of the earliest, Hypatia (born around 350–370 and died 415 CE), lived in Alexandria, which was then part of the Eastern Roman Empire. Like her father, she was a university professor and taught mathematics, philosophy, and astronomy. Considered the first female mathematician whose story is reasonably well known, her wisdom and scholarly work were celebrated in her own lifetime (Deakin 2007). A pagan and tolerant of other religions, she was murdered by early Christian leaders who regarded her philosophical views as heretical.

Through the ages, religious traditions often barred or discouraged women's intellectual aspirations. According to ancient Hebrew and early Christian teaching, women were disruptive to the intellectual process, and contact with them made men unfit for serious intellectual endeavor (Schiebinger 1993). In the Middle Ages, professors at such universities as Oxford and Cambridge were forbidden to marry; the requirement for celibacy lasted until the late 19th century (Schiebinger 1993). Women's scientific contributions in this time came almost exclusively from nuns. Hildegard von Bingen (1098–1179), known today as Saint Hildegard, was a Benedictine abbess in what is present-day Germany who authored theological, botanical, and medicinal texts (Engbring 1940). She also composed religious music and pioneered aspects of cosmology, and some regard her as the founder of natural history science in Germany. Hildegard's nine-volume work, *Physica*, recorded scientific and medicinal properties of plants, fish, reptiles, and other animals. Her *Causae et Curae* recorded findings about human anatomy, diseases and treatments, and connections to the natural world.

The church's domination of Western learning waned in the 17th century with the emergence of modern science and new institutions such as the Royal Society of London, Académie des sciences in Paris, and Akademie der Wissenschaften in Berlin. The admission of women to these academies would take another 300 years (Schiebinger 1993). It wasn't for lack of qualifications or effort on the women's part. Acclaimed German astronomer Maria Winkelmann (1670–1720) shared fully in her husband Gottfried Kirch's work. She discovered a comet, published many important works, and enjoyed a reputation of excellence in her field. Maria petitioned to assume her husband's position at the Berlin Academy of Sciences after Gottfried's death in 1710, but the academy's executive council refused in order to avoid the precedent of appointing a woman (Schiebinger 1993).

Not until the 18th century did a woman attain a teaching position at a European university. Laura Bassi (1711–1778) was born into a wealthy Italian family, married Giuseppe Veratti, a doctor of philosophy and medicine, and bore 12 children. She received a doctoral degree in philosophy from the University of Bologna in 1732, making her the world's second woman to receive a doctorate, behind Elena Cornaro Piscopia in 1678. Starting as a professor of physics in 1732, she became the highest salary earner at the University of Bologna. In 1776 at age 65 she was appointed the chair of experimental physics at the Bologna Institute of Sciences, with her husband serving as teaching assistant (Schiebinger 1993).

Maria Mitchell's (1818–1889) accomplishments as an astronomer, librarian, naturalist, and educator made her one of America's most celebrated women scientists of the 19th century. While still a teen she performed complex navigational computations for whaling expeditions and became the first librarian for the Nantucket Atheneum. She was best known for her work in astronomy, including the discovery of a comet in 1847, for which she was awarded a gold medal from King Frederick VI of Denmark. Mitchell was the first woman elected to the American Academy of Arts and Sciences, to join the American Association for the Advancement of Science, and to teach astronomy at the university level. Beloved by her students, she taught at Vassar until 1888 when failing health necessitated her retirement. Kohlstedt (1978) suggested that Mitchell may have been the only American woman to support herself in scientific employment, and she achieved international recognition in the 1850s. She was an activist for women's rights and promoted the involvement of women in science.

Why So Few?

These examples demonstrate that there have always been women with the passion and intellectual capacity to excel in science. That they rarely succeeded was a result of the social barriers to the serious pursuit of their interests and achievement of recognition. Those who succeeded did so through *"ingenuity, careful planning, and a willingness to defy social conventions"* (Rositter 1989, xii).

We have already noted that various religious traditions held women to be incapable of, and a distraction from, serious intellectual endeavor. Such biological determinism existed in the ancient world of Hippocrates, Aristotle, and Galen. Aristotle argued that women are *"colder and weaker than men . . . and do not have sufficient heat to cook the blood and thus purify the soul"* (Schiebinger 1993, 15). Biological determinism was central to the leading hypothesis in the 18th and 19th centuries that men's more powerful brains explained their greater intellectual prowess. Harvard doctor Edward Clark (1820–1877) argued that allowing women into universities would cause *"great cost to their reproductive development"* because *"exercising their brains caused women's ovaries to shrivel, nervous collapse, hysteria, and sterility"* (Schiebinger 1993, 16).

Clark's claim, although outrageous today, needs to be examined in the context of Western social and cultural norms that prevailed for much of history. Many societies maintained the place of women to be hearth and home, with reproduction and family care their highest calling. In this view, intellectual pursuit was not only unnecessary for women; it could do harm by distracting or detracting from their primary calling in the domestic sphere. The outward-looking, objective, intellectual sphere occupied by men sharply contrasted with the domestic domain of women. These norms created a steep uphill path for women, who had to defy cultural conventions to participate in serious intellectual activity.

Another reason for the scarcity of famous women scientists relates to gender-based differences in how scientific work was credited and acclaimed. Many of the women who rose to prominence in the sciences did so by working with a male mentor such as a father, husband, or brother. These collaborations provided women with access to labs, equipment, and scientific literature they wouldn't otherwise have had. The integration of their accomplishments with their mentor's provided a sort of protective cover from society's disapproving eyes, but they seldom received the recognition their contributions merited (Abir-am and Outram 1989). In the long history of male-dominated science, individuals and institutions alike were loath to acknowledge the accomplishments of women and give credit where deserved. For example, Pierre Curie often had to emphasize to colleagues the status of Marie Sklodowska Curie as a full and essential partner in the couple's research accomplishments. Even after Pierre's death at a young age, peers continued to underplay Marie's groundbreaking contributions to the study of radiation. Despite Marie Curie's distinction as the first person ever to win two Nobel Prizes, she was denied membership in the Académie des sciences in Paris. The reason given was that it was *"eminently wise to respect the immutable tradition against the election of women"* (Schiebinger 1993, 14).

The Golden Age of Nature Study

Having examined why so few women rose to fame as mainstream scientists, we now focus on the second historical track leading to the wildlife profession. This track, natural history, involves the investigation of plants, animals, minerals, and other aspects of the natural environment. Nature study was a very popular pastime among the leisure classes of Britain and some European countries in the 17th to 19th centuries and would give rise to scientific disciplines such as botany, ornithology, and geology. While pursued by both sexes, more is known about the natural history accomplishments of men (Norwood 1993). This seems a paradox because those centuries were a period of widespread and intense scientific endeavor by women, but these women remained largely anonymous.

The "golden age" refers to the period of about 1650 to 1850, when it was common in Britain for leisure-class women to engage in nature study. The socially imposed mandate for domesticity meant that girls and young women in middle- and upper-class families were educated at home, whereas young men typically attended universities. The education received differed greatly between the sexes. Universities such as Oxford and Cambridge provided a classical education based on the works of such Latin and Greek luminaries as Aristotle, Homer, and Virgil. Society considered an education in the classics to be an *"intrinsically masculine avocation"* that constituted *"the only worthwhile field of scholarship for men of gentle birth and good breeding"* (Phillips 1990, ix). In contrast, this study was deemed unbecoming of the fair sex, and the female mind unfit for grasping and debating the classics. There was also concern that women's morality could be easily corrupted by what they might read.

At play too was a social bias against science, considered intellectually trivial compared to the classics, which explored the fundamental truths of human existence. This diminished view of science gave women a social license to pursue it as a hobby. Such study was thought to encourage domesticity, godliness, and virtue and to fill idle time that might otherwise invite flightiness and moral wandering. Expressing the thinking of the time, Rev. John Bennett wrote that women could *"safeguard their womanliness, their beauty, and their social graces by dabbling gently in scientific matters, specifically natural history, geology, and botany"* (Phillips 1990, 179). It was also believed that this study of science, unlike the classics, did not endanger the more fragile female brain.

Thus, science and nature study became a popular pastime for leisure-class women during the 17th and 18th centuries, as many sought intellectual fulfillment in the study of botany, zoology, geology, astronomy, and other branches of natural history. A form of "science tourism" arose as women of means and unburdened by family obligations (often by choice) traveled to the Americas and other parts of the world to pursue their interests (Guelke and Morin 2002). Although nearly all these women were born into wealth, Jeanne Baret (1740–1807) was an exception with an extraordinary story (Ridley 2010). Born a peasant in France, Jeanne's family encouraged her interest in plants, and she built a reputation as an "herb woman," or practitioner in medicinal botany. Her live-in lover, Dr. Philibert Commerçon, was appointed to the botanist position aboard the French Navy's first expedition to circumnavigate the globe. Jeanne joined the expedition disguised as a man, bunked with Commerçon, and shared in his botanical work. The plan worked until the expedition reached Mauritius, where Jeanne's true sex was discovered, and the couple was put ashore. Eventually, Jeanne returned to France as the first woman to circumnavigate the globe and was awarded a pension for her services in collecting botanical specimens.

Botany, the Feminine Science

Botany was part of the education girls received at home, and personal study was regarded "*appropriate to a feminine ideal and thus acceptable for young women, wives, and mothers*" (Shteir 1989). Women's botanical activities included collecting, cataloging, drawing, and detailed observation. Some women branched out from collecting to author popular articles and books such as Priscilla Wakefield's (1751–1832) *Introduction to Botany* (1796) and Anne Pratt's (1806–1893) five-volume *Flowering Plants and Ferns of Great Britain* (1855). The German naturalist Amalie Dietrich (1821–1891) got her start supporting her husband's work. After fleeing what was an abusive relationship, she built a celebrated career as a collector of botanical, zoological, and ethnological specimens for the Museum Godeffroy in Hamburg. More than 30 plant and insect species were named after her by European naturalists (Ogilvie 1989).

Those examples were the exception, however, as most botanical accomplishments by women remained invisible. While many male botanists relied on the detailed studies, collections, and drawings completed by women, these contributions were seldom acknowledged in the lectures, scientific publications, and university textbooks comprising "men's" botanical work. We will see this pattern again in the section on women naturalists in the United States.

Science Goes Mainstream

In response to women's popular pursuit of science, some publishers began to produce periodicals and books targeting women. The *Ladies Diary* commenced publication in 1704 and was promoted as a compact volume that ladies could tuck into their handbags. The full title of this periodical was *The Woman's Almanac . . . Containing many Delightful and Entertaining Particulars peculiarly adapted for the Use and Diversion of the Fair Sex* (Phillips 1990, 98). This periodical had substantial science and mathematics content, including math puzzles indicating *"that stereotypes about the inability of women to understand and enjoy mathematics were less strongly believed in the 18th century than they are today"* (Perl 1979, 1).

Many women desired to expand their engagement but were denied entry to the professional and scientific academies. The British Association for the Advancement of Science, established in 1831 in York, initially excluded women from both membership and participation. After a few years it allowed women to attend lectures and other events, but the heavy turnout created resentment that women were monopolizing association activities. The first woman was admitted as a member in 1853.

End of the Golden Age

By the mid-19th century the British government saw economic advantage in providing applied and practical science education to the working classes, including girls and women. However, by around 1870, interest in science was in decline. Phillips (1990) attributed this to a growing sense that girls should not be denied the "superior," classical education provided to boys. And in fact, girls aspiring to higher education were disadvantaged by their science backgrounds; the universities favored applicants with a classical education.

Two additional forces spelled the end of women's golden age in science. As science matured from a leisure-time pursuit to a profession, women were excluded from science careers and paying jobs that society deemed to be men's domain. And by the early 20th century, women's education was being blamed for declining domesticity, falling birth rates, and other social problems. In response, "domestic science" subjects such as cookery, housekeeping, and household hygiene were introduced into public schools in lieu of science courses. Phillips (1990, 257) summed the demise of the golden age: *"Just as science was poised to become a respectable profession . . . women were finally excluded."*

American Women: Nature, Science, and Adventure

As noted in the previous section, many women of means traveled from Britain and Europe to America to study plants, birds, shells, and other natural history subjects. Often these travelers perceived American women as being isolated from nature, ignorant of the flora and fauna around them, and "*extremely domestic*" (Norwood 1993, 5). Despite these perceptions of visitors, America had its share of accomplished female botanists, malacologists, ornithologists, and others who pointed the way for today's women of wildlife. They worked within societal norms that accepted nature study as the "*middleclass women's salvation from aimless leisure*" (Norwood 1993). As in England, botany was deemed the most acceptable subject for women. An 1871 essay by naturalist Wilson Flagg in the *Atlantic Monthly* stated, "*Women cannot conveniently become hunters or anglers, nor can they without some eccentricity of conduct follow birds and quadrupeds to the woods . . . [T]he only part of natural history which they can pursue out of doors is the study of plants*" (Nicholson 2009, 40).

The lives of these women are explored in Marcia Bonta's books *Women in the Field: America's Pioneering Women Naturalists* (1991) and *American Women Afield: Writings by Pioneering Women Naturalists* (1995), and in Bronstein and Bolnick's (2018) examination of women who published in the *American Naturalist* during its first 50 years (1867–1916). Bonta noted the challenge of researching works by early women naturalists: "*Neither the women nor their friends and colleagues had thought their papers important enough to save*" (Bonta 1991, xii). Few of these women held professional positions; rather, they conducted the work on their personal time and turned over their notes and drawings to men to publish. Most were regarded as amateurs, regardless of the depth of their work or the scientific training they may have brought to it. Bonta (1991) found that many of these pioneering women shared the trait of not having families to care for. Of the 25 women featured in her book, 11 never married, 5 began careers after the early death of a husband, 2 married in middle age after completing much of their scientific work, and 6 married supportive men in a similar field, whereas 2 were discouraged in their work by spouses. Only 2 had children. Bonta also noted that these women tended to form mutual support groups, writing to one another about their work and the frustrations they encountered. Most were passionate and driven; their work was their life.

Home, Family, and Nature Study

The social norms prescribing women's lives in colonial times were just as strong in America as in Britain and Europe. Such activities as gardening, drawing and painting, keeping nature journals, and writing poems and essays were compatible with the "home and hearth" mandate imposed on colonial women. The colonial botanist Jane Colden (1724–1766) learned botany from her father, who described it as "*an amusement which may be made agreeable to the ladies, who are often at a loss to fill up their time*" (James et al. 1971, 357). Colden collected and described flora near her home in the lower Hudson River Valley of New York. She was an early adopter of Carl Linnaeus's classification system, and her work was widely respected by male botanists of the time. Her botanical work ceased when she married in 1759, and she died in childbirth at age 41.

Susan Fenimore Cooper (1813–1894), daughter of the novelist James Fenimore Cooper, remained unmarried in the family home, where her interest in nature and writing were encouraged (Norwood 1993). She studied botany and zoology as well as languages and the arts while the family lived abroad in Europe from 1826 to 1833. They resettled in Cooperstown, New York, in 1936, and Susan kept a journal about her explorations of local gardens, woods, and fields. It was published in 1850 as the highly popular book *Rural Hours*, making her the first woman among renowned US nature writers such as Ralph Waldo Emerson and Henry David Thoreau. In *Rural Hours* she reported declines in numbers of deer (*Odocoileus virginianus*), bear (*Ursus americanus*), beaver (*Castor canadensis*), otter (*Lontra canadensis*), fish, and game birds, and she advocated for laws to protect wildlife. While a pioneer in some respects, she fully endorsed the ideology of domesticity. In *Rural Hours* she wrote, "*Home, we may rest assured, will always be, as a rule, the best place for a woman; her labors and interests, should all centre there, whatever be her sphere of life*" (Cooper 1850, 161–62).

Maria Martin (1796–1863) is considered one of the best nature artists of the 18th century. She provided childcare and household work for her sick older sister, Harriet Bachman, and married her brother-in-law, John Bachman, after Harriet's death. She was taught to paint by John James Audubon, who used her illustrations in *The Birds of America* and other works. Mary Treat (1830–1923) commenced scientific studies through collaborations with her husband, entomology professor Joseph Burrell. Her initial focus on botany expanded to include ornithology and entomology, and she conducted detailed observation and experimentation on spiders, ants, wasps, birds, and insectivorous plants. Mary Treat worked locally in New Jersey, and her most popular work, *Home Studies in Nature*, encouraged women to study nature close to home. Her scientific output, including five books and 76 popular and scientific articles, enabled her to support herself after separating from her husband in 1874.

Animals and Adventure

While most of the early women naturalists conducted their work close to home, Martha Maxwell (1831–1881) ventured farther afield (Dartt 1879) She was a naturalist, artist, and taxidermist who pioneered the display of lifelike animal specimens in dioramas depicting their native habitats (Schantz 1943; Benson 1986). These displays promoted public awareness and interest in native wildlife. She is recognized as the first woman to collect and prepare her own specimens, and her taxidermy methods were well ahead of their time.

Like Martha Maxwell, Annie Montague Alexander (1867–1950) defied convention by pursuing an adventurous life dedicated to study and collection of animal specimens and fossils (Stein 2001; Slavicek 2004). A woman of means, she circumvented barriers against women in the field by financing the expeditions she wished to participate in. In fact, she bankrolled the establishment of the Museum of Vertebrate Zoology and the Museum of Paleontology, both at the University of California, Berkeley, thereby solidifying her role in decision making for the museums' operations and specimens (Stein 1996). Her impact is evidenced by the naming of at least 17 species in her honor.

Ornithology

Compared to botany, ornithology was a more difficult field for women to enter because they were "*considered too delicate to tramp around the field and shoot birds, which is how most ornithological field study was done in the 19th century*" (Bonta 1991, 181). Mostly, women observed and sketched birds close to home or studied skin collections in museums. Work was done on an amateur basis, and paid positions would remain scarce until the 1970s (Ainley 1989). Nevertheless, the personal efforts of women were significant, and some gained respect and recognition within the ornithological community.

Quaker, teacher, and illustrator Graceanna Lewis (1821–1912) studied skins at the Philadelphia Academy of Natural Sciences under renowned ornithologist John Cassin. She became a member of that academy and the Delaware County Institute of Science, and in 1869 she published *The Natural History of Birds*. Avid bird watcher Harriet Mann Miller (1831–1918, pen name Olive Thorne Miller) wrote articles for popular magazines and, later, books such as *Bird Ways* (1885). She also wrote for the Audubon Society's journal and advocated to stop the slaughter of birds for the plume trade (Merchant 2005). In 1903 Harriet Miller, Mabel Osgood Wright, and Florence Merriam Bailey were the first women elected to membership in the American Ornithologists' Union. Bailey's primary interest was bird research (Dunlap 1996a), but she believed in educating women and children about the value of birds and thought bird study should be taught in schools (Kofalk 1989).

Mabel Osgood Wright (1859–1934) wrote prolifically about children, nature, outdoor life, and birds. In 1895 she published *Birdcraft*, a field book on song, game, and water birds that is considered a prototype of today's field guides (Welker 1971). She served 11 years as associate editor for *Bird-Lore* and was the founding president of the Connecticut Audubon Society. In 1914 she established America's oldest private sanctuary for songbirds, Birdcraft Sanctuary.

Harriet Miller, Mabel Wright, and others were considered "dabblers" and "bird lovers" by some women ornithologists, such as Althea Rosina Sherman (1853–1943), who published their work in the *Auk*, *Wilson Bulletin*, *Condor*, and other scientific journals (Bonta 1991). Illustrator, teacher, writer, and self-taught ornithologist Althea Sherman studied avian behavior and life cycles for 38 years. Her more than 70 publications were used extensively by Arthur Cleveland Bent in his series, *Life Histories of North American Birds*. She belonged to 15 scientific societies, including the American Ornithologists' Union (AOU), but lamented, "*I have said and believed it, that no woman will ever be made a fellow of AOU. . . . No, man's nature must change before a woman is a fellow*" (Bonta 1991, 204). She attracted criticism in the scientific community by advocating protection for "good birds" but elimination for "bad species" such as predators and pests (Boyle 2009).

Cordelia Stanwood (1865–1968) was trained in teaching and art but took up birdwatching at age 39 following a nervous breakdown (Bonta 1991). For the next 50 years she completed detailed studies of birds, writing for both popular magazines and scientific journals. She has additional distinction as a pioneer in bird photography. She was willing to sit in blinds for hours to capture exceptional photos for her articles, and these pictures were in demand for books by others. Another self-taught naturalist, Amelia Laskey (1885–1973), made major contributions to the field of ornithology (Bonta 1991). Her interest awoke at age 43 when she attended her first bird club meeting in Nashville, Tennessee. Amelia joined the Tennessee Ornithological Society and soon was writing for their journal, the *Migrant*. Her study of mockingbirds (*Mimus polyglottos*) commenced in 1929 and ran for 30 years. After acquiring a federal banding permit in 1931, Amelia began long-term studies of the eastern bluebird (*Sialia sialis*), tufted titmouse (*Baeolophus bicolor*), and northern cardinal (*Cardinalis*

cardinalis); by 1934 she had banded 3,734 birds of 69 species. She pioneered the rehabilitation of injured birds, use of bluebird boxes in research, and monitoring of bird casualties at airports and television towers. In an early example of citizen science, she enlisted volunteers for her research on chimney swifts (*Chaetura pelagica*). Some of her banded birds were recovered in Peru in 1943, providing the first record of where chimney swifts migrate for the winter. Amelia Laskey published her long-term studies in prestigious journals and, in 1966, was elected a fellow of the AOU.

Conservation Activism

Women in the early 20th century were excluded from emerging professional fields such as forestry and ornithology (Phillips 1990). In contrast, the "progressive conservation crusade" of this period was largely driven by women who turned to social and political activism to advance their engagement in nature and environmental issues (Merchant 1984, 57). Women advocated for many causes, including preservation of public lands for wildflowers and wildlife; humane treatment of domestic, laboratory, and wild animals; and protection of forests, waterways, and birds (Norwood 1993).

Examples abound of women who advocated tirelessly for the preservation of natural places and resources. Laura Lyon White (1839–1916) founded the women's California Club and led a successful national movement to save the state's iconic redwood (*Sequoia giganteum* and *S. sempervirens*) groves. Women were also instrumental in creating forest reserves in Florida. In Pennsylvania, their lobbying efforts succeeded in creating the Pennsylvania Bureau of Forestry. A key activist among them, Mira Lloyd Dock (1853–1945), was the first woman appointed to the Pennsylvania Forest Commission (Rimby 2005).

Women were prominent in the crusade to protect birds against the ravages of the plume industry, which was driven by the demand for feathers to decorate women's hats and fashions. So great was the demand that snowy egrets (*Egretta thula*), great egrets (*Ardea alba*), and other species were being driven to extinction. In what Merchant (1984) termed the "Audubon movement," women led the formation of local Audubon clubs and state societies to advocate for the protection of wild birds. Harriet Hemenway (1858–1960) founded the Massachusetts Audubon Society and together with her cousin Minna Hall led a boycott of feathered fashions and advocated for the protection of birds. Their efforts paved the way for the 1913 passage of the Weeks-McLean Act, cosponsored by Massachusetts representative John Weeks and Connecticut senator George McLean, which outlawed the plume trade and other forms of market hunting (Souder 2013). Also known as the Migratory Bird Act, the law was a key landmark in the early conservation movement.

Women were active participants in the conservation congresses of the early 20th century, where according to Merchant (1984, 73), they drew on a *"trilogy of slogans"* to rationalize conservation activism as an appropriate calling for women. The first slogan, *"conservation of true womanhood,"* portrayed conservation as fundamental to the ideals and ethics defining a woman's life. *"Conservation of the home"* emphasized connections of abundant natural resources and a clean environment to the health and happiness of the family. *"Conservation of the child"* conveyed that women, as givers of life, should be concerned with the well-being of future generations. While many women conservationists also fought for women's rights, some anti-suffragists used the slogans to argue that the burden of voting would sap women's strength and detract from *"the great work Nature has given her to do"* (Merchant 1984, 76).

The mid-20th century saw the emergence of diverse environmental movements with women in leading roles. Margaret Thomas "Mardy" Murie (1902–2003) has been called the grandmother of the conservation movement in recognition of her tireless environmental work, including the 1950s campaign to protect 8 million acres of Alaskan wilderness in what is now the Arctic National Wildlife Refuge (Nicholson 2009). Rachel Carson (1907–1964) is best known for her groundbreaking book, *Silent Spring* (Carson 1962), which brought to light the science proving the perils of widespread use of agricultural pesticides. That contribution stands among the seminal works on ecology and human relationships to the natural world.

Women in the Emerging Field of Wildlife, 1930s–1970s

By the 1920s the progressive conservation movement—in which, as we have seen, women played a large role—was well established and realizing gains. However, its leaders realized that protective laws alone would not be enough to reverse the declines in America's wildlife resources (Organ et al. 2010). Aldo Leopold and colleagues put forward a broader vision for wildlife restoration in the 1930 document *American Game Policy* and proposed that the work be carried out by trained biologists (American Game Protective Society 1930). Thus was born the wildlife management profession, which subsequently gave rise to new university programs and curricula, US Cooperative Wildlife Research Units, and professional societies (Organ et al. 2010).

The premier society for wildlife management, The Wildlife Society (TWS), was founded in 1937 by the pioneering wildlife professionals of that time, who were all men. More than a decade passed before TWS had its first female member, Elizabeth Beard Losey (1912–2004), born Elizabeth Browne in East Orange, New Jersey. While earning her bachelor of arts degree in English, she developed a deep interest in waterfowl and became the first woman to receive a master of science degree in wildlife management and conservation from the University of Michigan. Despite this advanced degree, the Michigan State Game Division informed Elizabeth that *her hiring would be inappropriate because the attendant field work would involve long hours and possible overnight stays with male associates*" (Casselman 2006, 558). Instead, she became the first female wildlife biologist of the US Fish and Wildlife Service (FWS). She also taught wildlife ecology and game management courses at the University of Michigan and joined TWS in 1948. She left the FWS after her marriage in 1953 but continued to keep meticulous notes of birds and other natural history subjects.

For nearly half a century, the wildlife profession remained an unconventional career path for women. Many shared in the work of their husbands as unpaid assistants whose contributions remain anonymous. One such example is Joyce McTaggart Cowan (1912–2002), wife of famous Canadian wildlife biologist Ian McTaggart Cowan. Joyce's father was Kenneth Racey, an avid naturalist and collector of birds and mammals, who involved the entire family in his outdoor pursuits (Penn 2015). Joyce brought to her marriage a full set of field and lab skills that enabled her full (though unpaid) participation in Ian's wildlife and zoological work. Ian freely credited Joyce for her contributions, but as an uncredentialed participant, she did not receive professional recognition.

Other early women of wildlife obtained credentials that enabled them to work on a par with their husbands. Fran Hamerstrom (1907–1998) was the only female graduate student supervised by Aldo Leopold. She and her life partner, Fred Hamerstrom, completed groundbreaking work on prairie chickens (*Tympanuchus cupido*), northern harriers (*Circus cyaneus*), great horned owls (*Bubo virginianus*), Harris hawks (*Parabuteo unicinctus*), and other species. Her legacy includes a dozen books, more than 100 scientific papers, conservation milestones, and countless stories about her colorful personality (Box 2.1).

Elizabeth "Libby" Schwartz (1912–2013) earned a doctorate in zoology from the University of Missouri, where she met and married fellow biology student Charles Schwartz (*Columbia Tribune* 2013). Libby's research on box turtles (*Terrepene carolina*) and other species generated interest in the ecology and conservation of nongame animals. Charles and Libby had joint careers in the

BOX 2.1 Frances Carnes Flint Hamerstrom 1907–1998

Is *she* coming too?

Frances "Fran" Carnes Flint Hamerstrom was often described as flamboyant, spontaneous, daring, and eccentric, and she left an indelible impression on everyone she met. Personality aside, she was an outstanding researcher, biologist, and passionate conservationist. She and husband Frederick Nathan Hamerstrom Jr. ("Hammy") pioneered and perfected greater prairie chicken (*Tympanuchus cuipido pinnatus*) research and conservation in Wisconsin.

Fran (rhymes with "brawn") was born to a wealthy Bostonian family. Her father, an international criminologist, moved the family to Europe, where the young Fran learned five languages living in seven countries before returning to the United States in 1914 (Hamerstrom 1994). Fran's father, a strict disciplinarian, often disapproved of her actions. Her heiress mother loved to read and tried to make the best of any situation. A governess educated Fran and her brother Bertram in academics, languages, and social graces, and prepared Fran for her studies at the Milton Academy (Hamerstrom 1994). Later, according to Fran, she flunked out of Smith College in favor of birds, nature, ornithology, hunting, "*raising hell*," boys, and fashion (Mandula 1997; Corneli 2002; Kessler 2012).

A chance date (Corneli 2002) ignited Fran and Hammy's 68-year love affair and shared passion for hunting and wildlife research, lasting until Hammy's death. The steadfastly independent, outspoken, and spontaneous Fran delighted in shocking people with her personal stories. Hammy, in contrast, was quiet and reserved. Their lives were so intertwined that a former apprentice described them as a "*super organism*" (Gawlik and Anderson 1998, ii).

Wanting a life centered on hunting and wildlife management, Fran and Hammy enrolled in the Game Conservation Institute in Clinton, New Jersey (Kessler 2012). When this trade school failed to meet their goals, they left to study wildlife management under Paul Errington at Iowa State College, where Hammy obtained a master's and Fran an undergraduate degree (Mandula 1997). Subsequently, Hammy was accepted by Aldo Leopold at the University of Wisconsin for a PhD study on greater prairie chickens. Fran assisted with the project but independently initiated a behavioral study of color-banded chickadees (*Poecile* spp.), which earned her a master's degree and the designation of the only woman to obtain an advanced degree under Leopold (Johannsen 2017).

Settling in Plainfield, Wisconsin, the Hamerstroms commenced a half century of research on the greater prairie chicken. Their old farmhouse served as home, research lab, and lodging for countless scientists and thousands of "gabboons" (Fran's name for volunteers) who spent freezing mornings in blinds to record prairie chicken behavior (Kessler 2012).

The Hamerstroms' lifelong collaboration yielded 238 research papers and numerous reviews. They communicated with ornithologists worldwide (Mayr 1992), and Fran translated more than 100 ornithological articles and books into reviews for US scientists (Bildstein 1999). Together they created the first recovery plan for prairie chickens and established the Society of Tympanuchus Cupido Pinnatus, Ltd. (Kessler 2012). Devoted teachers, they contributed immensely to developing the next generation of field biologists through their mentoring and apprenticeships (Bildstein 1999)

Following Leopold's lead, Fran shared her knowledge in popular articles as well as scientific journals. Fran independently published more than 150 scientific papers, 12 books, and numerous popular articles, many focusing on greater prairie chickens, sharp-tailed grouse (*Tympanuchus phasianellus*), and northern harriers (*Circus cyaneus*) (Gawlik and Anderson 1998). Her popular works include hunting, wild game cooking, and personal stories. In one such book, *Is She Coming Too?*, Fran described a quail-counting field trip that Errington arranged for her and Hammy. The game warden assigned to guide them asked Hammy, "*Is* she *coming too?*" (Hamerstrom 1989, 84). Hammy's answer: "*Yes.*" Fran experienced many such incidents.

Fran's many awards included an honorary PhD from Carroll College in Waukesha, Wisconsin (Bildstein 1999), The Wildlife Society's Publication Award (1940, 1958), and the National Wildlife Federation Distinguished Service Award (1970) (Gawlik and Anderson 1998).

After Hammy's death, Fran traveled alone in her 80s to Saudi Arabia, Africa, and South America to study the hunting techniques of native peoples (Paulson 2013). While exploring the Amazon by dugout canoe (Hamerstrom 1994) at age 86, Fran broke her hip and had to be evacuated but returned to Peru the next year to continue her studies (Kessler 2012). Those of us who knew Fran never doubted that her final years would be filled with adventure. She lived life on her own terms. No doubt she would celebrate knowing that, yes, we are coming too!

Missouri Department of Conservation and traveled widely to produce books and films about wildlife. They received 23 awards for their 24 films. Their 1959 *The Wild Mammals of Missouri* is the best-selling book ever published by the University of Missouri Press. Libby completed the second revised edition of this book in 2001, a decade after Charles's passing.

Lucille Stickel (1915–2007) had a major influence on the developing field of wildlife management through her pioneering work in wildlife toxicology (Coon and Perry 2007). Her illustrious career at the FWS Patuxent Wildlife Research Center (where her husband also was employed) progressed from volunteer to junior biologist, to senior scientist, to director (1972–1980). Her work in pesticide contamination informed Carson's *Silent Spring* and earned her the distinction of the first woman to receive, in 1974, the Aldo Leopold Memorial Award from TWS (Box 2.2).

Breakthroughs in Field Behavioral Studies

Some women made their mark in the mid-20th century by inventing the study of animals in their natural habitats through prolonged and intensive observation. Perhaps "reinvent" is more accurate, since these methods mirrored the detailed observational studies carried out by women naturalists of preceding centuries.

Margaret Altmann (1900–1984) *"almost singlehandedly began the naturalistic study of the behavior of large animals in their normal habitat"* (Chiszar and Wertheimer 1988, 102). Margaret began life in Germany, earned a PhD in rural economics from the University of Bonn, and worked in farm management. She immigrated to the United States in 1933 and earned another PhD in animal breeding at Cornell University. She moved to Colorado in 1948, and for the next 36 years conducted detailed behavioral studies of moose (*Alces alces*), elk (*Cervus canadensis*), bison (*Bison bison*), and other large mammals in the field, often in rugged wilderness conditions.

Anne Innis Dagg (born 1933 in Toronto, Ontario) is credited with being the first to study animals in the wild in Africa (Peters 2019). Her path was set as a small child when she saw her first giraffe (*Giraffe giraffa*) at the Brookfield Zoo. Dagg graduated as a gold medalist in biology at the University of Toronto and stayed to complete a master's degree in genetics. She used her prize money to book passage alone to Africa, where against great odds she carried out detailed field studies of giraffe on private lands near Kruger National Park. Back in Canada she completed a doctorate in animal behavior at the University of Waterloo. Her book *The Giraffe: Its Biology, Behaviour, and Ecology* (1976) remains the definitive work on the species. Despite a strong academic record, Dagg was denied tenure at the University of Guelph and the University of Waterloo. She expanded her research interests to encompass feminist topics, including sexism in academia. Dagg has published extensively on giraffes, camels (*Camelus dromedarius*), primates, and various Canadian wildlife species, and is the author of 20 books. She has more than 60 refereed scientific papers on mammalian behavior, sociobiology, feminism, and gender bias in academia. She has sometimes been called "the Jane Goodall of giraffes," but because Dagg's pioneering

BOX 2.2 Lucille Farrier Stickel (1915–2007)

Listed in *American Men of Science*, she was among the highest ranking career women in the federal government.

—US NATIONAL WILDLIFE REFUGE SYSTEM, 2008

Lucille Stickel was not only a pioneer in the field of wildlife toxicology but also the first woman to attain the level of senior scientist in the US Civil Service System and head a major federal fish and wildlife agency. She directed the Patuxent Wildlife Research Center from 1973 until 1981, when she retired (Perry 2016). Stickel's research on DDT's impact on wildlife and food chains was integral to Rachel Carson's development of *Silent Spring* (US National Wildlife Refuge System 2008).

Stickel grew up in Hillman, Michigan, and was free to roam the fields, woods, and lakesides near her home. Like many biologists, these activities stimulated her interest in the natural world. Stickel graduated Phi Beta Kappa from Eastern Michigan University with a BA in 1936 and earned her MS (1938) and PhD (1949) in zoology from the University of Michigan (Coon and Perry 2007). In 1974 her alma mater, Eastern Michigan University, conferred upon her an honorary PhD (Wayne 2011).

Lucille and Bill Stickel married while both were attending graduate school at the University of Michigan. Bill was offered a wildlife biologist position in 1940 with the Civil Service Commission in Washington, DC, but transferred to the Patuxent Wildlife Research Center in 1941 (Coon and Perry 2007). Although Lucille was offered paid positions at Patuxent, she chose instead to volunteer there until 1943. At that time the United States was recovering from the Depression, and Lucille believed that family men were in greater need of jobs. In 1943 she was offered and accepted the position of junior biologist at the Patuxent Wildlife Research Center (Coon and Perry 2007).

Research on environmental contaminants increased under Stickel's leadership at Patuxent, with the number of publications on environmental contaminants averaging 30 per year in the 1970s (Perry 2016). She hired approximately 40 research scientists, who produced more than 1,000 scientific papers and books during that period, and some of whom attained leadership roles in the US Fish and Wildlife Service and the US Geological Survey (Coon and Perry 2007). Under her directorship, three Research Natural Areas were designated, and 712 hectares (1,760 acres) were transferred from the US Department of Agriculture to Patuxent (Dunkel et al. 1989). Studies of migratory birds on Patuxent lands during this period focused on forest fragmentation, population modeling, and development of statistical methodologies (Perry 2016).

Stickel's personal research focused predominately on toxic effects of DDT and other contaminants on wildlife populations. She published 44 scientific papers in this subject area (US National Wildlife Refuge System 2008; Perry 2016). Other contributions, including literature such as her "Wildlife Abstracts 1952–55/A Bibliography and Index of the Abstracts in Wildlife Review Numbers 67–83" (Stickel 1957), formed a cornerstone in the foundation of wildlife management. Lucille also studied small mammal populations and home ranges, and she and Bill conducted research on black rat snakes (*Pantherophis obsoletus*) and box turtles (*Terrapene carolina*) (US National Wildlife Refuge System 2008).

Stickel spent her entire professional career (1943–1981) at Patuxent. She was the first woman to receive the Aldo Leopold Memorial Award from The Wildlife Society (1974). She also received the Federal Women's Award in 1968, given by the US Civil Service Commission, as well as the Distinguished Service Award in 1973 from the Department of the Interior (Coon and Perry 2007) and the Society of Environmental Toxicology and Chemistry's Rachel Carson Award in 1998. Stickel appeared in the 2009 exhibit of the Michigan Women's Historical Center, *Resourceful Women: 30 Who Worked to Preserve Michigan's Water, Woods, and Wildlife*, and she was inducted into the Michigan Women's Hall of Fame in 2014 (Michigan Women Forward n.d.).

Together Lucille and Bill Stickel amassed 78 years of research at Patuxent. In 1989 the Biochemistry and Wildlife Pathology Laboratory building at Patuxent was renamed the Stickel Laboratory in their honor (Perry 2016).

---

work commenced a decade earlier, some have suggested that Jane Goodall (born 1934) is "the Anne Dagg of chimpanzees."

Like Dagg, Goodall's desire to work in Africa stemmed from a childhood fascination with animals (Gerber 2017). She began her field studies of chimpanzees (*Pan troglodytes*) with no higher education other than a secretarial certificate verifying her typing skills. She became a public sensation when her pioneering work on chimpanzees in the wild was featured in publications and films of the National Geographic Society. Goodall's impact stems from her novel field methods and discoveries that revolutionized primate science, and her enduring commitment to protecting chimpanzees and their habitat. Her successes paved the way for other women pioneers in primate biology and conservation, such as the groundbreaking research and conservation work of Dian Fossey (1932–1985) on mountain gorillas (*Gorilla beringei*), and the tireless work on orangutans (*Pongo pygmaeus*) by researcher and conservationist Biruté Galdikas (born 1946 in Wiesbaden, Germany).

Margaret Morse Nice's (1883–1974) paper on song sparrows (*Melospiza melodia*) was the first long-term field study of any free-living wild animal, leading Konrad Lorenz to deem her the actual founder of ethology (Bonta 1991, xiii). Although Nice's research produced many publications, rather than working in a professional position, she instead chose to support her husband and family in the traditional role of wife and mother. Her invention of the technique of color-banding birds opened the world of ornithology to the science of bird behavior (Dunlap 1996b) (Box 2.3).

The 1970s: Barriers Fall, Doors Open

In the United States, the decade of the 1970s saw a significant increase in the numbers of women hired in wildlife positions of federal agencies and other employers. The change was largely driven by new laws to protect civil rights and equal opportunity in employment.

Equal Employment Opportunity

Passage of the Civil Rights Act of 1964 (Public Law 88-352) was a critical step in removing barriers to women's employment. The act established the Equal Employment Opportunity Commission (EEOC), and Title VII of the act prohibited employment discrimination based on race, color, religion, sex, or national origin (EEOC 1964). While its enforcement powers were initially weak, the law signaled the end of agency policies that barred women from field and other male-dominated positions. Before this law, only rarely had women been hired in these jobs. Clare Marie Hodges (1890–1970) became the first paid female ranger for the National Park Service in 1918, when men were in short supply because of World War I (Kaufman 2006). In 1913 the US Forest Service (FS) broke new ground by hiring Hallie M. Daggett (1878–1964) as a fire lookout in Klamath National Forest (Holsinger 1983). Such hires were exceptions to the rule that women need not apply for field positions with the federal agencies.

Title VII was given teeth through its amendment in 1972 as the Equal Employment Opportunity Act (Public Law 92-261), which authorized the EEOC to sue in federal courts when there is reasonable cause to believe that employment discrimination has occurred based on race, color, national origin, sex, religion, age, disability, political beliefs, and marital or familial status. This act empowered women to fight back against discrimination based on their sex.

BOX 2.3 Margaret Morse Nice (1883–1974)

I have always felt that she, almost single-handedly, initiated a new era in American ornithology and the only effective countermovement against the list chasing movement. —MILTON B. TRAUTMAN, 1977

Margaret Morse Nice's college-educated mother and college professor father provided their seven children with a rich intellectual environment (Dunlap 1996b), including travel throughout the United States and to Europe (Contreras and Trautman 1997; Trautman 1977). One of Nice's most prized Christmas gifts was *Bird-Craft*, a book containing color illustrations of birds (Contreras and Trautman 1997), which helped to arouse her early interest in bird research. Her attention to detail was so great that she was able to utilize data she collected in her preteen years on bird migration, behavior, and fledgling success with data collected in her 70s for research purposes (Dunlap 1996b; Contreras and Trautman 1997). Ironically, her parents held the traditional view of women's roles being wives and mothers and discouraged Nice from seeking a professional career (Dunlap 1996b).

Nice obtained a BA in French at Mount Holyoke College, then traveled abroad developing skills in German, French, Italian, and Latin languages (Dunlap 1996b), which proved critical when she published her defining work on life history of the song sparrow, "Zur Naturgeschichte des Singammers," in the German periodical *Journal für Ornithology* (Nice 1933; Nice 1934; Contreras and Trautman 1997). This groundbreaking work, later published in the United States as "Studies in the Life History of the Song Sparrow" in the *Transactions of the Linnaean Society of New York*, brought Nice international recognition (Nice 1937).

Nice obtained a master of art degree in 1926 at Clark University on work she had accomplished in 1915 studying the food habits of bobwhite quail (*Colinus virgianus*) (Ogilvie 2018). She published her research results in the *Journal of Economic Entomology I* in 1910 (Contreras and Trautman 1997; Ogilvie 2018). While at Clark she met and married Leonard Blaine Nice, who received a PhD in physiology. Blaine's first job was at Harvard (Trautman 1977), but he held faculty positions at several other universities through his career. Margaret chose the traditional role of running the household over a professional career. However, she did so on her own terms and continued conducting research when her children's language development skills aroused her interest in child psychology. This research yielded 13 scientific papers (Dunlap 1996b).

Although Blaine's career moves, Margaret's domestic responsibilities, and the care of the couple's five daughters constrained her ability to conduct field studies of wild birds, the ever-persistent Margaret made time for research. She developed new techniques to study bird behavior, including her pioneering use of colored celluloid leg bands that permitted the identification of individual birds. Her groundbreaking methodology proved critical in the field of avian behavior (Dunlap 1996b) and opened a whole new opportunity for Nice to study birds in most any setting.

Margaret was successful in carving out a productive career for herself while fulfilling her familial responsibilities. Blaine was supportive of her research, helped fund her travel to meetings, and cared for the children while she conducted fieldwork (Dunlap 1996b), the entire family sometimes traveling and camping with her on research trips (Contreras and Trautman 1997).

Margaret's publications were sometimes unfairly judged because she lacked a PhD and professional employment. She expressed her frustrations with that, the burdens of housework, the lack of access and time for field opportunities, and the fact that she was considered a "housewife" even though she was trained as a biologist (Dunlap 1996b). Yet Margaret Nice published approximately 250 articles in scientific journals (including seven book-length works) and more than 3,000 reviews of other biologists' work.

Margaret received honorary PhDs from Mount Holyoke in 1955 and Elmira College in 1962. She was named the second woman Fellow of the American Ornithologists' Union and was awarded the AOU Brewster Medal in 1942 (Bonta 1995). Margaret held honorary memberships in ornithological societies of Britain, Finland, Germany, Switzerland, and the Netherlands (Trautman 1977), and her research was respected by such luminaries as Ernest Mayer, Konrad Lorenz (Dunlap 1996b), and Nikolaas Tinbergen (Trautman 1977).

Margaret remained active in the profession most of her life, slowing down in her 80s because of failing eyesight and other health issues (Trautman 1977). She died at age 90, and her autobiography, *Research Is a Passion with Me* (Nice, 1979), was published posthumously (Bonta 1995).

A famous and far-reaching example was the legal challenge brought by Gene C. Bernardi, a research sociologist who worked for the FS Pacific Southwest Forest and Range Experiment Station from 1968 to 1972 (Forest History Society 2011). In 1972 she filed a formal complaint of sex discrimination against the FS after being passed over for a promotion and pay raise. This escalated to a class-action suit alleging discrimination, filed in 1973 on behalf of all women in the US Department of Agriculture. The suit was resolved in 1979 by a consent decree stipulating that FS staffing in the Pacific Southwest Region must, within five years, be in line with the civilian labor force, with women in more than 43 percent of jobs in each series and pay grade. It also required training and opportunities to move more women into higher administrative positions.

The FS consent decree and other legal challenges showed the agencies that the courts were serious about enforcing equal employment opportunity laws. Not only were barriers removed, but women and other underrepresented groups were actively recruited through affirmative action policies and practices that sought to correct the dark legacy of employment discrimination. The change was most visible in the federal agencies, which actively recruited women into entry-level jobs and supported their career advancement. Slower but similar trends commenced in the state agencies, universities, and other employers. By the 1990s, the face of the wildlife profession had noticeably changed.

Emerging Leaders

Because most women began their careers in entry-level positions, years passed before they emerged in leadership positions of the agencies. The first female director of the US Fish and Wildlife Service, Mollie Beatty (1947–1996), was appointed in 1993 by President Bill Clinton (Gup 1996). Tragically, her tenure was short, as she died from a brain tumor just three years into the director's job. But her watch included the addition of 15 new national wildlife refuges to the system, completion of more than 100 conservation plans, and the reintroduction of the gray wolf (*Canis lupus*) into the Northern Rockies. Named in her honor, the Mollie Beatty Wilderness comprises 3.2 million hectares (8 million acres), or 40 percent, of the Arctic National Wildlife Refuge in Alaska.

The FS appointed its first woman to preside over the national wildlife program in 2005, the agency's centennial year. Anne Zimmermann's program responsibilities included wildlife, fish, watersheds, air, soils, and rare plants for the 154 national forests (covering about 76.2 million hectares, or 188.3 million acres) managed by the agency, and she served in that position for nine years. Anne's education included bachelor of science degrees in forestry and wildlife from Virginia Tech University and a master of science from Purdue University. The first woman appointed as FS chief, the agency's highest position, was Abigail R. "Gail" Kimbell. She had a forestry background and served from 2007 to 2009. In 2018, Vicki Christiansen was sworn in as the 19th person, and second woman, to hold this position.

The first woman elected to The Wildlife Society Council, the governing body of TWS, was Judith L. Tartaglia, a FS biologist. Tartaglia's council term spanned 1992–1995. The first woman president of TWS, Diana Hallett, was elected in 1999 for successive terms as vice president, president-elect, president (2001–2002), and past president. Beginning her career as the first woman game manager with the Wisconsin Department of Natural Resources (DNR), she returned to her home state of Missouri and became the first woman hired as a research biologist for the Missouri Department of Conservation (Gjestson 2013). She retired in 2003 as the resource science chief administrator for the Missouri DNR. A decade passed before the second woman, Winifred Kessler, was elected to serve as TWS president (2012–2013). She received the Aldo Leopold Memorial Award in 2017, only the second woman so honored since Lucille Stickel in 1974. The third woman elected, Carol Chambers, served as TWS's 75th president in 2020–2021.

It took time and perseverance to get this far, but the path ahead is bright for women of wildlife.

Reflections on Allies

The role of allies is an important and recurring theme in this book. But with few exceptions (usually a supportive father, husband, or the rare collaborator), the women profiled in this chapter had no allies in their fields of endeavor. That makes their contributions to science and conservation even more remarkable, considering the powerful social and cultural forces opposing them. The authors of this chapter feel a personal connection to these women's stories, having entered the wildlife profession when formidable social barriers remained. But fortunately, we both had the benefit of allies who offered encouragement, support, and opportunities to follow our passion.

One of us, Winifred Kessler, emphasized allies in her 2018 Aldo Leopold Memorial Award presentation (Kessler 2019). She highlighted the importance of unconventional faculty men who, against dire warnings from their male peers, gave her a solid start as their graduate student or research assistant. Equally important to Kessler was her bonding with the rare women who, like her, were forging their way in a male-dominated profession. These allies have remained an instrumental source of encouragement, support, and inspiration throughout her career.

The other, Selma Glasscock, experienced similar beneficial support from a select group of male professors, colleagues, and employers. Did she experience prejudice within the male-dominated profession? Yes, definitely. While not always overt, the demeanor of many revealed a deeply held conviction that women did not belong in the wildlife field, in any capacity. Like Kessler, Glasscock found and bonded with supportive women and men. This included her master's degree advisor, a young woman only slightly older than Glasscock herself. Those individuals, together with family, became Glasscock's foundation of support for navigating the rocky path to a wildlife career.

We encourage you, dear readers, to seek out allies who can support your journey toward personal and professional goals. And just as important, we hope you will become an ally to others who are seeking their place in this rewarding profession.

Lastly, we must stress that determination is key to any success. No matter the odds, if you are not determined to put your entire being into achieving your goals, then the greatest support team cannot get you there.

Discussion Questions

1. We've described historical barriers. Do these barriers exist today? What barriers do you see now?
2. Select three of the women discussed in this chapter (for example, the women featured in Boxes 2.1, 2.2, and 2.3), conduct your own search for historical information and pictures, and contrast the barriers they experienced and how they overcame them.
3. Identify women leaders in the wildlife profession you can talk with. How did they become involved in wildlife? Who were their mentors?
4. Describe three approaches that you can take to support women in the wildlife profession.

Activity: Traits for Success

Learning Objectives: Participants will learn more about the founding women of wildlife through mini research projects with a focus on their pathways and the unique traits that facilitated success. They will create and discuss lists based on their individual research. Participants will also reflect on their own strengths and weaknesses and develop a personal plan for success.
Time: Outside of class or group activity: 1–2 hr.
Large group or class activity: ~45–60 min.
Description of Activity:
Materials needed: Access to the internet.
Process:
Part 1: Individual Research (variable but likely 1–2 hr.)
Participants will be assigned or will select one of the women described in this chapter to further investigate and define traits the individual possessed that likely pushed them to succeed. Questions to investigate include:

1. How did character traits (i.e., communication skills, ability to interact/work with others, perseverance, creativity, attention to detail, organization, curiosity, integrity) support their success?
2. What roadblocks or challenges did they encounter, and how did they overcome them?
3. Who were their mentors and how did mentoring help these women overcome adversity? Did they then become mentors themselves?
4. Was it hard work for them to pursue their interests? Was self-sacrifice required to meet their goals?
5. What was their educational background? Did they receive professional training, or were they self-taught?
6. How did they rise above others of their time and receive recognition for their accomplishments?

Part 2: Table Comparison (15–20 min.)

After everyone has time for personal research, participants will form small groups of two or three, preferably with people who researched different women. Each group member will share what they learned from their research. The group will then create a list of accomplishments and shared traits that led to these women standing apart from other men and women of their time.

Part 3: Large Group or Class Comparison (depends on the number of small groups; estimate 2–4 min. per group)

After the small groups have met and created their list, provide each with time to report out and share their list with the larger group. As a large group or class, create another master list of similarities and differences in the women presented.

Part 4 (10–20 min.)

After the large group shares and creates the list, have everyone participate in a discussion.

Group or Class Debriefing and Discussion Questions

1. How did you feel learning about these women?
2. What did you learn?
3. What was one of the most striking differences in the women discussed?
4. What trait or characteristic was most notable for you or surprised you the most?
5. Was there a trait that you never thought of before as beneficial to a scientist or wildlife professional?

6. How would you summarize the role of mentoring in success for this group of women?

7. How would you summarize the role of education in success for this group of women?

8. How would you use the information you learned or your experience from this activity in your life as a wildlife professional, student, or citizen?

Part 5: Personal Reflection (10 min.)

Each person will now spend some time in quiet reflection writing notes and answering the following prompts:

1. What are some of your personal strengths or traits that will help facilitate your success?

2. What are one or two personal weaknesses that you will need to overcome to be successful?

3. Which one or two personal traits will you strengthen or improve, and how will you accomplish this?

Literature Cited

Abir-am, P. G., and D. Outram. 1989. *Uneasy Careers and Intimate Lives: Women in Science, 1789–1979*. New Brunswick, NJ: Rutgers University Press.

Ainley, M. G. 1989. "Field Work and Family: North American Women Ornithologists 1900–1950." In *Uneasy Careers and Intimate Lives: Women in Science, 1789–1979*, edited by P. G. Abir-am and D. Outram, 60–76. New Brunswick, NJ: Rutgers University Press.

American Game Protective Association. 1930. *A Proposed American Game Policy to be Discussed at the Seventeenth Annual American Game Conference, New York City, December 1–2, 1930*. New York: Jacques.

Benson, M. 1986. *Martha Maxwell, Rocky Mountain Naturalist*. Lincoln: University of Nebraska Press.

Bildstein, K. L. 1999. "In Memoriam: Frances Hamerstrom, 1907–1998." *Auk* 116(4): 1122–24.

Bonta, M. M. 1991. *Women in the Field: America's Pioneering Women Naturalists*. College Station: Texas A&M University Press.

Bonta, M. M. 1995. *American Women Afield. Writings by Pioneering Women Naturalists*. College Station: Texas A&M University Press.

Boyle, B. 2009. "Sherman, Althea Rosina." In *The Biographical Dictionary of Iowa*. University of Iowa Press Digital Editions. http://uipress.lib.uiowa.edu/bdi/.

Bronstein, J. L., and D. I. Bolnick. 2018. "'Her Joyous Enthusiasm for Her Life-Work . . .': Early Women Authors in The American Naturalist." *American Naturalist* 192(6): 655–63.

Carson, R. 1962. *Silent Spring*. Boston: Houghton Mifflin.

Casselman, T. 2006. "Elizabeth Beard Losey, First Female Member of the Wildlife Society." *Wildlife Society Bulletin* 34(2): 558.

Chiszar, D., and M. Wertheimer. 1988. "Margaret Altmann: A Rugged Pioneer in Rugged Fields." *Journal of History of Behavioral Sciences* 24(1): 102–6.

Clark, E. H. 1873. *Sex in Education or, a Fair Chance for Girls*. Boston: James R. Osgood. The Albert M. Greenfield Digital Center for the History of Women's Education. http://greenfield.brynmawr.edu/items/show/51.

Columbia Tribune. 2013. "Pioneering Conservationist Libby Schwartz Dies at Age 101." September 21. https://www.columbiatribune.com/article/20130921/News/309219892.

Contreras, A., and M. B. Trautman. 1997. "Margaret Morse Nice (1883–1974)." In *Women in the Biological Sciences: A Bibliographic Sourcebook*, edited by L. S. Grinstein, C. A. Biermann, and R. K. Rose. Westport, CT: Greenwood Press.

Coon, N. C., and M. C. Perry. 2007. "Lucille Stickel, 1915–2007." *Journal of Wildlife Management* 71(8): 2827–28.

Cooper, S. F. 1850. *Rural Hours: By a Lady*. New York: George P. Putnam.

Corneli, H. M. 2002. *Mice in the Freezer, Owls on the Porch: The Lives of Naturalists Frederick & Frances Hamerstrom*. Madison: University of Wisconsin Press.

Dartt, M. 1879. *On the Plains and among the Peaks; or, How Mrs. Maxwell Made Her Natural History Collection*. Philadelphia: Claxton, Remsen, and Haffelfinger. https://archive.org/details/onplainsamongpea00thom/page/n8.

Deakin, M. A. B. 2007. *Hypatia of Alexandria: Mathematician and Martyr*. Amherst, MA: Prometheus Books.

Dunkel, F., R. N. Smith, and H. J. O'Connor, eds. 1989. "Patuxent Wildlife Research Center 50th Anniversary: Wildlife Conservation through Scientific Research 1939–1989." *Fish and Wildlife News*, February–March. https://prd-wret.s3-us-west-2.amazonaws.com/assets/palladium/production/s3fs-public/atoms/files/pwrc50.pdf.

Dunlap, J. 1996a. "Florence Merriam Bailey (1863–1948) Ornithologist, Nature Writer." In *Notable Women in the Life Sciences: A Biographical Dictionary*, edited by B. F. Shearer and B. S. Shearer, 27–31. Westport, CT: Greenwood Press.

Dunlap, J. 1996b. "Margaret Morse Nice (1883–1974) Ornithologist." In *Notable Women in the Life Sciences: A Biographical Dictionary*, edited by B. F. Shearer and B. S. Shearer, 304–9. Westport, CT: Greenwood Press.

Engbring, G. M. 1940. "Saint Hildegard, Twelfth Century Physician." *Bulletin of the History of Medicine* 8:770–84.

Forest History Society Library and Archives. 2011. "Gene Bernardi Papers, 1971–1991." Durham, NC: Forest History Society. https://foresthistory.org/research-explore/archives-library/fhs-archival-collections/.

Gawlik, D. E., and R. K. Anderson. 1998. "In Memoriam: Frances Hamerstrom." *Journal of Raptor Research*. 32(4): ii–iv.

Gerber, T. 2017. "Becoming Jane." *National Geographic* 232:30–51.

Gjestson, D. 2013. *The Gamekeepers: Wisconsin Wildlife Conservation from WCD to CWD*. Madison: Wisconsin Department of Natural Resources.

Guelke, J. K., and K. Morin. 2002. "Gender, Nature, Empire: Women Naturalists in Nineteenth Century British Travel Literature." *Transactions of the Institute of British Geographers*. 26(3): 306–26.

Gup, T. 1996. "Woman of the Woods." *Washington Post*, July 1.

Hamerstrom, F. 1989. *Is She Coming Too? Memoirs of a Lady Hunter*. Ames: Iowa State University Press.

Hamerstrom, F. 1994. *My Double Life: Memoirs of a Naturalist*. Madison: University of Wisconsin Press.

Holsinger, R. 1983. "A Novel Experiment: Hallie Comes to Eddy's Gulch." *Women in Forestry* 5(2). www.webpages.uidaho.edu/winr/Daggett.htm.

James, E. T., J. W. James, and P. S. Boyer. 1971. *Notable American Women 1607–1950*. Cambridge, MA: Harvard University Press.

Johannsen, C. 2017. "Fran Hamerstrom." Filmed August 12, 1996, in Plainfield, WI, YouTube video, 14:39. www.youtube.com/watch?v=ZWgoW63lhTw.

Kaufman, P. W. 2006. *National Parks and the Woman's Voice*. Albuquerque: University of New Mexico Press.

Kessler, W. B. 2012. "An Irrepressible Pioneer Makes Her Mark: Profile of Frances Hamerstrom: 1907–1998." *Wildlife Professional* 6(3): 38–39.

Kessler, W. B. 2019. "Dying Wolf Moments: Inspirations and Reflections on a Wildlife Career." *Wildlife Professional* 13(2): 28–32.

Kofalk, H. 1989. *No Woman Tenderfoot: Florence Merriam Bailey, Pioneer Naturalist*. College Station: Texas A&M University Press.

Kohlstedt, S. G. 1978. "Maria Mitchell and the Advancement of Women in Science." *New England Quarterly* 51:39–63.

Mandula, B. 1997. "Frances Carnes Flint Hamerstrom (1907–)." In *Women in the Biological Sciences: A Biobibliographic Sourcebook*, edited by L. S. Grinstein, C. A. Biermann, and R. K. Rose, 196–210. Westport, CT: Greenwood.

Mayr, E. 1992. Preface to *Journal of Raptor Research* 26(3): 106–7.

Merchant, C. 1984. "Women of the Progressive Conservation Movement: 1900–1916." *Environmental Review* 8(1): 57–85.

Merchant, C. 2005. *The Columbia Guide to American Environmental History*. New York: Columbia University Press.

Michigan Women Forward. N.d. "Lucille Farrier Stickler." Accessed January 1, 2022. https://miwf.org/timeline/lucille-farrier-stickel/.

Nice, M. M. 1933. "Zur Naturgeschichte des Singammers." *Journal für Ornithologie* 81:552–95.

Nice, M. M. 1934. "Zur Naturgeschichte des Singammers." *Journal für Ornithologie* 82:1–96.

Nice, M. M. 1937. "Studies in the Life History of the Song Sparrow." *Transactions of the Linnaean Society of New York* 4:1–246.

Nice, M. M. 1979. *Research Is a Passion with Me*. Toronto: Consolidated Amethyst.

Nicholson, K. 2009. "Wanted: Female Role Models." *Wildlife Professional* 3(1): 40–42.

Norwood, V. 1993. *Made from This Earth: American Women and Nature*. Chapel Hill: University of North Carolina Press.

Ogilvie, M. B. 1989. "Marital Collaboration: An Approach to Science." In *Uneasy Careers and Intimate Lives: Women in Science, 1789–1979*, edited by P. G. Abiram and D. Outram, 104–25. New Brunswick, NJ: Rutgers University Press.

Ogilvie, M. B. 2018. *For the Birds: American Ornithologist Margaret Morse Nice*. Norman: University of Oklahoma Press.

Organ, J. F., S. P. Mahoney, and V. Geist. 2010. "Born in the Hands of Hunters: The North American Model of Wildlife Conservation." *Wildlife Professional* 4(3): 22–27.

Paulson, E. H., ed. 2013. *Hamerstrom Stories*. Stevens Point, WI: R. Schneider.

Penn, B. 2015. *The Real Thing: The Natural History of Ian McTaggart Cowan*. Victoria: Rocky Mountain Books.

Perl, T. 1979. "The Ladies' Diary or Woman's Almanac, 1704–1841." *Historia Mathematica* 6:36–53.

Perry, M. 2016. "The History of Patuxent: America's Wildlife Research Story." *Circular 1422*. Washington, DC: US Department of Interior, US Geological Survey. www.researchgate.net/publication/313818651_Patuxent's_Development_The_People_and_the_Projects.

Peters, D. 2019. "Pioneering Biologist Anne Innis Dagg Gets Her Due." *University Affairs*, June 26. www.universityaffairs.ca/pioneering-biologist-Anne-Innis-Dagg-gets-her-due.

Phillips, P. 1990. *The Scientific Lady: A Social History of Women's Scientific Interests 1520–1918*. London: Weidenfeld and Nicolson.

Ridley, G. 2010. *The Discovery of Jeanne Baret: A Story of Science, the High Seas, and the First Woman to Circumnavigate the Globe*. New York: Broadway Paperbacks.

Rimby, S. 2005. "Better Housekeeping Out of Doors: Mira Lloyd Dock, the State Federation of Pennsylvania Women, and the Progressive Era." *Journal of Women's History* 17:9–34.

Rositter, M. W. 1989. Foreword to *Uneasy Careers and Intimate Lives: Women in Science, 1789–1979*, edited by P. G. Abiram and D. Outram, xi–xii. New Jersey: Rutgers University Press.

Schantz, V. S. 1943. "Mrs. M. A. Maxwell, a Pioneer Mammalogist." *Journal of Mammalogy* 24(4): 464–66.

Schiebinger, L. 1993. "Women in Science: Historical Perspectives." In *Women at Work: A Meeting on the Status of Women in Astronomy; Proceedings of a Workshop Held at the Space Telescope Science Institute, Baltimore, Maryland, 8–9 September 1992*, edited by C. M. Urry, L. Danly, L. E. Sherbert, and S. Gonzaga, 11–19.

Shteir, A. B. 1989. "Botany in the Breakfast Room: Women and Early Nineteenth Century British Plant Study." In *Uneasy Careers and Intimate Lives: Women in Science, 1789–1979*, edited by P. G. Abiram and D. Outram, 31–43. New Brunswick, NJ: Rutgers University Press.

Slavicek, L. C. 2004. *Annie Montague Alexander: Naturalist and Fossil Hunter*. Broomall, PA: Chelsea House.

Souder, W. 2013. "How Two Women Ended the Deadly Feather Trade." *Smithsonian Magazine*, March. www.smithsonianmag.com/science-nature/how-two-women-ended -the-deadly-feather-trade-23187277/.

Stein, B. R. 1996. "Women in Mammalogy: The Early Years. Museum of Vertebrate Zoology, University of California, Berkeley." *Journal of Mammalogy* 77(3): 629–41.

Stein, B. R. 2001. *On Her Own Terms, Annie Montague Alexander and the Rise of Science in the American West*. Berkeley: University of California Press.

Stickel, L. F. 1957. *Wildlife Abstracts 1952–1955: A Bibliography and Index of the Abstracts in Wildlife Review Numbers 67–83*. Washington, DC: US Department of the Interior, Fish and Wildlife Service.

Trautman, M. B. 1977. "In Memoriam: Margaret Morse Nice." *Auk* 94(3): 430–41.

US Equal Employment Opportunity Commission (EEOC). 1964. Title VII of the Civil Rights Act. www.eeoc.gov/statutes/title-vii-civil-rights-act-1964.

US National Wildlife Refuge System. 2008. "Lucille Farrier Stickel: Research Pioneer." www.fws.gov/refuges/about/conservationheroes/stickelLucille_07232012.html.

Wayne, T. K. 2011. *American Women of Science since 1900*, 2:892–893. Santa Barbara, CA: ABC-CLIO.

Welker, R. H. 1971. "Wright, Mabel Osgood." In *Notable American Women, 1607–1950: A Biographical Dictionary*, edited by E. T. James, J. W. James, and P. S. Boyer, 3:682–84. Cambridge, MA: Harvard University Press.

CHAPTER

3

Lisette P. Waits, Lubia N. Cajas-Cano, and Kerry L. Nicholson

Who Are We Now?

Shaping Future Generations of the
Wildlife Profession

Women and minorities are substantially underrepresented among wildlife faculty and leadership. Surveys completed in 2020 indicate that only 27% of full professors in wildlife are women, and only 15% are Hispanic, Asian, African American, or Native. At the department head level, only 13% are women, and all current department heads are Caucasian. Our students need more diverse role models.

Before the turn of the 20th century, there were few laws or regulations imposed on the use of wild animals in North America. The first efforts to manage wildlife in the United States started in 1885 with the establishment of the US Department of Agriculture's Division of Economic Ornithology and Mammalogy. Its primary objectives were to study bird populations and the damage they did to agricultural crops. Many of the men managing wildlife at this time were trained as foresters, zoologists, or biologists. Wildlife management was established as a university degree in North America in 1934 when Aldo Leopold formalized a program at the University of Wisconsin. Currently, at least 400 universities in North America offer a wildlife science degree program (TWS n.d.[a]). Women and people of color have been students in the wildlife field since the beginning, albeit sporadically and in low numbers. The long struggle for acceptance into the wildlife field for women and minorities advanced with the help of pioneers, such as those described in Chapter 2. This has laid the foundation for where we are today.

Colleges and universities in North America have struggled to increase the racial, ethnic, and gender diversity of their students and faculty in scientific fields. Despite longstanding efforts, large sectors of the population remain underrepresented, and diversity remains an elusive goal. Universities, as training grounds for the next generation of wildlife professionals, are key in this effort. The lack of gender, racial, and ethnic diversity among faculty and students has frequently been highlighted as an area of weakness and concern for wildlife and other natural resource education programs (Kern et al. 2015; Sharik et al. 2015; Arismendi and Penaluna 2016; Cho et al. 2017; Gharis et al. 2017; Bal and Sharik 2019; Batavia et al. 2020).

The lack of diversity at the university level leads to monocultural workforces. A diverse workforce is critical to improve performance in organizations because people with diverse perspectives generate new ideas, promote innovation, lead to better problem solving (Woolley et al. 2010; Østergaard et al. 2011; Rice 2011), increase financial performance or value (Campbell and Mínguez-Vera 2008; Isidro and Sobral 2015; Perryman et al. 2016), and increase the probability that science will have a high impact (Maes et al. 2012; Freeman and Huang 2015; Valantine and Collins 2015; Nielsen et al. 2017). Barriers to diversifying workforces include lack of employment opportunities, gender gaps in salary, stymied career advancement, and perceived double standards for women and people of color. These hamper the development of the field of wildlife sciences (Hunter et al. 1990; Lidestav and Egan Sjölander 2007; Balcarczyk et al. 2015), limit the performance of organizations and deprive organizations of employment of women's leadership (Anderson 2020), and may lead to problems with retention of women and minorities in land grant institutions (Gumpertz et al. 2017) or with their recruitment to replace retiring tenure-track faculty (McChesney and Bichsel 2020).

Understanding the diversity and trends among university students (the future workforce) and among university faculty (our role models and mentors) is critical to efforts to increase diversity in the wildlife profession in North America (Taylor 2018). This chapter has three primary goals. The first is to review findings from earlier assessments of diversity in natural resource education and the natural resource workforce. These studies make it clear how much change is needed. The second is to present some of our own original data collected to evaluate changes in gender, racial, and ethnic diversity in university wildlife programs for students, faculty, and leaders over a 15-year period (2005–2020) and to identify ways that current university wildlife programs foster diversity. Finally, based on our findings, we reflect on the implications of our results, the challenges for women and people of color, and the opportunities for university programs and individuals to increase diversity and foster positive change.

What Are Gender and Ethnic Diversity Levels?

Women are substantially underrepresented in all ranks among natural resources faculty, although this is slowly beginning to change (Gumpertz et al. 2017). For example, only 15% of full professors in fisheries and 18% of tenured faculty in forestry are women (Kern et al. 2015; Arismendi and Penaluna 2016). The proportion of women faculty (27%–29%) is only slightly higher among the younger generation at the untenured ranks in fisheries and forestry (Kern et al. 2015; Arismendi and Penaluna 2016). The representation of people of color is very low among fisheries faculty, ranging from 8% to 11% depending on rank (Arismendi and Penaluna 2016).

Lack of diversity within universities is mirrored in the natural resource and environmental workforce of North America (Taylor 2007, 2014; Kern et al. 2015; Sharik et al. 2015; Arismendi and Penaluna 2016). For example, US Census Bureau data indicated that agriculture and natural resources professions ranked lowest in the percentage of minorities and second lowest in the percentage of women in the workforce (Sharik et al. 2015). In US federal fisheries sciences and management jobs, 91% of employees are Caucasian and 26% are women (Arismendi and Penaluna 2016). An evaluation of gender diversity in the USDA Forest Service research and development branch (Kern et al. 2015) found a similar low proportion of female employees (25%). In Natural Resources Canada, a 2014–2015 study found that women make up only 19% of science research jobs and 26% of science and technology employees (NRC 2015). No minorities were employed at 35% of the most widely recognized private environmental and natural resource organizations and 19% of natural resource government agencies (Taylor 2007).

In the United States the demand for jobs in zoology and wildlife was projected to increase by 7.6% between 2016 and 2026 (BLS 2016). In forestry it was projected to increase by 5%, and for fish and game positions 4.3% (BLS 2016). International demand for professionals trained in these fields has likely increased as interest in global environmental challenges grows (Edge 2016). Workforce diversity must advance in these sectors as the job force expands.

Wildlife conservation and management faces increasingly complex environmental challenges driven by rising global population growth rates. Diversity will be key in leading to better solutions to these challenges. By improving diversity in the profession, support broadens to the benefit of financial contributions and commitment to conservation initiatives and policies (Lopez and Brown 2011; Leisher et al. 2016). Diversifying the profession brings to the fore ideas among disciplines, countries, and groups that would otherwise be ignored.

Increasing our knowledge base with new concepts and perspectives enhances the potential for innovative solutions (Aslan et al. 2014; Newer 2018).

Recognizing the value of diversity, multiple North American natural resources organizations have developed diversity, equity, and inclusion position statements and initiatives to this end (e.g., DOI 2012, NRC 2019b, Diaz 2016, and TWS 2021). The importance of gender diversity for the protection of natural resources and biodiversity is also being reviewed and discussed at a global scale as part of the International Union for the Conservation of Nature's Environment and Gender Information Platform (IUCN 2020). All these efforts are steps in the right direction and bode well for the future of diversity in the workforce if the plans are proactively implemented.

Evaluation of the Changes

Data Collection Methods

To determine whether gender and racial diversity grew over a 15-year period (2005–2020), we compiled information from educational databases that track student demographics and developed and implemented surveys of university wildlife programs to collect faculty demographic data, student historical data, and information on scholarships and support programs. For a breakdown of student demographics by type who enrolled and graduated with wildlife degrees (i.e., baccalaureate undergraduate [BS], master's [MS], doctor of philosophy [PhD]), we queried the USDA Food and Agricultural Education Information System (Table 3.1). We used demographic records only from public and national land grant institutions with natural resource and conservation colleges in the United States. These colleges offered BS, MS, or PhD programs in wildlife biology or wildlife, fish, and wetlands science and management.

We used most of the terminology provided the Integrated Postsecondary Education Data System to define our racial groupings: African American, Asian, Caucasian, Latino, Native, international, two+ races, and unknown. Latino includes Hispanic and non-Hispanic; Native includes Native American, Native Alaskan, Native Hawaiian, or other Pacific Islanders; two+ races are those from two or more of these defined groups and was available in 2015, 2019, and 2020; international are neither citizens nor nationals of the United States; unknown are those who declined to answer. This database does not monitor gender statistics beyond male/female.

For faculty demographics and student history, we surveyed 50 wildlife degree programs in North America using the National Association of University

TABLE 3.1 US universities that provide history of students and demographic data of faculty in the wildlife field or scholarships from the National Association of University Fish and Wildlife Programs (NAUFWP) and USDA Food and Agricultural Education Information System (FAIES, www.faeis.cals.vt.edu/) student demographic data. Note 2020 graduation information was incomplete, so we used 2019 graduation data; 2020 enrollment data was complete.

*Indicates universities reporting diversity scholarship or support programs for wildlife students.

NAUFWP	FAIES	Both NAUFWP and FAIES
Alabama A&M	Auburn University	Iowa State University
Cornell University*	Clemson University	Oregon State University*
Tennessee Tech University*	Humboldt State	Pennsylvania State University*
University of Arizona*	McNeese State University	Texas Tech University
University of Minnesota	Michigan State University	University of Alaska Fairbanks
University of Montana	Michigan Technological University	University of Alberta
University of Tennessee*	Mississippi State University	University of Florida*
	North Carolina State, Raleigh	University of Idaho*
	Ohio State University	University of Nebraska–Lincoln
	Purdue University	Virginia Polytechnic Institute and State University
	South Dakota State University	
	Sul Ross State University	
	Tarleton State University	
	Texas A&M University	
	University of Illinois at Urbana-Champaign	
	University of Delaware	
	University of Maine	
	University of Missouri	
	University of Nevada, Reno	
	University of Rhode Island	
	University of Vermont	
	University of Wisconsin–Madison	
	University of Wisconsin–Stevens Point	
	West Virginia University	

Fish and Wildlife Programs (NAUFWP n.d.) email LISTSERV (Table 3.1). We specifically requested data for 2005, 2015, and 2020 for the following categories: department head or wildlife program lead, faculty (by type: assistant, associate, full), by gender (male, female, transgender) and racial background. Assistant professors are the youngest group, followed by associate, and full professors are the most senior faculty. In addition, we asked for the earliest dates that women graduated from undergraduate and graduate programs in wildlife and the dates women became faculty and department heads in these programs. We also asked survey respondents to provide and describe information on scholarships and support or mentoring services for underrepresented students.*

The Change in Student Demographics

In the 15-year period we examined, there was a large increase in women enrolling in wildlife programs (Figure 3.1A). For all three degrees, by 2020 there was an increase to nearly 50:50 male:female (M:F) balance, with substantial changes in the ratios of BS and PhD students from a 60/40 M:F ratio in 2005 (Figure 3.1A). In fact, women outnumbered men at the BS and MS degree level by 2020.

Transgender students were not represented in the 2015 FAEIS national database that we used; however, our survey of NAUFWP department heads indicated that the number of students in wildlife identifying as transgender increased from zero in 2005 to two in 2015, though this information was not reported in 2020. We did not expect many students to openly identify themselves as transgender, thus we believe that these two students provide evidence of an increase in social acceptance for gender identification conversations (see Chapter 8 for more on this topic). Sex ratios of students graduating with wildlife degrees mirrored enrollment patterns. By 2015 the percentage of women graduating with BS degrees exceeded the percentage of men, and by 2019 the percentage of women graduating with MS degrees exceeded the percentage of men. However, more men (54.5%) than women (45.5%) earned a PhD in 2019, indicating there will continue to be fewer women than men available for future faculty positions.

In 2019, Caucasians made up 61% of the US population (Figure 3.2). During the 15-year evaluation period, we saw the Caucasian student population enrollment decrease and the racial diversity increase within each degree. However, racial diversity among undergraduate and graduate students lags far behind the diversity levels of the country as a whole. According to the US population census in 2019, Latino and African American students are strikingly underrepresented in enrollment in the wildlife field; international students are overrepresented (Figure 3.2). Awarded degrees generally follow enrollment patterns with a five-year lag, suggesting that most of the students enrolled did complete their degrees. In 2015, for example, 11.3% of the students enrolled in BS programs for wildlife were Hispanic, and in 2020 11.8% of graduating students were Hispanic.

* This survey was approved by the University of Idaho Institutional Review Board (protocol #16-063) and certified as exempt under category 2 at 45 C.F.R. 46.101(b)(2).

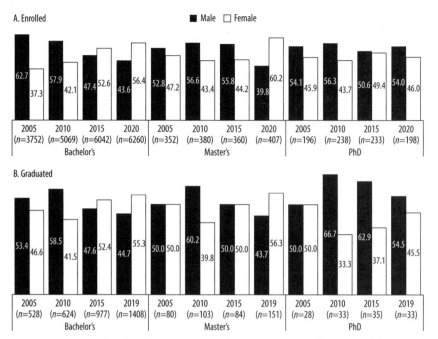

FIGURE 3.1. Sex ratios of wildlife students (undergraduate, master's, and PhD) in 2005, 2010, 2015, and 2020 who were enrolled (A) and graduated (B); *n* is the number of students registered.

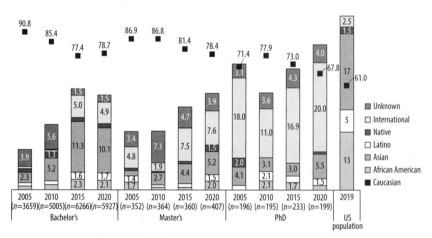

FIGURE 3.2. Percentage of wildlife students by ethnic diversity enrolled (undergraduate, master's, and PhD) in 2005, 2010, 2015, and 2020. Ethnic diversity of US population in 2019 is provided for comparison. The Integrated Postsecondary Education Data System (NCES n.d.) provided the terminology to define groupings. International is combined with two+ races; *n* is the number of students registered; groups with <1% are not labeled.

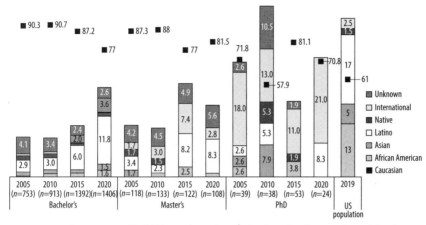

FIGURE 3.3. Percentage of wildlife students by ethnic diversity that graduated (undergraduate, master's, and PhD) in 2005, 2010, 2015, and 2020. Ethnic diversity of the US population in 2019 is provided for comparison. The Integrated Postsecondary Education Data System (NCES n.d.) provided the terminology to define groupings. International is combined with two+ races; n is the number of students registered; groups with <1% are not labeled.

We see stochastic changes among the races except Caucasians (Figure 3.3) at the PhD level, where the sample sizes are much smaller. For example, Asian Americans did not receive PhDs in 2015 or 2020, and Latinos did not receive PhDs in 2015 (Figure 3.3). The proportion of international students graduating by degree program fluctuated over time but was always highest at the PhD level. We see the increase of international students in wildlife majors as beneficial to our programs because of their diverse cultural backgrounds and global perspectives.

The Change in Faculty Demographics

Similar to recent studies of faculty in fisheries, agriculture, and forestry (Kern et al. 2015; Arismendi and Penaluna 2016; Cho et al. 2017), increases in gender diversity were observed at all faculty levels, but women are still substantially underrepresented in all positions, particularly as full professors and department heads, the most senior faculty members and leaders. Women assistant, associate, and full professors increased by 4%, 15%, and 11%, respectively, from 2005 to 2015 and increased again by 8% for assistant and full professors but lost some ground (−4%) at the associate level between 2015 and 2020 (Figure 3.4). Department heads briefly saw an increase of 6% in women from 2005 to 2015 but lost those leaders by 2020 (Figure 3.4). At all levels, the ratio was still well below a M:F gender-balanced 50:50 ratio. There were no transgender faculty reported in the survey.

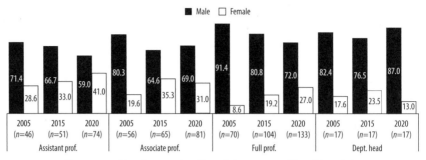

FIGURE 3.4. Sex ratios of professors (prof.) and department (dept.) heads in 2005, 2015, and 2020 at universities with fish and wildlife programs; *n* is the number of faculty employed at each level.

The lack of female leadership and retention at the upper levels is a concern. We discovered that 47% of the universities in our survey had never hired a woman at the department head or program lead level. However, there was a brief increase in hiring women as department heads between 2005 and 2015, based on our survey data. In 2014 when coauthor Lisette Waits became the first woman to head a department at the University of Idaho, she was also the only one who attended meetings of the NAUFWP. She was so surprised by this that one year she introduced herself as "Lisette Waits, representing the University of Idaho and my gender." However, since 2015 the number of women department heads attending the meetings has risen. Thus, the drop from 2015 to 2020 may just be a stochastic effect of the small sample size (17). To evaluate this, we looked at the gender ratio for all 42 NAUFWP members in 2021 and found 24% were women. The gender ratio of regional chairs for NAUFWP is 50:50 (NAUFWP n.d.), and the organization elected its first woman as president in 2019. So overall, we conclude that there has been continuous progress in the percentage of women in leadership positions in wildlife programs. In addition, the proportion of women graduating with PhDs in 2010 (33%) is very similar to the proportion of women at the assistant professor (33%) levels in 2015, suggesting that most qualified women were retained over this period and did progress from graduate school to faculty positions. We see a similar trend with 37% women graduating with PhDs in 2015 and 41% women at the assistant professor level in 2020. This is a positive trend; however, it is concerning to see a drop at the associate professor level in 2020 compared to 2015.

There were fewer gains and some losses in racial diversity among faculty over this 15-year period. Racial diversity, excluding international faculty, declined in assistant professor hires over time, which is concerning for future diversity moving up the faculty ranks. The percentages of racial diversity among full professors did increase during the period, but it decreased among associate

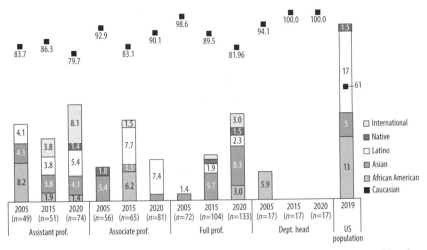

FIGURE 3.5. Percentage of ethnic diversity of professors (prof.) and department (dept.) heads in 2005, 2015, and 2020 at universities with fish and wildlife programs. Ethnic diversity of the US population in 2019 is provided for comparison. The Integrated Postsecondary Education Data System (NCES n.d.) provided the terminology to define groupings. International is combined with two+ races; n is the number of students registered; groups with <1% are not labeled.

professors from 2015 to 2020, which reflects the negative impacts of the drop in racial diversity among assistant professors from 2005 to 2015. There was very little racial diversity (1 in 17) among department heads in 2005 and none in 2015 or 2020 (Figure 3.5).

The largest increases in faculty racial diversity were observed for Latino faculty at all ranks. There was also a large increase in Asian American faculty at the full professor level. There was an overall decline in African American faculty by 2020 in all ranks. Native American faculty were either completely absent or below 2% in all years. We also observed an increase in the proportion of international faculty at the assistant professor and full professor ranks; however, there are few or no international faculty present in the associate professor ranks. The growth at the assistant professor levels is an encouraging trend that should increase the diversity of knowledge and perspectives in teaching and mentoring of wildlife students (Figure 3.5). Overall, the low proportion of racial diversity among the ranks of assistant and associate professors and among PhD students is concerning for the future of diversity in wildlife programs at the faculty level. Proactive efforts are clearly needed; otherwise, the progression of change will be slow. We hope to see a reversal in this alarming downward trend in the future and more diverse role models in leadership positions, because it is critical to increasing diversity among students and assistant professors.

When Did It All Start?

Nineteen universities responded to our survey regarding the history of hiring women faculty and graduating students (Table 3.2). We would expect that the earliest women graduates would come from the universities that have the longest history of offering wildlife degrees. However, tracking down this information was more difficult than expected. Many universities did not and still do not have wildlife programs that are distinctly separate from biology, ecology, or zoology programs. We also discovered that many institutions have reorganized departments or changed names over time. Additionally, few schools tracked demographic information. Therefore, the information we report here may not reflect "the first" and may vary depending on how we define the degree awarded.

Oregon State University (OSU) was one of the first universities that offered a degree in wildlife, and the first BS was awarded to a woman in 1939. In 1938, Margaret Whipple was in the first cohort of MS degrees awarded in fish and game management, and she worked with Oregon Department of Fish and Wildlife on a ring-necked pheasant (*Phasianus colchicus*) research project. The first woman to graduate with a PhD from the Department of Fisheries and Wildlife at OSU was likely Anne Christine Kristiansson Barton, who worked with spring Chinook salmon (*Oncorhynchus tshawytscha*) and graduated in 1980 (Table 3.2). The first PhD from OSU on terrestrial wildlife study was completed in 1984 by Marcia Wilson. She studied comparative ecology of northern bobwhite (*Colinus virginianus*) and scaled quail (*Callipepla squamata*).

The earliest women tenure-track faculty we identified were Winifred Kessler (1976, University of Idaho) and Marylin D. Bachmann (1979, Iowa State University). Most of the universities we surveyed did not hire their first woman to a tenure-track position until after 1992, and 32% did not hire a woman to the faculty until 2002 or later. As of 2019, Auburn University had not yet hired a woman to a full-time wildlife faculty job.

Patricia Werner was the first woman to head a fish and wildlife program, in 1992, at the University of Florida. All other female department heads in our survey were hired after 2004. Forty-seven percent of the universities that we surveyed have yet to hire one.

Scholarships and Support Programs for Underrepresented Students

Ten universities answered our questions regarding scholarships and support programs for underrepresented students in wildlife (Table 3.1). We were

TABLE 3.2 Survey results on earliest dates for women students, faculty and department heads in wildlife programs. "None" indicates that data were available, but no woman has ever served in that role. NA indicates information was not available as of 2019.

* US Geological Survey (USGS) Fish and Wildlife Cooperative Research Unit (COOP) faculty
† only quarter time

Name of University	Name of Department	Year of First Woman as				
		Faculty	Dept. Head	BS	MS	PhD
Auburn University	School of Forestry and Wildlife Science	2015[†]	None	1979	1985	2006
Iowa State University	Natural Resource Ecology and Management	1979	2011	1970	1974	1993
Louisiana State University	School of Renewable Natural Resources	2005	None	1969	1977	2001
Missouri State University	Biology	1993	2005	NA	NA	NA
Oregon State University	Fisheries and Wildlife	2002	2014	1939	1938	1980
Pennsylvania State University	Ecosystem Science and Management	1988	None	1983	1972	1985
Purdue University	Forestry & Natural Resources	2004	None	1971	1975	1998
Texas Tech University	Natural Resources Management	1990	None	1971	1977	1984
University of Arizona	School of Natural Resources & the Environment	2002*	2007	1979	1975	1989
University of Florida	Wildlife Ecology and Conservation	1992	1992	1977	NA	NA
University of Georgia	Warnell School of Forestry & Natural Resources	1992	2009	NA	NA	NA
University of Idaho	Fish and Wildlife Sciences	1976	2014	NA	NA	NA
University of Minnesota	Fisheries, Wildlife, and Conservation Biology	1985	2006	1978	1976	1992
University of Montana	Wildlife Biology	1990	None	1966	1954	1997
University of Nebraska–Lincoln	School of Natural Resources	1987	2012	1990	1983	1997
University of Tennessee	Forestry, Wildlife and Fisheries	1999	None	NA	NA	NA
University of Wisconsin–Stevens Point	College of Natural Resources	2007	2015	NA	NA	NA
Virginia Polytech Institute and State University	Fish and Wildlife Conservation	1993	None	NA	1972	1996
Washington State University	School of the Environment	1996	None	1962	NA	1983

encouraged to find all of the universities provided at least one support program. Five reported specific scholarships, professional societies, and training programs that targeted underrepresented students. These programs included the National Science Foundation Research Experience for Undergraduates programs, Doris Duke Conservation Scholars Program (DDCSP), and Minorities in Agriculture, Natural Resources and Related Sciences (MANRRS). All of the schools had some form of support programs for Native American, Hispanic, migrant-worker, and multicultural students at the college or university level, but not necessarily at the level of the department or major. Two of the universities noted support from the DDCSP, launched in 2013, which currently provides scholarships and mentoring for underrepresented undergraduate students interested in career paths related to conservation at nine universities in the United States (Doris Duke Charitable Foundation n.d.). In general, the scholarship programs reported by universities were small (<$5,000 per scholarship per year), and our findings highlight the need for larger, more targeted scholarships to recruit and retain underrepresented students in wildlife.

Reflecting on These Findings

Looking to the Future

There are many challenges for women and minorities in wildlife, but our survey demonstrates that we have made impressive strides. For example, we are currently graduating an approximately equal proportion of men and women from wildlife programs, providing the opportunity for gender equity in the academic and nonacademic wildlife workforce. In another positive step for diversity and female leadership, the NAUFWP, has elected its first woman president and added diversity and inclusion initiatives to its 2019 strategic plan. These steps indicate a purposeful focus on increasing diversity throughout US universities that have fish and wildlife programs. However, we have more work to do before the student population and our workforce reflect the racial and gender diversity of the general population.

Dramatic shifts in US and Canadian population demographics should provide opportunities to increase racial and ethnic diversity in the wildlife workforce if we can lower and remove current barriers, improve the educational and workplace climate, and provide the mentoring support needed. For example, the Census Bureau estimates that by 2050, there will be no clear racial or ethnic majority and approximately 45% of the US population will be people of color (US Census Bureau 2011).

Many actions are needed to accomplish our goals of achieving greater diversity in the wildlife profession, and other chapters in this book cover this topic in more detail. Studies seeking to understand the underrepresentation of women in science, technology, engineering, and mathematics (STEM) have highlighted the important fact that women are as dedicated to their profession as men (Xu 2008), and some might argue more so given the hurdles they have had to overcome. There is a significant and increasing amount of information on the factors and barriers that drive the underrepresentation of women in STEM and natural resource fields. Here we emphasize three focal areas that are key to increasing diversity in academia: (1) highlighting diverse role models and increasing mentoring, (2) improving culture and climate, and (3) investing in proactive programs.

Role Models and Mentoring

Role models from underrepresented groups are crucial to increasing diversity in wildlife because they demonstrate that individuals from these groups belong and can succeed in the profession. Good mentoring establishes a positive workplace culture. Faculty and students who receive regular mentoring are more likely to succeed in their academic pursuits (Blau et al. 2010; Reybold et al. 2012; Mervis 2016). Yet women receive less mentoring and are less frequently selected as role models than men, especially in male-dominated fields (Baugh and Scandura 1999; Feeney and Bernal 2010). Interestingly, women in ecological fields value mentoring and a positive workplace environment more than men do (McGuire et al. 2012; Hooker et al. 2017).

Identity-based motivation theory (Oyserman 2014) suggests that women and some minorities may find STEM incompatible with their identity. However, this can be overcome with role models of the same underrepresented group (Oyserman 2014). For example, female students in STEM are more likely to attend class and ask questions when their professors are women (Solanki and Xu 2018), and female instructors positively influenced the likelihood of women taking future STEM courses and graduating with a STEM degree (Carrell et al. 2010). Creating a more diverse pool of faculty and students in wildlife, including those with international backgrounds, will help recruit and educate the next generation to think globally. Students should have mentors who can understand their diverse cultures, academic backgrounds, and needs to increase retention and graduation rates.

Coauthor Waits has seen this dynamic at the University of Idaho, where she joined the exclusively male wildlife faculty in 1997. The department finally hired a second woman to the faculty in 2001. Now women occupy five out of six wild-life tenure-track positions there, probably the highest proportional representation of women in wildlife faculty in North America. These women are playing important leadership and mentoring roles: three are full professors, and Waits is the department head. Many female undergraduate and graduate students have commented that having so many successful women as senior faculty and as department head empowers them. This fish and wildlife department has observed large increases in the percentage of women enrolling in undergraduate (27% to 41%) and graduate study (41% to 61%) in wildlife from 2005 to 2015. In contrast, there have been smaller increases in the proportion of female students in fisheries where the department has no female tenure-track faculty among seven male faculty and only one female federal USGS COOP unit faculty, who retired in 2017.

Others have highlighted the importance of identifying and celebrating women and other underrepresented role models in the wildlife profession (e.g., Smith and Kaufman 1996; Nicholson et al. 2008; Nicholson 2009). See Boxes 3.1 and 3.2 to celebrate two exceptional women who've pushed beyond barriers and taken and made the most of the opportunities they've crossed. Some organizations like the US Geological Survey (USGS) have worked to accomplish this with special publications highlighting women in STEM careers (Aragon-Long et al. 2018), and groups like The Wildlife Society Women of Wildlife (TWS 2022) were formed to celebrate the contributions of women and to promote networking. Additionally, TWS's recurring feature in the *Wildlife Professional*, Wildlife Vocalizations, highlights and connects the members of their society (TWS n.d.[b]). Other organizations are highlighting the importance of diversity among keynote speakers at scientific meetings and developing sets of recommendations for improving gender equity at conferences (Sardelis et al. 2017). While role models are critical for inspiring young people from underrepresented groups to pursue wildlife careers, strong mentoring programs are critical for retaining those individuals in academic programs and in the workplace (Mervis 2016; Kohl et al. 2017; Hansen et al. 2018). Women and underrepresented students need mentoring to increase retention and to gain valuable leadership, time management, and other professional skills that improve the likelihood of success and persistence in academic careers.

One valuable way to expose students and faculty in wildlife and other natural resource fields to diverse role models and mentors is through diversity-focused professional societies such as the Minorities in Agriculture, Natural Resources and Related Sciences (MANRRS), Society for Advancement of Chi-

BOX 3.1 Women Are Hunters and Anglers, Too

According to the first woman dean of the College of Natural Resources at the University of Wisconsin–Stevens Point, Christine Thomas, being a woman in the wildlife profession is a double-edged sword. When she entered the wildlife field in the late 1980s, the lack of female peers made her stick out like a parrot among ravens. Sometimes this was advantageous for gaining access to new projects, people, and positions. But the overall lack of peers, support, and encouragement for women to participate in the wildlife field was a concern for her.

Thomas's game-changer occurred in 1985 when her dean, Dan Trainer, asked her to attend a Forest Service–sponsored conference on women in natural resources, held in Dallas. She was floored by what she heard about sexual discrimination and harassment women faced in the natural resource field. The conference exposed her to some wonderful female role models, and she decided that to make a difference in the world, she needed to get her PhD.

As a new assistant professor looking for her niche, two colleagues asked for her help developing a project about women and hunting. They wished to investigate reasons for the low representation of women in the hunting and angling community. Were women even interested? If they were, what prevented their participation? The project she developed was to host a national workshop, "Breaking Down the Barriers to Participation of Women in Angling and Hunting," with Frances Hamerstrom (see Chapter 2) delivering the keynote speech. The 65 participants identified 21 barriers, ranging from complex social pressures to simple things such as clothing and equipment that fit women and met their particular needs, for example, outdoor hygiene issues such as menstruation. The lack of opportunities to learn outdoor skills was a key barrier, and it changed the trajectory of Thomas's career. As a result of this conference, she and her colleagues established Becoming an Outdoors-Woman (BOW). This weekend program provides outdoor skills training for women in a nonthreatening atmosphere. Originally conceived as a one-time event with a budget of $150 and all borrowed equipment, the 100-person workshop filled in 1 day. Based on that initial success, groups like the National Shooting Sports Foundation and Safari Club International stepped up with funding to support additional workshops. Today, BOW has expanded from Wisconsin to 38 US states and six Canadian provinces, clearly answering the initial question, "Do women want to learn about hunting and fishing?" with a resounding yes!

BOX 3.2 Open Doors Open Opportunities

Sometimes necessity and unforeseen opportunities open doors we never dream existed. Your life can unexpectedly change because of personal circumstances or events that occur at much larger scales, and women should not shy away from opportunities when they come their way. This is how Merav Ben-David has lived her life. Ben-David was born and raised on a family farm in a small village south of Tel Aviv, Israel, among the youngest of a large extended family of 21 cousins. Her position in the family dramatically changed at age 14 after the tragic loss of her father, the main caretaker of the family farm. Out of necessity she assumed his duties including managing the farm, directing the hired hands, and maintaining the family income, which was a very unusual role for a young woman in this society. Through this unforeseen opportunity, she quickly learned how to make informed decisions and assert her opinion. Both skills would serve her well for the rest of her personal and professional life.

Another unforeseen opportunity that profoundly changed Ben-David's life happened on a mountaintop. She had finished her master's degree in zoology in 1988 at the Tel Aviv University on marbled polecats (*Vormela peregusna*). After graduating she struggled to make a living, so she went to Kenya to be a tour guide, yet she still longed to effect real change in the wildlife profession. While hiking in the Abruzzo Mountains outside of Rome during a post-conference trip, she met a university professor from Fairbanks, Alaska, and casually began discussing her future goals. This professor was so impressed by her tenacity that he offered her a PhD project in southeast Alaska focused on the seasonal diet of mink and marten. Completing her PhD in 1996 jumpstarted her academic career, and she went on to become a faculty member at the University of Wyoming. There she rose through the academic ranks from assistant to full professor and in 2017 assumed the position of department head of zoology and physiology for three years.

During her time as a professor, Ben-David initiated several studies on polar bears that gave her unique experiences jumping through bureaucratic government hoops and facing the stark reality of the dramatic negative impacts of climate change on polar bear populations. This experience reinforced her passion to effect changes for the environment and wildlife. Another door opened to help her reach this goal during the 4th annual Wyoming Women's March on Equality in 2020, when Ben-David announced her candidacy for the upcoming US Senate seat. She ran an impressive campaign supported by many and inspiring other scientists to fight for change at the political level. If she had won, she would have been the first woman US senator in the 150 years since Wyoming granted women the right to vote. Ben-David has inspired us all by continuing to push the barriers on female leadership in science and politics.

canos/Hispanics and Native Americans in Science (SACNAS), Native American Fish and Wildlife Society, and working groups and other programs within professional societies like the Native Peoples' Wildlife Management Working Group, the Inclusion, Diversity, Equity and Awareness Working Group (formerly the Ethnic and Gender Diversity Working Group) of The Wildlife Society, and the Strategies for Ecology Education, Diversity, and Sustainability (SEEDS) program of the Ecological Society of America (ESA) (Mourad et al. 2018). These organizations help build support networks and encourage cooperation and coordination, key for any workforce to succeed. We encourage university wildlife programs to consider adding local student chapters for MANRRS, SACNAS, or both encouraging students and faculty from underrepresented groups to join the relevant national organizations, and providing financial support to attend their meetings.

Culture and Climate

Wildlife and other natural resources fields have suffered from a history of overt sexism, harassment, and exclusion. Women experience sexual harassment and discrimination in natural resource and ecology fields at a much higher rate than men (Clancy et al. 2014; Hooker et al. 2017). Establishing codes of conduct and reporting structures for sexual harassment and gender and ethnic bias in the workplace, in the classroom, at scientific meetings, and in fieldwork is critical for improving the culture and climate for women and underrepresented groups in wildlife and all academic fields (Mansfield et al. 2019). Some universities have begun to incorporate this topic into introductory wildlife courses, which is a positive and proactive step that all should consider.

Some have attributed the lower proportion of female professors to the inflexibility of the US workplace, configured around a male career model, that forces women to choose between work and family (Wolfinger et al. 2008). Women generally put more time into household chores and the care of children and elderly parents than men (Schiebinger and Gilmartin 2010; McGuire et al. 2012; Loison et al. 2016; Hooker et al. 2017), and they are more likely to take a break from science or work part time to raise a family (Hooker et al. 2017; Shauman 2017). This affects the number of women who continue to a tenure-track position and their success as faculty. Some suggest that the extra time commitments of women, particularly women with children, decreases their professional productivity, disadvantaging them in an increasingly competitive academic job market (Adamo 2013). Women in natural resources careers are also more likely to have a spouse in the same field or profession (Primack and

O'Leary 1993). Although careers of men benefit from having a partner in academia, women's careers suffer (Sweet and Moen 2004; Loison et al. 2016). Often, women follow their spouse and accept temporary or non-tenure-track positions, or even drop out of the profession to live in the same location and preserve the family structure (Ceci and Williams 2011).

One cultural shift recommended for addressing work-life balance challenges for women is changing to a model where granting organizations permit grantees to use a certain percentage of their grant award funds to pay for childcare, eldercare, or family-related expenses, which would make it easier for women to travel to give invited lectures or attend scientific meetings and conferences (Smith et al. 2015). Another innovative idea is to fund "extra hands" to all young investigators with families. The "extra hands" would allow investigators to maintain research efforts without major interruption by hiring technicians, administrative assistants, or postdoctoral fellows, thus increasing productivity in the early critical years of their careers. Dean Laurie Glimcher pioneered a version of this idea, the Primary Caregiver Technical Assistance Supplements. In another example from Germany, the Christiane Nusslein-Volhard Foundation (2022) supports talented young graduate student and postdoc parents in the natural sciences and medicine by providing financial assistance for household chores and childcare. Gender equity task forces at universities have suggested other ideas for improving work-life balance. Examples include no-cost grant extensions; adjusting the length of time to work on grants to accommodate child-rearing; supplements to hire postdocs to maintain momentum during family leave; reduction in teaching responsibilities for women with newborns; delaying the time clock for promotion or tenure after childbirth, medical leave, or major caregiving responsibilities; grants for retooling after leaves of absence; couples hiring (spousal accommodation); childcare to attend professional meetings; and part-time tenure track jobs that transition to full time (Smith et al. 2015; Hooker et al 2017).

To mitigate the negative effects of implicit bias against women and minorities in science, some have suggested that grant review committees, search committees, and promotion review committees should include statements that describe the concept of implicit bias and emphasize the commitment of organizations to gender equality and diversity in all its forms (Smith et al. 2015; Hooker et al 2017). Also, some have proposed that improved metrics could minimize gender biases in STEM. For example, Cameron et al. (2016) found that men researching in ecology used self-citation more often than women, but if self-citations and non-research-active years were excluded, there are no gender differences in research performance. Also, research performance metrics that rely on quality as well as quantity of publications can benefit female job candidates because women's impact factors are equal to or higher than men's in STEM (Symonds et al. 2006; Goulden et al. 2011; Bendels et al. 2018).

Proactive Programs

Many countries have launched initiatives and funding to address gender inequity in STEM fields. In the United States, the first broad efforts at gender equality in education were initiated with the passage of Title IX in 1972. This law prohibits sex discrimination in education or any federally funded program. It states, *"No person in the United States shall, on the basis of sex, be excluded from participation in, be denied the benefits of, or be subjected to discrimination under any education program or activity receiving federal financial assistance."* In the United Kingdom, higher education institutions have been strongly encouraged to join the Athena SWAN Charter of 2005, which contains six principles that actively address gender inequalities and the unequal representation of women in STEM. In 2018, the American Association for the Advancement of Science started a similar program modeled after the SWAN charter titled STEM Equity Achievement (SEA) Change. The SEA Change program *"supports systemic, structural institutional transformation in support of diversity and inclusion, especially in colleges and universities, enabling success in education and research missions by ensuring that the full range of talent can be recruited, retained, and advanced in science, technology, engineering, medicine, and mathematics."* Universities voluntarily enroll in the program and then receive awards for meeting specific diversity and inclusion targets and metrics. Also in the United States, the NSF launched the ADVANCE program in 2001 to promote women and other underrepresented groups in STEM through organizational changes at universities, and it has invested more than $270 million in this effort.

Natural Resource Canada has launched two efforts of note on gender diversity since 2018. First is the Charter for Gender Equality (NRC 2019b) that outlines concrete measures for detecting bias, recruiting, and supporting women in reaching their potential. Second is the Equal by 30 campaign (NRC 2018), launched in May 2018 under the banner of the Clean Energy Ministerial, an international framework that asks companies and governments to endorse principles, then take action to make gender equality central to the transition to a clean energy future by promoting equal pay, opportunity, and leadership for women. Canada has also started a diversity in STEM reentry program for women and Indigenous people with STEM degrees who have been out of the workforce for five or more years (NRC 2019a).

Earlier we mentioned the DDCSP as one innovative US program for recruiting and retaining diverse students. There are currently nine US universities with DDSCP funding, and each program is slightly different but includes the key components of intensive mentoring, research training, and internship placement with potential natural resource employers. At the University of Idaho, the DDCSP program is led by a woman on the faculty from an under-represented ethnic group who also recruits a graduate student from an under-represented group to assist with mentoring and recruitment. Undergraduates in the DDSCP are supported and mentored for two years as part of a four- to five-student cohort that includes biweekly mentoring and professional development meetings, a summer of field research mentored by faculty and graduate students, presentations of their research project at professional meetings, and a summer internship with a natural resource employer. The University of Idaho DDSCP program has been very successful at recruiting and retaining diverse students in natural resources and currently has a 100% graduation rate, with most students (~70%) continuing in the natural resources conservation field or to graduate school.

Another diversity-enhancing student-focused program is SEEDS, of the ESA. This program is designed to increase participation in the field of ecology for underrepresented minority undergraduate students, through activities such as mentoring, field trips, leadership development, and research fellowships (Mourad et al. 2018). SEEDS was created in 1996 by a consortium led by the Institute of Ecosystem Studies in partnership with the ESA and the United Negro College Fund, and the first phase focused efforts on faculty development and student support at historically Black colleges and universities. In 2002, the ESA assumed full management of SEEDS and expanded it to engage students from all backgrounds, including Asian American, Hispanic, Native American, and Caucasian students (Mourad et al. 2018). The program has been very successful. For example, 80% of SEEDS alumni completed at least one degree in an ecology-related field, and the completion rate for underrepresented minority students was 85%. In addition, 71% of working SEEDS alumni had careers in ecology (Mourad et al. 2018).

A proactive program at the postdoc and faculty level is the Women Evolving Biological Sciences (WEBS) symposia that were developed in 2007 to provide early career female biologists with increased access to career development, mentoring, and networking opportunities that provide participants with a sense of community and empowerment (Katz 2008; Horner-Devine et al. 2016). The effort is focused on the transition from graduate student to postdoctoral scholar to permanent faculty or researcher. Preference is given to promising biologists considered at risk of leaving academia because of lack of mentoring, career support, career satisfaction, and peer networks, and to applicants from underrepresented groups. The WEBS program has distinctive elements that create a counterspace, a setting that promotes positive self-concepts among individuals from underrepresented groups (Solorzano 1998; Case and Hunter 2012) and helps women establish a national network and peer cohort of biologists. WEBS symposia include panel discussions and personal reflections from participants and senior women scientists, as well as interactive skill-development workshops. Participants report that the WEBS symposia had strong, positive effects on them and reduced negative feelings about themselves and their profession by more than 50%. The successful WEBS program model has been adapted to advance diversity in neuroscience (see BRAINS: Broadening the Representation of Academic Investigators in Neuroscience) and engineering (LATTICE: Launching Academics on the Tenure-Track: an Intentional Community in Engineering) and seems a great model to consider for wildlife.

These types of proactive programs that include targeted diversity recruitment, scholarships and funding for underrepresented groups, professional mentoring, and initiatives that address culture and climate in the workplace through diversity training and inclusive curricula need to be incorporated in all academic wildlife programs. This can be achieved through new initiatives, grant writing to organizations like the groups highlighted above, and departmental fundraising.

Conclusions and How to Be an Ally

Our goal for this chapter was to review the change in gender and ethnic diversity of students and faculty in US university wildlife programs from 2005 to 2020 to evaluate our progress. We found great improvement and a shift to equality of gender diversity at the student level, but our results highlight much need for increases in gender diversity at the faculty and department head level and in racial diversity at all levels in university wildlife programs. Thus, our chapter also explored approaches for supporting and enhancing diversity in university programs and the wildlife profession into the future. Women and people of color are tomorrow's leaders, and we must better prepare our future workforce. Below are some key takeaway messages on how to foster positive change and be an ally of women and underrepresented minorities in the wildlife profession:

- Changing the demographics of diversity in the wildlife profession starts with increasing the diversity among undergraduate and graduate students.
- Diverse role models and strong mentoring is critical for recruiting and retaining women and ethnic minorities in the wildlife profession.
- Increasing the diversity among department heads and faculty role models is a critical next step.
- More scholarships and funding for programs that recruit, support, and retain underrepresented students, postdocs, and faculty are needed.
- Underrepresented students need support of underrepresented peers, and joining diversity-focused programs in STEM and natural resources can greatly improve retention.
- We must find ways to support childcare, eldercare, and flexible work schedules to make it easier for women to progress professionally without sacrificing family obligations.
- University wildlife programs should all establish codes of conduct, curricula, and reporting structures that create a culture and expectation of diversity, equity, and inclusion.

Resources

ADVANCE program: www.nsf.gov/crssprgm/advance/

Broadening the Representation of Academic Investigators in Neuroscience (BRAINS): www.brains.washington.edu

Christiane Nusslein-Volhard Foundation: www.cnv-stiftung.de/en/goalsl

Inclusion Diversity Equity and Awareness Working Group of The Wildlife Society: https://wildlife.org/dei/

International Union for the Conservation of Nature Gender and the Environment Resource Center: https://genderandenvironment.org/

Launching Academics on the Tenure-Track: an Intentional Community in Engineering (LATTICE): https://advance.washington.edu/about/national/lattice

Minorities in Agriculture, Natural Resources and Related Sciences: www.manrrs.org/

National Association of University Fish and Wildlife Programs: www.naufwp.org

Native American Fish and Wildlife Society: https://nafws.org/

Native Peoples' Wildlife Management Working Group: https://wildlife.org/npwmwg/

Primary Caregiver Technical Assistance Supplements: www.niaid.nih.gov/grants-contracts/research-supplements#A4

Society for Advancement of Chicanos/Hispanics and Native Americans in
   Science (SACNAS): www.sacnas.org/
STEM Equity Achievement (SEA) Change: https://seachange.aaas.org/
Strategies for Ecology Education, Diversity, and Sustainability: https://esa
   .org/seeds/

Discussion Questions

1. Did you identify with any of the challenges highlighted in the chapter?
   Please share your experiences.
2. Which ideas for addressing these challenges to support women and
   people of color seemed most important to you and why?
3. Can you identify any "firsts" in your social arena? For instance, the first
   in your family to attend college? Your teacher is the first to _____.
   Your friend was the first female to_____. What kinds of barriers do you
   think this person had to overcome or is still tackling?
4. Can you identify an important mentor in your life? What are the
   qualities that you are looking for in a mentor? Have you had experiences with bad mentors? What are some steps you could take to be a
   better mentor for others?

Activity: Forming Groups (or Packs, Prides, Pods, or Parliaments)

Learning Objectives: Participants will explore ways in which people form
groups and discuss considerations when intentionally creating groups.
Time: 20 min.
Description of Activity
Materials needed: Before the activity, create animal cutouts that can be
attached via a lanyard or double-sided tape to each participant (see link in
"References and Resources" below for free shapes available on the web).
Animal cutouts can be in different sizes and on different colored paper.
There should be some duplicates of each as well as a diversity of animals,
sizes, and colors of the badges for the group. Each participant will only
receive one badge.
Process:
Part 1 (5 min.): Once everyone has their animal attached to their shirt, instruct
   them to form groups without talking. After forming the first group, repeat

and have them create new groups again without talking. Repeat this at least four times. There are no criteria on how to form group; the participants simply create groups silently for this activity. Remind participants to record who ends up in their group during each trial.

Part 2: Debriefing and Discussion Questions (10–15 min.)

1. How would you describe the composition of the groups that formed during this activity?
2. What did you learn?
3. What was most surprising about this exercise?
4. Have you experienced real-life experiences similar to the results of this activity?
5. How would you use the information you learned or your experience from this activity in your life as a wildlife professional, student, or citizen when intentionally creating groups?

References and Resources

This activity is modified from Fowler, S. M. 2006. "Training across Cultures: What Intercultural Trainers Bring to Diversity Training." *International Journal of Intercultural Relations* 30(3): 401–11.

Animal shapes are available as free clip art but also offered in sites with open access like Wikimedia Commons. For example, a selection of animal shapes like bat, bear, beetle, crow, frog, hippo, octopus, etc. can be found at https://commons .wikimedia.org/wiki/Category:Animal_shapes_by_Trisorn_Triboon.

Literature Cited

Adamo, S. A. 2013. "Attrition of Women in the Biological Sciences: Workload, Motherhood, and Other Explanations Revisited." *BioScience* 63(1): 43–48. doi: 10.1525/bio.2013.63.1.9.

Anderson, W. S. 2020. "The Changing Face of the Wildlife Profession: Tools for Creating Women Leaders." *Human-Wildlife Interactions* 14(1): 15.

Aragon-Long, S. C., V. R. Burkett, H. S. Weyers, S. M. Haig, M. S. Davenport, and K. L. Warner. 2018. *A Snapshot of Women of the U.S. Geological Survey in STEM and Related Careers* (Reston, VA: US Geological Survey). http://pubs.er.usgs.gov/publication /cir1443.

Arismendi, I., and B. E. Penaluna. 2016. "Examining Diversity Inequities in Fisheries Science: A Call to Action." *BioScience* 66 (7): 584–91. doi: 10.1093/biosci/biw041.

Aslan, C. E., M. L. Pinsky, M. E. Ryan, S. Souther, and K. A. Terrell. 2014. "Cultivating Creativity in Conservation Science." *Conservation Biology* 28(2): 345–53.

Bal, T. L., and T. L. Sharik. 2019. "Web Content Analysis of University Forestry and Related Natural Resources Landing Webpages in the United States in Relation to Student and Faculty Diversity." *Journal of Forestry* 117(4): 379–97. doi: 10.1093/jofore/fvz024.

Balcarczyk, K. L., D. Smaldone, S. W. Selin, C. D. Pierskalla, and K. Maumbe. 2015. "Barriers and Supports to Entering a Natural Resource Career: Perspectives of Culturally Diverse Recent Hires." *Journal of Forestry* 113(2): 231–39. doi: 10.5849/jof.13-105.

Batavia, C., B. E. Penaluna, T. R. Lemberger, and M. P. Nelson. 2020. "Considering the Case for Diversity in Natural Resources." *BioScience* 70(8): 708–18. doi: 10.1093/biosci/biaa068.

Baugh, S. G., and T. A. Scandura. 1999. "The Effect of Multiple Mentors on Protege Attitudes toward the Work Setting." *Journal of Social Behavior & Personality* 14(4): 503–21.

Bendels, M. H. K., R. Müller, D. Brueggmann, and D. A. Groneberg. 2018. "Gender Disparities in High-Quality Research Revealed by Nature Index Journals." *PLoS ONE* 13(1): e0189136. doi: 10.1371/journal.pone.0189136.

Blau, F. D., J. M. Currie, R. T. A. Croson, and D. K. Ginther. 2010. "Can Mentoring Help Female Asistant Professors? Interim Results from a Randomized Trial." *American Economic Review* 100 (2): 348–52. doi: 10.1257/aer.100.2.348.

Cameron, E. Z., A. M. White, and M. E. Gray. 2016. "Solving the Productivity and Impact Puzzle: Do Men Outperform Women, or Are Metrics Biased?" *BioScience* 66(3): 245–252. doi: 10.1093/biosci/biv173.

Campbell, K., and A. Mínguez-Vera. 2008. "Gender Diversity in the Boardroom and Firm Financial Performance." *Journal of Business Ethics* 83(3): 435–51.

Carrell, S. E., M. E. Page, and J. E. West. 2010. "Sex and Science: How Professor Gender Perpetuates the Gender Gap." *Quarterly Journal of Economics* 125(3): 1101–44. doi: 10.1162/qjec.2010.125.3.1101.

Case, A. D., and C. D. Hunter. 2012. "Counterspaces: A Unit of Analysis for Understanding the Role of Settings in Marginalized Individuals' Adaptive Responses to Oppression." *American Journal of Community Psychology* 50(1–2): 257–70. doi: 10.1007/s10464 -012-9497-7.

Ceci, S. J., and W. M. Williams. 2011. "Understanding Current Causes of Women's Underrepresentation in Science." *Proceedings of the National Academy of Sciences* 108(8): 3157. doi: 10.1073/pnas.1014871108.

Cho, A., D. Chakraborty, and D. Rowland. 2017. "Gender Representation in Faculty and Leadership at Land Grant and Research Institutions." *Agronomy Journal* 109(1): 14–22. doi: 10.2134/agronj2015.0566.

Christiane Nusslein-Volhard Foundation (2022). http://www.cnv-stiftung.de/en/goalsl.

Clancy, K. B. H., R. G. Nelson, J. N. Rutherford, and K. Hinde. 2014. "Survey of Academic Field Experiences (SAFE): Trainees Report Harassment and Assault." *PLoS ONE* 9(7): e102172. doi: 10.1371/journal.pone.0102172.

Diaz, B. 2016. "Mainstreaming Gender in National Biodiversity Strategy of Mexico and Action Plan 2016–2030." Knowledge Base: Best Practices. NBSAP Forum. Last modified November 16. http://nbsapforum.net/knowledge-base/best-practice /mainstreaming-gender-national-biodiversity-strategy-mexico-and-action.

Doris Duke Charitable Foundation. N.d. "Doris Duke Conservation Scholars Program." Accessed May 2019. www.ddcf.org/funding-areas/environment/.

Edge, W. D. 2016. "Learning for the Future: Educating Career Fisheries and Wildlife Professionals." *Mammal Study* 41(2): 61–69.

Feeney, M. K., and M. Bernal. 2010. "Women in STEM Networks: Who Seeks Advice and Support from Women Scientists?" *Scientometrics* 85(3): 767–90. doi: 10.1007/ s11192-010-0256-y.

Freeman, R. B., and W. Huang. 2015. "Collaborating with People like Me: Ethnic Coauthorship within the United States." *Journal of Labor Economics* 33(S1): S289–S318. doi: 10.1086/678973.

Gharis, L. W., S. G. Laird, and D. C. Osborne. 2017. "How Do University Students Perceive Forestry and Wildlife Management Degrees?" *Journal of Forestry* 115(6): 540–47. doi: 10.5849/jof-2016-080r3.

Goulden, M., M. A. Mason, and K. Frasch. 2011. "Keeping Women in the Science Pipeline." *Annals of the American Academy of Political and Social Science* 638(1): 141–62. doi: 10.1177/0002716211416925.

Gumpertz, M., R. Durodoye, E. Griffith, and A. Wilson. 2017. "Retention and Promotion of Women and Underrepresented Minority Faculty in Science and Engineering at four Large Land Grant Institutions." *PLoS ONE* 12(11): e0187285. doi: 10.1371/journal. pone.0187285.

Hansen, W. D., J. P. Scholl, A. E. Sorensen, K. E. Fisher, J. A. Klassen, L. Calle, G. S. Kandlikar, N. Kortessis, . . . and M. E. Shea. 2018. "How Do We Ensure the Future of Our Discipline Is Vibrant? Student Reflections on Careers and Culture of Ecology." *Ecosphere* 9(2): e02099. doi: 10.1002/ecs2.2099.

Hooker, S. K., S. E. Simmons, A. K. Stimpert, and B. I. Mcdonald. 2017. "Equity and Career-Life Balance in Marine Mammal Science?" *Marine Mammal Science* 33(3): 955–65.

Horner-Devine, M. C., J. W. Yen, P. N. Mody-Pan, C. Margherio, and S. Forde. 2016. "Beyond Traditional Scientific Training: The Importance of Community and Empowerment for Women in Ecology and Evolutionary Biology." *Frontiers in Ecology and Evolution* 4(119). doi: 10.3389/fevo.2016.00119.

Hunter, M. L. J., R. K. Hitchcock, and B. Wyckoff-Baird. 1990. "Women and Wildlife in Southern Africa." *Conservation Biology* 4(4): 448–451.

International Union for the Conservation of Nature (IUCN). 2020. Gender and the Environment Resource Center. https://genderandenvironment.org/.

Isidro, H., and M. Sobral. 2015. "The Effects of Women on Corporate Boards on Firm Value, Financial Performance, and Ethical and Social Compliance." *Journal of Business Ethics* 132(1): 1–19.

Katz, S. J. 2008. "WEBS: Practicing Faculty Mentorship." *BioScience* 58(1): 15. doi: 10.1641/ B580105.

Kern, C. C., L. S. Kenefic, and S. L. Stout. 2015. "Bridging the Gender Gap: The Demographics of Scientists in the USDA Forest Service and Academia." *BioScience* 65(12): 1165–72. doi: 10.1093/biosci/biv144.

Kohl, M. T., S. J. Hoagland, A. R. Gramza, and J. A. Homyak. 2017. "Professional Diversity: The Key to Conserving Wildlife Diversity." In *Becoming a Wildlife Professional*, edited by S. E. Henke and P. R. Krausman, 188–95. Baltimore: John Hopkins University Press.

Leisher, C., G. Temsah, F. Booker, M. Day, L. Samberg, D. Prosnitz, B. Agarwal, E. Matthews, D. Roe, . . . and D. Wilkie. 2016. "Does the Gender Composition of Forest and Fishery Management Groups Affect Resource Governance and Conservation Outcomes? A Systematic Map." *Environmental Evidence* 5(1): 6. doi: 10.1186/s13750-016-0057-8.

Lidestav, G., and A. Egan Sjölander. 2007. "Gender and Forestry: A Critical Discourse Analysis of Forestry Professions in Sweden." *Scandinavian Journal of Forest Research* 22(4): 351–62. doi: 10.1080/02827580701504928.

Loison, A., S. Paye, A. Schermann, C. Bry, J.-M. Gaillard, C. Pelabon, and K.-A. Bråthen. 2016. "The Domestic Basis of the Scientific Career: Gender Inequalities in Ecology in France and Norway." *European Educational Research Journal* 16(2/3): 230–57. doi: 10.1177/1474904116672469.

Lopez, R., and C. Brown. 2011. "Why Diversity Matters: Broadening Our Reach Will Sustain Natural Resources." *Wildlife Professional* 5(2): 20–27.

Maes, K., J. Gvozdanovic, S. Buitendijk, I. R. Hallberg, and B. Mantilleri. 2012. "Women, Research and Universities: Excellence without Gender Bias." *League of European Research Universities*. www.leru.org/publications/women-research-and-universities -excellence-without-gender-bias#.

Mansfield, B., R. Lave, K. Mcsweeney, A. Bonds, J. Cockburn, M. Domosh, T. Hamilton, R. Hawkins, . . . and C. Radel. 2019. "It's Time to Recognize How Men's Careers Benefit from Sexually Harassing Women in Academia." *Human Geography* 12(1): 82–87.

McChesney, J., and J. Bichsel. 2020. *The Aging of Tenure-Track Faculty in Higher Education: Implications for Succession and Diversity*. College and University Professional Association for Human Resources. www.cupahr.org/surveys/research-briefs/.

McGuire, K. L., R. B. Primack, and E. C. Losos. 2012. "Dramatic Improvements and Persistent Challenges for Women Ecologists." *BioScience* 62(2): 189–96. doi: 10.1525/bio.2012.62.2.12.

Mervis, J. 2016. "Mentoring's Moment." *Science* 353 (6303): 980. doi: 10.1126/science.353.6303.980.

Mourad, T. M., A. F. Mcnulty, D. Liwosz, K. Tice, F. Abbott, G. C. Williams, and J. A. Reynolds. 2018. "The Role of a Professional Society in Broadening Participation in Science: A National Model for Increasing Persistence." *BioScience* 68(9): 715–21. doi: 10.1093/biosci/biy066.

National Association of University Fish and Wildlife Programs (NAUFWP). N.d. Accessed May 2021. www.naufwp.org.

National Center for Education Statistics (NCES). N.d. "Definitions for New Race and Ethnicity Categories." Integrated Postsecondary Education Data System. https://nces.ed.gov/ipeds/report-your-data/race-ethnicity-definitions.

Natural Resources Canada (NRC). 2015. "10 Key Facts on Canada's Natural Resources; Key Facts and Figures Supporting Gender-Based Analysis at NRCan; Women in Science and Technology at NRCan (presentation)." www.nrcan.gc.ca/sites/www.nrcan.gc.ca/files/files/pdf/10_key_facts_nrcan2015_e.pdf.

Natural Resources Canada (NRC). 2018. "The Equal by 30 Campaign." Last modified December. www.nrcan.gc.ca/21638.

Natural Resources Canada (NRC). 2019a. "Diversity in STEM: Re-entry Program." Last modified April. www.nrcan.gc.ca/careers/21701?fbclid=IwAR2CMGtKzVOh-GrFd_Rm4ieEaZXi1qZIb8_USQvgOcVPGUwpjiBzIDAHLog.

Natural Resources Canada (NRC). 2019b. "Gender-Based Analysis Plus." Last modified April. www.nrcan.gc.ca/plans-performance-reports/dp/2019-20/21792.

Newer, R. 2018. "Meet the 'Brave Ones': The Women Saving Africa's Wildlife." Future. BBC. Last modified 2019. http://www.bbc.com/future/story/20180926-akashinga-all-women-rangers-in-africa-fighting-poaching.

Nicholson, K. L. 2009. "Wanted: Female Role Models." *Wildlife Professional* 3(1): 40–42.

Nicholson, K. L., P. R. Krausman, and J. A. Merkle. 2008. "Hypatia and the Leopold Standard: Women in the Wildlife Profession 1937–2006." *Wildlife Biology in Practice* 4:57–72.

Nielsen, M. W., S. Alegria, L. Börjeson, H. Etzkowitz, H. J. Falk-Krzesinski, A. Joshi, E. Leahey, L. Smith-Doerr, A. W. Woolley, and L. Schiebinger. 2017. "Opinion: Gender Diversity Leads to Better Science." *Proceedings of the National Academy of Sciences* 114(8): 1740. doi: 10.1073/pnas.1700616114.

Østergaard, C. R., B. Timmermans, and K. Kristinsson. 2011. "Does a Different View Create Something New? The Effect of Employee Diversity on Innovation." *Research Policy* 40(3): 500–509. doi: 10.1016/j.respol.2010.11.004.

Oyserman, D. 2014. "Identity-Based Motivation: Core Processes and Intervention Examples." In *Motivational Interventions*, vol. 18 of *Advances in Motivation and Achievement*, 213–42. Bingley, UK: Emerald Group.

Perryman, A. A., G. D. Fernando, and A. Tripathy. 2016. "Do Gender Differences Persist? An Examination of Gender Diversity on Firm Performance, Risk, and Executive Compensation." *Journal of Business Research* 69(2): 579–86. doi: 10.1016/j.jbusres.2015.05.013.

Primack, R. B., and V. O'Leary. 1993. "Cumulative Disadvantages in the Careers of Women Ecologists." *BioScience* 43(3): 158–65. doi: 10.2307/1312019.

Reybold, L. E., S. D. Brazer, L. Schrum, and K. W. Corda. 2012. "The Politics of Dissertation Advising: How Early Career Women Faculty Negotiate Access and Participation." *Innovative Higher Education* 37 (3): 227–42. doi: 10.1007/s10755-011-9200-1.

Rice, C. 2011. "Scientific (E)quality." *Interdisciplinary Science Reviews* 36(2): 114124. doi: 10.1179/030801811X13013181961356.

Sardelis, S., S. Oester, and M. Liboiron. 2017. "Ten Strategies to Reduce Gender Inequality at Scientific Conferences." *Frontiers in Marine Science* 4(231). doi: 10.3389/fmars.2017.00231.

Schiebinger, L., and S. K. Gilmartin. 2010. "Housework Is an Academic Issue." *Academe* 96(1). www.aaup.org/article/housework-academic-issue#.YdSonmjMKHs.

Sharik, T. L., R. J. Lilieholm, W. Lindquist, and W. W. Richardson. 2015. "Undergraduate Enrollment in Natural Resource Programs in the United States: Trends, Drivers, and Implications for the Future of Natural Resource Professions." *Journal of Forestry* 113(6): 538–551. doi: 10.5849/jof.14-146.

Shauman, K. A. 2017. "Gender Differences in the Early Employment Outcomes of STEM Doctorates." *Social Sciences* 6(1): 24.

Smith, F. A., and D. M. Kaufman. 1996. "A Quantitative Analysis of the Contributions of Female Mammalogists from 1919 to 1994." *Journal of Mammalogy* 77(3): 613–28.

Smith, K., P. Arlotta, F. M. Watt, and S. L. Solomon. 2015. "Seven Actionable Strategies for Advancing Women in Science, Engineering, and Medicine." *Cell Stem Cell* 16(3): 221–24. doi: 10.1016/j.stem.2015.02.012.

Solanki, S. M., and D. Xu. 2018. "Looking beyond Academic Performance: The Influence of Instructor Gender on Student Motivation in STEM Fields." *American Educational Research Journal* 55(4): 801–35. doi: 10.3102/0002831218759034.

Solorzano, D. G. 1998. "Critical Race Theory, Race and Gender Microaggressions, and the Experience of Chicana and Chicano Scholars." *International Journal of Qualitative Studies in Education* 11(1): 121–36. doi: 10.1080/095183998236926.

Sweet, S., and P. Moen. 2004. "Coworking as a Career Strategy: Implications for the Work and Family Lives of University Employees." *Innovative Higher Education* 28(4): 255–72. doi: 10.1023/B:IHIE.0000018909.62967.b5.

Symonds, M. R. E., N. J. Gemmell, T. L. Braisher, K. L. Gorringe, and M. A. Elgar. 2006. "Gender Differences in Publication Output: Towards an Unbiased Metric of Research Performance." *PLoS ONE* 1(1): e127. doi: 10.1371/journal.pone.0000127.

Taylor, D. E. 2007. "Diversity and the Environment: Myth-Making and the Status of Minorities in the Field." In *Equity and the Environment*, vol. 15 of *Research in Social Problems and Public Policy*, 89–147. Bingley, UK: Emerald Group.

Taylor, D. E. 2014. *The State of Diversity in Environmental Organizations: Mainstream NGOs, Foundations & Government Agencies.* Green 2.0 Working Group. www.diversegreen.org/the-challenge/.

Taylor, D. E. 2018. "Racial and Ethnic Differences in the Students' Readiness, Identity, Perceptions of Institutional Diversity, and Desire to Join the Environmental Workforce." *Journal of Environmental Studies and Sciences* 8(2): 152–68. doi: 10.1007/s13412-017-0447-4.

US Bureau of Labor Statistics (BLS). 2016. "Employment Projections, Occupational Separations and Openings." US Department of Labor, Bureau of Labor Statistics. Last modified January 30, 2018. Accessed April 4, 2019. www.bls.gov/emp/tables/occupational-separations-and-openings.htm#top.

US Census Bureau. 2011. *Overview of Race and Hispanic Origin: 2010.* 2010 Census Briefs. Washington, DC: US Census Bureau. www.census.gov/prod/cen2010/briefs/c2010br-02.pdf.

US Department of the Interior (DOI). 2012. Diversity and Inclusion Strategic Plan. https://edit.doi.gov/sites/doi.gov/files/migrated/pmb/eeo/whoweare/upload/Diversity-and-Inclusion-Strategic-Plan-Department-of-the-Interior-3-16-2012.pdf.

Valantine, H. A., and F. S. Collins. 2015. "National Institutes of Health Addresses the Science of Diversity." *Proceedings of the National Academy of Sciences* 112(40): 12240. doi: 10.1073/pnas.1515612112.

The Wildlife Society (TWS). N.d. (a). "Where to Get Your Degree." Accessed January 2, 2022. https://wildlife.org/next-generation/career-development/where-to-get-your-degree/.

The Wildlife Society (TWS). N.d. (b). Wildlife Vocalizations. Accessed January 2, 2022. https://wildlife.org/tag/wildlife-vocalizations/.

The Wildlife Society (TWS). 2021. *Diversity Equity and Inclusion TWS Vision within the 2019–2023 strategic Plan and Beyond.* https://wildlife.org/wp-content/uploads/2021/08/20210802-DEI-Vision-FINAL.pdf.

The Wildlife Society (TWS). 2022. "Diversity, Equity, and Inclusion." https://wildlife.org/dei/.

Wolfinger, N. H., M. A. Mason, and M. Goulden. 2008. "Problems in the Pipeline: Gender, Marriage, and Fertility in the Ivory Tower." *Journal of Higher Education* 79(4): 388–405. doi: 10.1080/00221546.2008.11772108.

Woolley, A. W., C. F. Chabris, A. Pentland, N. Hashmi, and T. W. Malone. 2010. "Evidence for a Collective Intelligence Factor in the Performance of Human Groups." *Science* 330(6004): 686–88. doi: 10.1126/science.1193147.

Xu, Y. J. 2008. "Gender Disparity in STEM Disciplines: A Study of Faculty Attrition and Turnover Intentions." *Research in Higher Education* 49(7): 607–24. doi: 10.1007/s11162-008-9097-4.

CHAPTER

4

Evelyn H. Merrill, Patricia L. Kennedy, Susan K. Skagen, and Kathy A. Granillo

Personal and Institutional Barriers to Success

Challenges and Solutions for Women

Why Aren't There More Women Leaders in Science?

How could a Nobel Prize winner not be a distinguished full professor? In 2018 Associate Professor Donna Strickland became only the third woman to win the Nobel Prize in Physics. When asked why she wasn't a full professor at Canada's University of Waterloo she said, *"I never applied,"* and commented, *"I have been treated as an equal by my male peers"* (Stack 2018). The bottom line is that her eminence in science is not reflected in her professional stature. Was she so enamored with her research that she was not in tune with academic advancement, or is this further evidence of institutional problems related to mentoring and advancement of women?

We have come a long way from the era when a woman had to quit the workplace once she married or could not be hired by the same institution that hired her husband because of nepotism rules. The number of women with science, engineering or health (SEH) doctorates has more than doubled since 1997 (Foley et al. 2019). Currently, 35% of the SEH doctorate holders are women, compared to 23% twenty years ago. Growth in female employment varied by employment sector. The largest gain was in the US government, increasing from 19% of all federally employed SEH doctorates in 1997 to 35% in 2017. Similarly, the share of women employed in four-year educational institutions increased from 25% in 1997 to 38% in 2017. To what degree gender currently impacts a woman's ability to enter and advance through the SEH workplace is debatable, but gaps between men and women still exist, particularly in salaries and representation in leadership positions (Homyack et al. 2014).

To understand what contributes to this discrepancy, we lean on the framework of Holmes (2015) and Risman and Davis (2013), who characterize potential barriers as individual, interactional, and institutional (Figure 4.1). Individual factors are personal qualities reflecting physical capabilities or personality traits and career and personal choices. These barriers may change with education, networking, role models, or advocacy for women. Interactional barriers arise from interpersonal relations. These can be positive, for example, the presence of a strong mentor-mentee relationship, or negative, in the case of sexual discrimination or harassment. For example, the early career trajectory of one of the authors, Susan Skagen, was dramatically and positively changed by the intervention of Nina Leopold Bradley, Aldo Leopold's eldest daughter. Based on her prior life experience, Bradley encouraged Skagen to undertake her own PhD project rather than act as volunteer for her then-husband's proposed field work; she then helped her secure a fellowship to start her PhD program. Institutional barriers include formal policies and informal practices that may reflect influences of current or historical people in top positions.

We begin this chapter with a discussion of individual and interactional issues focusing on implicit biases, stereotypes, and overt sexual harassment. We

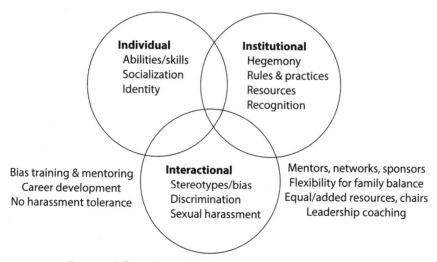

FIGURE 4.1. Framework for understanding barriers and solutions to gender parity. Redrawn combining Holmes (2015) and Risman and Davis (2013).

take a broader perspective than women in wildlife or natural resources because most of the published literature on these topics is on science, technology, engineering, and mathematics (STEM) careers. We then focus our attention to comparing research universities (hereafter referred to as academia) and US federal government workplaces to explore issues of institutional barriers in more detail, highlighting how their differences may influence gender parity from recruitment to advancement. Finally, we end with suggestions for generating a more inclusive workplace environment.

Gender Bias and Stereotypes

Most workplaces that hire wildlife biologists have substantially reduced overt forms of gender discrimination that impede a woman's ability to pursue a career. However, discrimination resulting from gender bias is an ongoing problem in society in general and the wildlife profession in particular.

Implicit bias (or **implicit social cognition**) refers to attitudes or stereotypes unconsciously affecting our understanding, actions, and decisions. These unconscious thought patterns are not generally accessible through introspection and can have the unintended effect of conferring advantage to some groups of people and disadvantages to others. Implicit biases about gender roles are formed in childhood. For example, Bian et al. (2017) studied young children to assess when perceptions of intellectual brilliance emerge. At age 5, children seemed not to differentiate between boys and girls in expectations of "really, really smart"—childhood's version of adult brilliance. But by age 6, girls lumped more boys into this category and steered away from games intended for the "really, really smart." These perceptions are reinforced by parents and teachers, although not necessarily consciously. As children grow, peers are also likely to influence these biases. Youth with peer groups who encourage, endorse, or exemplify high math and science achievement are more likely to take math and science classes (Wang and Degol 2017). Girls may be more susceptible to peer social influences in these areas than boys.

These gender-stereotypical beliefs influence behavior through adolescence and adulthood, including choice of academic majors and careers. At an early age, boys prefer working with objects, whereas girls prefer working with people (Wang and Degol 2017). Young women may avoid STEM career paths because they perceive STEM as antithetical to their communal goals of helping and collaborating. As Diekman et al. (2010) point out, it is ironic that STEM fields help many people but are commonly regarded as individualistic, as in the stereotype of the "the lone scientist or naturalist." Within STEM fields, women are more likely to choose careers that emphasize community or are perceived to be more people oriented, such as biomedicine and psychology. Helping the environment and love of the natural world has also been shown to motivate women to choose careers in natural resources (Angus 1995). One of the authors, Patricia Kennedy, grew up on the South Side of Chicago, where she had little experience with wildlife, so her passion was directed at improving the lives of companion animals. There were always dogs and cats in the house, and she worked at a local vet clinic in high school, which cemented her desire to become a veterinarian. As a first-year biology major at Colorado College, faculty mentors expanded Kennedy's awareness of the natural world and helped her channel her interests into the conservation of wildlife. A future wildlife biologist was born.

Numerous studies have affirmed that people link men with more traits that connote physical and mental strength and view women as physically weaker and more emotional, which is assumed to limit mental stamina. Natural resource jobs that involve fieldwork are frequently in remote locations, can be physically demanding, and pose safety issues. For many of these jobs, gendered inequalities are perpetuated by using masculine traits as the point of reference for what a desirable employee should look like (often referred to as **hegemonic masculinity**). In male-dominated fields, like wildland firefighting, competencies are linked to physical capabilities for managing risk and emotional capabilities for suppressing emotional distress (Eriksen et al. 2016). Equality tends to equate to women becoming like men, and femininity is considered a "problem." However, there are many ways to approach field-based problems that do not necessarily involve physical strength. For example, during shorebird surveys along the Texas coast in the early 1990s, coauthor Skagen had a field technician who was newly pregnant. She was frightened that a venomous snake bite would harm her fetus and so was reluctant to walk in some of the survey sites. Skagen told the technician to prioritize her baby's well-being and not to jeopardize her own health. As a result, the technician relaxed and found creative ways to complete all the surveys. Giving field technicians permission up front to place their safety and well-being ahead of data collection empowers staff to find ways around uncomfortable or unsafe situations while doing their job. In these workplaces, programs to promote diversity are not likely to be effective until differences are respected and hegemonic masculinity is no longer viewed as the gold standard (Eriksen et al. 2016).

There are few women in high-ranking positions in all workplaces. Overall, the higher the rank in STEM, the less likely it is to be occupied by a woman (Amon 2017). Challenges for women reaching leadership positions are complex, but gender stereotyping is one factor influencing the slow rate of advancement of women in STEM. Women are often associated with communal qualities, which convey a concern for the compassionate treatment of others. In contrast, men are associated with qualities that convey assertion, authority, and control, traits commonly associated with successful scientists. As a result, women find themselves in a double bind. If they are highly communal, they may be criticized for not being authoritative and competent enough. But if they are highly assertive, they may be criticized for lacking communal qualities (Eagly and Carli 2007). It's a struggle to appear both competent and warm. People's divergent expectations of leaders result in resistance to women's leadership.

Although few people in society explicitly state that men are better at science or math than women, subtler versions of this message persist and are reinforced by the media. For example, in television programs popularly watched by middle-school-aged students, male scientists appear 1.6 times as frequently as female scientists (Eddy and Brownell 2016). In contrast, representations of women and men in science in the *New York Times* from 2011 to 2018 were more accurate and less fraught with gender stereotypes (Mitchell and McKinnon 2019).

Training for a Career in STEM

Although retention and graduation rates are quite favorable for women in wildlife biology and many STEM fields, there is growing evidence that women and men experience their undergraduate programs differently based on gender stereotypes they acquired in childhood. High numerical representation of women in the undergraduate population does not mean there are no gender inequalities in their academic environment. Gender disparities in STEM classrooms can influence women's academic experiences and ultimately impact retention of women in STEM majors (Cundiff et al. 2013; Eddy and Brownell 2016; Wang and Degol 2017).

Three important predictors of STEM retention in undergraduate programs are a student's sense of belonging, self-efficacy (confidence in one's ability to complete a task), and disciplinary identity. Although the data are equivocal, there is evidence that women in STEM majors at the beginning of their college career report significantly lower academic self-efficacy compared with their male peers, particularly on study skills, test-taking skills, and coping with test anxiety. One of the major sources of women's self-efficacy is recognition from others. Evidence suggests women in STEM classes may not be gaining as much recognition as men by their peers and instructors (Zeldin and

Pajares 2000), and thus as women advance through their college degrees, their self-efficacy does not always improve (Eddy and Brownell 2016).

Regular lack of recognition can cumulatively lead to women believing they do not belong in the university environment. Studies suggest only women with a heightened sense of belonging consider graduate school, whereas a sense of belonging does not influence men's decision to pursue graduate school (Eddy and Brownell 2016). What graduate disciplines they choose is influenced by the identity of others in the discipline. Although students of both genders equally identified with fields in science, white women identified more with biology than did white men. Men and women of color did not differ from each other in the type of science they identified with (Cundiff et al. 2013; Eddy and Brownell 2016). This strong discipline identity may partially explain the high rates of white women graduating in wildlife biology.

Finally, closing the gender gap in STEM educational experiences is jeopardized by the persistence of sexual harassment in our nation's colleges and universities. In 2018, the National Academies of Sciences, Engineering, and Medicine thoroughly examined research on sexual harassment conducted over the past 30 years (NASEM 2018). Its analysis of the literature sadly indicates that depending on the institution, about 20%–50% of women experienced sexual harassment from faculty or staff during their STEM undergraduate and graduate education. As they note, it is especially discouraging that while extensive resources are being invested in efforts to attract and retain women in STEM programs, women are often bullied or harassed out of career pathways in these fields. These behaviors can be aggravated for women working in the field, both students and nonstudents, who may face these social dangers beyond the myriad of environmental and physical safety concerns in a field situation. As women scientists concerned about the well-being of our field employees and colleagues, we join the call of Rinkus et al. (2018) for an increased dialogue among conservation professionals on gendered considerations for safety in fieldwork. Little emphasis is placed on avoiding social dangers by using the buddy system, avoiding or leaving areas where danger is detected, and trusting your intuition and judgment. Of more than 100 safety training courses at the US Department of Interior, for example, none explicitly address how women can identify and respond to unsafe social situations in the field (see "Resources" at the end of the chapter). Luckily, woman can avail themselves of private courses in self-defense that provide tips on how to assess danger, how to "be a bad target," how to identify pre-attack indicators and diffuse potentially dangerous encounters, evasion tactics, and pre-positioning moves (e.g., FunctionalSelfDefense, see "Resources").

Hiring and Advancement

The gender ratios of STEM positions suggests a "leaky pipeline" exists where women have achieved parity with men in training, yet at subsequent career stages continue to be less well represented. Hewlett and Luce (2005) suggest this is more adequately portrayed as a "highway off ramp" whereby women exit by making active choices rather than passively pursuing, retaining, and advancing along career paths. This is consistent with recent literature that provides strong evidence that the reason for underrepresentation in some disciplines is not an anti-woman bias in hiring but that fewer women apply (Glass and Minnotte 2010; Ceci and Williams 2011; Williams and Ceci 2015). As one of the few documented examples, Shauman (2017) tracked members of nine groups of PhD students who earned their degrees between 1995 and 2013. Although most doctorates (>94%) entered the workforce, women were less likely to do so than men, and if they entered the labor market they were more likely to work part time, a pattern across the disciplines in STEM fields that did not change over the time of the study. Importantly, her results demonstrate that the presence of children and being in a dual-earner couple were strong influences on women's decisions to enter the labor market and if they entered it, to only work part time.

Institutional Support for Women

Institutions influence how women are recruited and advance by determining conditions for recruiting, enacting rules for advancement, and providing resources for success (Uriarte et al. 2007). These influences are social and organizational and can foster or hinder women's careers (Fox and Mohapatra 2007). We contrast institutional differences between academia and the US federal system (Table 4.1) to explore potential reasons for them and how they contribute to gender parity.

The Academic Workplace

In 1973, 55% of US doctorates in the biological sciences secured tenure-track positions within six years of completing their PhDs, compared to only 10%–15% of PhDs more recently (Shauman 2017). Severe budgets cuts in higher education have resulted in fewer tenure-track positions. Even though the number of academic jobs in the life and health sciences is not keeping pace with the production of PhDs (Langin 2019), women are pursuing and obtaining academic positions in the wildlife and natural resources fields.

TABLE 4.1 Comparison of characteristics of positions within academia and US federal government and potential effects on women.

	Academia	US Federal Government	Potential Impact on Gender Parity
History in diversity programs	Since 2000s: external programs (e.g., NSF ADVANCE)	Affirmative action, 1961 Agency responses to lawsuits, ~1980s Government-wide diversity strategic plan, 2011	Lower demographic inertia in government due to earlier and agency-wide programs
Institution structure	Loosely decoupled across individual universities	Hierarchical for personnel decisions Independence in choosing research topics	Flexibility but less coordination across universities than in government to address gender parity
Recruitment	Individualistic Potential to accommodate partner	Widespread access to all job listings and standard application process through USAJOBS Recruitment of STEM students K–12 and college	Easier to find all job listings for federal positions Active recruitment by agencies to increase participation in STEM professions
Salaries	Negotiated at recruitment Merit raises based on performance	Standardized by grade and time/performance and research grade evaluations	Less discrepancy in government salaries for time in position
Evaluation	Tenure/promotion 1/time at ~6 yrs. High initial risk for long-term security	Research: RGEG: person-in-job 4-yr. evaluations by panel of peers based on publication record, creativity, originality, impact Management agencies: advancements may be based on transfers to new job in new locale	Government job security sooner than academia, transparent evaluation process in government research May be hard to advance to highest levels in management agencies without moving
Reward metrics	Metrics defined for tenure but less so for advancement Priority for publication and grant funding Teaching and service secondary Demonstrated leadership role for full professorships	Research: well defined in RGEG Publications primary but also reward service to management and scientific communities Management agencies: metrics are individualized	In government, metrics defined throughout career More expectation of service to management community through impact of research and activities
Time allocation	High time investment Flexible hrs. Sabbatical leave	Some flexibility Not expected to work >40hrs./wk.	More predictable hours fostering more stable life-family balance in government

Kern et al. (2015) describe academic institutions as networks of independent universities that are loosely coupled, where research is decentralized and investigator led, unlike the government. Hiring and tenure decisions are made in most cases based on recommendations from peer review (Table 4.1). This may provide greater flexibility within institutions to overcome gender bias and become allies for women. Nonetheless, most mandates to diversify faculty composition followed the US federal equal opportunity laws in the 1970s or were motivated more by recommendations from external federal funding organizations, such as the National Science Foundation (NSF) in the United States and the Natural Sciences and Engineering Council (NSERC) in Canada (Morimoto et al. 2013; Stepan-Norris and Kerrissey 2016).

Hiring and Advancement in Academia

Disparities between the proportions of women graduates and women academics may be explained by attrition due to demographic inertia (Shaw and Stanton 2012; Bakker and Jacobs 2016), continuing biases (Martin 2012; Moss-Racusin et al. 2012), or free choice or choices dictated by circumstances (Ceci and Williams 2011). Demographic inertia relates to historical inequalities that have existed for women and will take some time to change even if all barriers to them were removed. Therefore, reaching parity between men and women within a reasonable time frame will require not only vigorous hiring of qualified women as men retire but changes in policies that influence retention and career progression (Thomas et al. 2015). It is clear that women at times also have to "lean in" and advocate for themselves (Sandberg and Scovell 2013).

The evidence that unconscious bias against women still exists is equivocal. For example, Moss-Racusin et al. (2012) found faculty gave identical applicants for a lab manager higher scores if they had male names. In contrast, Williams and Ceci (2015) presented faculty in the United States with hypothetical summaries of men and women applying for tenure-track assistant professorships. The applicants had identical scholarly outputs but marital status and number of children, for example, were varied systematically. Both men and women faculty preferred female applicants in most fields 2:1 over identically qualified men. Williams and Ceci (2015) concluded that efforts to combat formerly widespread sexism have succeeded in the academic setting, assuming women apply as frequently as men (Kaminski and Geisler 2012).

A factor in hiring is that women (40%–44%) are more likely than men (29%–36%) to have academic partners (Schiebinger et al. 2008; Girod et al. 2011). Dual-career hiring in academia increased from 3% in the 1970s to 13%

in 2000. The "trailing partner," who often is still usually a woman, received lower salary and start-up and accepted a temporary or non-tenure-track position (Schiebinger et al. 2008). Advice on when to reveal dual partnership when applying for a job is mixed (Kaplan 2010; Morton and Kmec 2018). Although waiting until a job is offered provides the greatest opportunity, revealing a need for spousal accommodation at an interview gives the institution more time to find a satisfactory situation. When institutions do not make dual-faculty accommodations, faculty are reported to have lower satisfaction and retention (Costa and Kahn 2000). In contrast, dual couples who were equal in their positions felt a high level of institutional commitment (Zhang and Kmec 2018). This satisfaction finally came for one of the authors, Evelyn Merrill, who spent 13 years seeking a full-time faculty position as part of a dual-career couple.

An assistant professor typically is hired in a tenure-track position with duties of teaching, research, and academic committees or professional service. Unlike in government, an assistant professor has six years to demonstrate their ability to be highly productive in these areas, at which point they are granted tenure or lose their position. After this extended, highly demanding early period, advancement criteria to full professor are less well defined but continue to focus on strong research productivity with increased emphasis on leadership and broad-scale recognition (Table 4.1).

What constitutes productivity and how do women fare? Publications remain the gold standard for judging productivity in academia, yet women are reported to have fewer (Smith and Kaufman 1996; Symonds et al. 2006; Addessi et al. 2012; McGuire et al. 2012; Sugimoto and Blaise 2013). This claim has been debated, but evidence supports it even when time to tenure decision (Lee et al. 2013; Cameron et al. 2016), academic rank, and institution prestige (NRC 2010) are accounted for in the comparison. In the three journals of The Wildlife Society from 1937 to 2006, only 18% of articles had women contributing, which increased to 41% by 2000–2006 (Nicholson et al. 2008). At the same time, because women published less, it was claimed that articles by women were of higher impact (Goulden et al. 2011; Duch et al. 2012). We found mixed support for higher citation rates as an index of publication impact. Instead, citation rate depended on the particular citation metric, the age cohort of the women, and how the comparisons were made (Peñas and Willett 2006; Tower et al. 2007; Esarey and Bryant 2018). If women have had a higher citation rate than men (Powell et al. 2009), it seems to be disappearing (van Arensbergen et al. 2012; Pautasso 2013; but see Cameron et al. 2016). Men are still found in the prestigious first and last author positions (West et al. 2013), and men are more likely to self-cite than women (Cameron et al. 2016; Fox et al. 2016).

An investigator's prestige is also based on number and amounts of successful grants. Studies show that women in medicine, science, and engineering have been awarded fewer grants than men at least initially (first four years) in their careers (Bornmann et al. 2007; Sheridan et al. 2017), but determining the reasons is difficult. Women in academia submitted fewer grants than men did, but their success rate was on par with that of men. The amount of funding awarded to faculty in Canada's federal NSERC Discovery Grant program from 2013 to 2019 was found to differ by stage of career but not gender (NSERC 2019), whereas the median funding awarded to women from NSF was 57% lower ($59,000 versus $137,000) than awards to men in 2010 (NSF 2011).

Academic institutions invest heavily in recruiting faculty (e.g., costs of interviews, start-up funding, lab space), so ensuring their retention is key. Most universities are recognizing the value of effective mentoring for retaining faculty (Montgomery 2017) by helping young faculty design courses, recruit students, navigate grant submission and publication, and form collaborative networks. Men generally have better academic networks (Kyvik and Teigen 1996; Fuchs et al. 2001), and extent of research collaboration indeed appears to influence academic performance (Lee and Bozeman 2005). Woman who had mentors had more publications in peer-reviewed journals and greater career satisfaction than women with no mentors (Levinson et al. 1991). Approaches to mentoring are diversifying from a hierarchical system whereby the department head assigns a senior faculty to small-committee mentoring. Hierarchical mentoring typically promotes organizational norms but not diverse networking (Higgins and Kram 2001). The effectiveness of mentoring also hinges on senior faculty having enough time and training to foster a mentee. In contrast, a committee provides varied perspectives on professional development and issues such as time management and work-life balance (Trube and vanDerveer 2015). Engagement in professional networks, for example, within The Wildlife Society working groups, also develops connections that can provide access to broader resources, facilitate collaboration, share teaching tools, and assist in navigating obstacles. Regardless of the type of mentoring, its positive effects, including improvement in self-esteem, were noted in all studies (Gardiner et al. 2007; Laver et al. 2018).

Work-Life Balance in Academia

Tenure pressures over an extended period put higher demands on faculty's time and work-life balance. Several studies show that women on average take one to four years longer to obtain tenure than men. This has been attributed to women having more stop-start cycles where they take breaks to attend to family responsibilities, agreeing to more teaching and service duties, and engaging in fewer supporting and collaborative networks (Cameron et al. 2016). The view that the tenure clock competes with the biological clock is a prevailing theme in the literature on academic mothers. Joecks et al. (2014) reported that giving birth at an earlier career stage correlated to lower work productivity than giving birth after tenure, whereas they did not find a similar relationship for men. In contrast, Harris et al. (2019) found mixed results, with outcomes depending on the number of children and how far apart they were born. An initial one-time birth had a short-term positive impact but a negative effect on productivity over time (Hunter and Leahey 2010). Women scientists with preschool but not school-age children were actually more productive than their childless counterparts (Fox 2005).

Although women with children expressed a desire to work fewer hours than male researchers, in reality they worked similar hours (Sieverding et al. 2018). By the time women are in their PhD programs, they already recognize that if they enter academia and choose to have families, they will face exacting time demands, which itself may contribute to fewer women faculty (Sax et al. 2002). Decisions to leave academia (exit the highway) have been attributed largely to maternity and differences in family responsibilities, either by choice or constraint (Ceci and Williams 2011; Whittington 2011; McGuire at al. 2012). In the late 1990s, Carr et al. (1998) reported that women with children on the faculty were much more likely to self-report greater obstacles in academic careers, perceived slower career progress, and were less satisfied than men. No differences were reported for women and men faculty without children. In a survey of tenure-track mothers early in their careers who were followed until mid-career, women reported they had made academic choices in light of their family priorities and expressed hesitancy to pursue promotion because they prioritized service or teaching to gain flexibility in work-life balance (Ward and Wolf-Wendel 2016).

Institutions are beginning to follow policies promoting a family-friendly workplace that address a range of family issues, including eldercare and parenting, and these will be crucial to promoting diversity in academia. In the face of high productivity demands, flexible scheduling policies (e.g., being able to work at home, control over scheduling contact hours) improve work-family balance of all employees and increase positive work attitudes (McCampbell 1996; Scandura and Lankau 1997). When women elect to have children or when families are faced with care of aging parents, accommodation for parental/eldercare leave also contributes to a supportive environment. The benefits of such flexibility are being recognized. For example, across 205 research universities in the United States and Canada, about 60% now have some kind of paid parental leave. A key will be how accommodation for leave is factored into tenure decisions because leave can have long-term implications on productivity if additional resources are not provided after time off (O'Brien and Hapgood 2012).

Full Professorship and Leadership in Academia

Most of the women-in-science literature has focused on barriers to women early in their career (Goulden et al. 2011; Kulp et al. 2019). There is an increasing number of women reaching full professor in the wildlife field. Nevertheless, in 2019 men across fields were still about twice as likely as women to advance to the full professor rank at doctoral-granting institutions. When they achieved full professorship, women were paid only 81% of their male peers' salaries (AAUP 2019), which forced the courts to address discrimination or establish proceedings for arbitration. For example, coauthor Merrill was among a cohort of full professors who recently received a settlement of back pay from her university to rectify historical salary discrepancies. Unlike promotion at tenure, which follows a set schedule and is based on potential, achieving full professorship in academia has no time limit. Gumpertz et al. (2017) reported that women across fields including natural resources were less likely to be promoted to full professor in six to eight years than were men. Gender differences in advancement have been attributed to a lack of clear process or access to information (Fox and Colatrella 2006; Gardner and Blackstone 2013). Women also appear more reluctant to apply for promotion because of career choices involving more service, teaching, and administration (Macfarlane and Burg 2019). These duties diminish time for serving on prestigious scientific committees and editorial boards and running professional societies, which demonstrate the required distinction and leadership.

US Federal Government Workplace

Wildlife jobs are available to women across a wide spectrum of government agencies (federal, state, and local) in the United States. For example, the US federal government has more than 70,000 jobs in natural resource management and biological sciences (the 0400 job family, Table 4.2). The majority of wildlife biologist (0486 series) jobs are found in the US Departments of the Interior and Agriculture and include both research and management positions.

TABLE 4.2 Number of federal jobs in natural resources management and biological sciences (job family GS 0400) across government bureaus (www.federaljobs.net/Occupations/gs-0400_jobs.htm).

[a] Other agencies plus cabinet level positions
[b] Both management-oriented and research positions

Department (or Bureau)	Interior	Agriculture	Health & Human Services	Environmental Protection Agency	Commerce	Army	Other[a]	Total
General biological sciences 0401	3,674	4,522	3,672	1,039		3,100	4,404	20,411
Biological science technician 0404	1,820	5,081	171				809	7,881
Microbiology 0403; other (0405, 0414, 0415)		872	2,249	349		40	404	3,914
Ecology 0408, zoology 0410	637	363		159	90		185	1,434
Physiology 0413, genetics 0440	33	348	147		24	66	183	801
Plant science (0421, 0430, 0434, 0435, 0437)	142	1541				15	65	1,763
Range/fire/soils (0454–0459, 0470, 0471)	2,111	7,508				20	77	9,716
Forestry 0460, forestry technician 0462	1,692	16,728					232	18,652
Fish & wildlife administration	247				78		22	347
Fish biologist 0482	958	319			931		151	2,359
Wildlife refuge management 0485	595							595
Wildlife biologist 0486[b]	1,069	1,057			108		77	2,311
Animal science 0487	1,064	1,079			24		150	2,317
Totals	14,042	39,418	6,239	1,547	1,255	3,241	6,759	72,501

Government agencies currently employ 10.3% of doctoral scientists working in the biological, agricultural, and environmental life sciences (hereafter life sciences; total >200,000) and even more (28.1%) of doctoral scientists in the subdiscipline of natural resources and conservation (NCSES 2018, Table 12.1). Of doctoral scientists in the life sciences employed in government agencies, women now comprise 43.4% (NCSES 2018, Table 43), an increase compared with 28.9% in 2001 (NCSES 2001, Table 37).

In the 1990s and 2000s, workplace diversity arose as a formalized management concept in government agencies and emphasized recruitment and retention of women and minorities through training programs, family-friendly policies, and mentoring (Pitts 2009). By 2009, nearly 90% of federal agencies had institutional management plans promoting diversity, which, though variable, had positive effects on job satisfaction and work performance (D'Agostino 2015; Pitts 2009). A supportive workplace climate and greater inclusion of employees seem to be the necessary ingredients for improvements in performance (Sabharwal 2014). Today, all federal departments and agencies have programs to increase diversity, some of which include outreach efforts to schools and universities to encourage internships with federal agencies. These efforts have been part of a broader effort by the federal government to encourage citizens nationwide to enter the STEM professions, with some programs targeting natural resources (Box 4.1). Coincident with these efforts has been a doubling of women doctorates in the US science, engineering, and health workforce in the past 20 years (Foley et al. 2019).

Hiring and Advancement in the Federal Government

In contrast to hiring practices in academia, the 1996 implementation of the online federal recruitment and hiring system, USAJOBS (see also the activity at the end of the chapter) overcame one barrier to hiring women because it made information on federal jobs accessible to all potential applicants with internet access. Prior to its use, the recruitment of permanent hires was often through informal networks, disadvantageous for women with less access to such networks (Glass and Minnotte 2010; Schiebinger et al. 2011–2018). Before internet access, finding out about open science positions nationwide was often difficult and often required periodic visits to federal buildings with postings.

The disparity in entry-level salaries between men and women still seen in academia is lower in federal and state agencies in large part because of hiring processes (Homyack et al. 2014). For example, gender gaps in median annual

BOX 4.1 Online Policies, Programs, and Tools for Recruiting Women and Minorities to STEM Professions

---

Government-Wide Diversity and Inclusion Strategic Plan, by the US Office of Personnel Management, in response to the 2011 Executive Order 13583:
  www.opm.gov/policy-data-oversight/diversity-and-inclusion/reports /governmentwidedistrategicplan.pdf.

A collaborative hiring excellence campaign of the US Office of Personnel Management, focusing on strategic recruitment as well as diversity and inclusion:
  www.opm.gov/blogs/Director/hiring-excellence-campaign/

Outreach to science teachers pre-K through college from the USGS Office of Science Quality and Integrity:
  usgs.gov/science-support/osqi/yes/resources-teachers/

Opportunities for students and recent graduates:
  www.science.gov/STEM_Opportunities.html and www.doi.gov /pathways

Snapshots of 15 women in USGS wildlife biology science and science support positions, showcasing a diversity of STEM careers (Aragon-Long et al. 2018):
  https://pubs.usgs.gov/circ/1443/cir1443.pdf.

---

salaries of full-time women life scientists with doctorates in the federal ($0.90/$1; $104,000 for women vs. $115,000 for men) or state ($0.96/$1; $77,000 for women vs. $80,000 for men) government are within 10%, whereas in academia the gap is almost three times as large ($0.78/$1; $72,000 for women vs. $92,000 for men, NCSES 2018, Table 55). The reduced discrepancy in salaries for civilian, white-collar, federal employees results from pay standardization through the General Schedule (GS) classification and pay system, with pay grades established prior to hiring based on job difficulty, responsibility, and required qualifications. Indeed, the Federalpay.org website (n.d.) identifies the set range of salaries for an entry-level wildlife biologist, who typically starts at a GS-5 or GS-7, as well as for a wildlife research biologist with a doctorate, who generally starts at a GS-11.

The system for evaluation and promotion of woman scientists in federal service may alleviate some of the aforementioned difficulties facing women in tenure-track positions in academia, but they pose their own challenges. For most management agency biologists and administrators, advancing to a higher GS level often requires taking a new position, which most frequently is at a new location. Thus, salary increases are tied to career changes and mobility, which can disrupt the environment for children, eldercare, and careers of partners in dual-earning couples. For research positions that range from GS-11 to GS15+ (a GS-13 is equivalent to tenured faculty; Kern et al. 2015), scientists are promoted in accordance with the Research Grade Evaluation (RGE) process (US Office of Personnel Management 2006; see also "Resources"). The process is well defined and standardized across federal agencies and science centers and is less encumbered. Researchers can attain a GS-15 or an even higher nonexecutive scientific level (ST) based on high research productivity and impact without requiring relocation. Unlike in the tenure system in academia, federal scientists are evaluated every four years by panels of peers in similar disciplines from across the country and rarely within the local institution. Attributes that are assessed focus broadly on the scientist's demonstrated originality and creativity; scientific productivity (e.g., number and quality of publications); and stature, impact, and recognition at national and international levels. The RGE process may alleviate some implicit bias due to preferences of masculine over feminine approaches in science because the research position is classified with the scientist's expertise and productivity in mind. This approach is termed a "*person-in-job*" concept in which the responsibilities depend on "*the interplay between the research situation . . . and the individual qualities of the incumbent*" (OPM 2006, 21). Therefore, there is a broad range of models for success in the federal government.

Despite an evaluation process that is designed to level the playing field, federal women scientists still have lower representation at the higher grade levels. Among US Geological Survey (USGS) RGE scientists evaluated in 2013, women lagged behind men both in number and grade levels, with women representing about 25% of the total number (approx. 1,400). Median grade level of men was GS-14 and women GS-13. Representation of women declined with each successive level from GS-12 to GS-15 (37%, 28%, 21%, and 16%, respectively) and especially at the very highest level, senior scientist, when women were fewer than 5% (unpublished data, V. Burkett). The same pattern of representation held (in 2014) for the non-RGE series as well, with GS-12 to GS-15 levels at 39%, 36%, 27%, and 19%, respectively. As in academia, the relative roles of demographic inertia and ongoing bias underlying these patterns are particularly understudied for federal agencies. As of the writing of this chapter, a newly formed RGE women's peer group within the USGS offers specific guid-

FIGURE 4.2. As a refuge manager, coauthor Kathy A. Granillo (*right*) took advantage of Fish and Wildlife Service agreements that facilitate creating a diverse, more inclusive workplace. She and her staff mentored young people brought on for internships through an agreement with the Student Conservation Association and in the Fish and Wildlife Directorate Fellows Program. The fellows program started in 2014 and brings on 60-80 students per year to conduct specific research projects for fish and wildlife programs, including those on National Wildlife Refuges. When these students successfully complete the program, they are eligible for placement in a permanent position. Courtesy of the US Fish and Wildlife Service.

ance, provides informal and formal mentoring, and discusses approaches for dealing with remaining implicit bias and other topics (Figure 4.2)

How organizational structure differs between academic and government research situations may also influence attainment of gender diversity goals. Although in theory, hierarchies in academia can preserve patriarchal structures (see above), the top-down government model can also lead to greater leadership influence on the "*demographic composition of scientists*" (Kern et al. 2015). A fine-scale comparison of gender and rank between research scientists in a federal agency (USFS Research and Development) and universities by Kern et al. (2015) revealed a 9% greater representation of women among federal scientists than among university faculty representing similar fields of expertise. This difference was attributed in part to the long, albeit painful, history of workforce diversification efforts in the US Forest Service (FS 2015; Kern et al. 2015), which began 20 years earlier than programs the federal funding agencies eventually imposed on academic systems (e.g., NSF ADVANCE). The higher representation could also be at least partially attributable to working conditions, pay, or benefits favorable to women described here.

There are several tracks to attaining administrative leadership positions in the federal government, including avenues in public administration and business and industry, often in combination with productive research careers (DOI 2017). For example, Sally Jewell, the second woman secretary of the interior, 2013–2017, held a BS in mechanical engineering, a professional background in the oil and banking industries, and was CEO of Recreational Equipment, Inc. Similarly, Dr. Marcia McNutt, the first woman director of the USGS, 2009–2013, came from an eminent background as a marine geophysicist at Stanford University and CEO of the Monterey Bay Aquarium Research Institute. Scientists in research grade positions can achieve science leadership roles through superior research productivity. In the past 40 years, women have increasingly assumed leadership roles as US Forest Service district rangers and forest supervisors, National Park superintendents, and National Wildlife refuge managers. Today, women comprise 23% of the unit leaders and 28% of the assistant unit leaders in the USGS's long-standing Cooperative Fish and Wildlife Research Unit program (1935–present) (USGS n.d.[a]). In this program, leadership roles were filled exclusively by men until 1988; from 1935 to 2006, fewer than 3% of 238 unit leaders and 10% of 257 assistant unit leaders were women (Goforth 2006).

A new challenge emerging for scientists in federal research analogous to the increasing use of non-tenure-track instructors in academia is the workforce flexibility strategy (Nica 2018). Just as numbers of women in science and engineering positions are increasing, the 2015–2020 USGS Bureau Workforce Plan indicates that fewer research positions in the future will be permanent. Instead, the plan outlines a major change in strategy for maintaining scientific capacity via a flexible workforce that could expand or contract with changing circumstances. This proposed workforce includes a mix of permanent and other-than-permanent hires, including temporary and term appointments, scientist emeriti, unpaid interns, volunteers, and contractors (USGS n.d.[b]). As of 2014, the USGS had 8,413 federal employees, of which about 80% were in permanent career positions and 15% in permanent research positions. This overall number represents a decline of nearly 20% from the prior decade, in response to uncertainties in annual budgets and resources, the "staggering pace" of technological change, and shifts in science and management priorities. Indeed, a growing trend is for science within USGS to be conducted by postdoctoral appointees, generally hired into federal service as term positions not to exceed four years, matching the duration of the funded project. The downside for employees (of continually needing to look for permanent jobs and feeling undervalued) is acknowledged in the workforce planning document but not adequately addressed. It is not clear whether these job trends will have a gender bias, but the paucity of permanent jobs will likely discourage some women from pursuing careers in wildlife biology. At the same time, these non-permanent positions might provide career flexibility for other women seeking part-time employment.

Work-Life Balance in the Federal Government

Work-life balance issues within the federal government are gradually improving as agencies seek to enable *"employees to balance their responsibilities at work and at home"* (White House 2014). Workplace flexibility and work schedules, long customary in university faculty positions, are increasingly possible for government scientists as well, albeit in a highly structured manner. There are work-life programs available to all federal employees such as worksite health and wellness, Employee Assistance Programs, workplace flexibilities, and telework (OPM n.d.). Work-life balance issues within agencies can improve with increased representation of women in upper leadership positions. For example, through strong leadership, USGS *"strives to support all staff throughout their life decisions and transitions, allowing them to raise a family, take care of elderly relatives, or pursue further education"* (Aragon-Long et al. 2018). However, in contrast to academia, the only formal spousal accommodation program in the federal government that we know of is for military spouses (USAJOBS n.d.).

Conclusions

We have focused our chapter on reviewing the evidence behind the forces that shape the interactions between personal and institutional barriers and on indicating solutions to women in the workplace. The most serious barriers to training for a position in STEM and advancing in the workplace are implicit biases about gender, feelings of low self-efficacy, complications arising from childcare and being a member of a dual-career couple, and pervasive sexual harassment. Nevertheless, we take away from our review that women are making strides in STEM disciplines broadly, and we expect similar conclusions for women in the wildlife profession. When looking at cross-sectional studies of the percentage of women at the various stages of a career—from working toward an advanced degree to a leadership role—it remains clear that the percentage of women in different cohorts of the workforce declines, which gives us pause. Whereas studies show patterns, they often are inadequate to examine the causes. A larger cohort of women is now in advanced positions in our field; more research on the applicability of STEM findings to natural resource fields is certainly warranted. We suggest that more studies addressing women's choices in critical transition periods are needed to disentangle the factors that may be barriers or choices in determining career paths. Nevertheless, we believe our review has pointed to several important insights.

Institutions with different histories, organizational structure, and policies provide different options for women that may lead them along different trajectories. We have attempted to bring these to the forefront in contrasting the workplace in academia and the US federal government. In government, a readily accessible network of jobs and standardized hiring practices may level the playing field at entry for women, whereas in academia, job entry depends more on negotiating a path that can have long-term implications for salary and access to resources. Once on board in the US federal government, advancement in management positions can demand geographic mobility, and job insecurity from political constraints on budgets plagues nonpermanent positions. In academia, high demands to show uninterrupted productivity early in a career may require short-term sacrifices for long-term job security. Emphasis on grant acquisition and publication still plays a vital role in securing tenure and promotion. Why women may not submit as many publications and grants or pursue full professorships as readily remain unresolved issues.

We argue that to at least some degree the decline in women in upper-level positions in both academia and government still reflects the historical lag in overcoming past biases. But we are quick to suggest there is evidence of improved practices in recruiting and retention, mentoring, and leadership training. Women's choices in how they pursue their careers define the future direction for women in our profession. At the same time, the workplace will not be truly inclusive until individuals step up, and the institution has zero tolerance of documented sexual harassment in the classroom, office and field settings. Based on our experience, no matter what path women choose, they will need to step up to advocate for themselves.

How to Be an Ally

Being an ally in academic and government workplaces starts with understanding your own privilege and biases, no matter what their source, and acknowledging there is a hierarchically developed system of power within institutions (Stephens 2020). A goal of this chapter was to point out some of the bias in institutional structures that may impose obstacles to women achieving career goals in academia and government agencies. Within institutions, you can be an ally of women, minorities, and LGBTQ+ communities. We offer suggestions on what you can do personally.

First, you can own the situation and be an agent of change. As an advisor, mentor, or supervisor, start a conversation about what is entailed in allyship, perhaps by citing current authors and organizations involved in this topic (TWS 2020). Second, ensure you are listening deeply, not speaking for but supporting the chosen paths of an individual for their own career goals, and not just adhering to how things are currently done within the institution. As an ally, promote opportunities and networking to a coworker, and for yourself, broaden the audiences with which you engage. As a supervisor or department head, monitor your workplace and take responsibility for a situation you view as unjust. Right a transgression in the present even if it does not lead to closure or lasting change. Institutions, whether academic or governmental, are slow to change but are making progress (Bishop et al. 2021; Haubold and Ross-Winslow 2021). Inroads can be imposed on institutions from the top, but effective change starts with self-reflection and a willingness to create room for diversity (Box 4.2).

Resources

FunctionalSelfDefense. N.d. http://www.functionalselfdefense.org/awareness-prevention.

Project Implicit. N.d. https://implicit.harvard.edu/implicit/.

US Geological Survey (USGS). N.d. *445-2-H, Occupational Safety and Health Program Requirements*, Appendix 14. https://www.usgs.gov/survey-manual/445-2-h-occupational-safety-and-health-program-requirements.

US Geological Survey (USGS). N.d. Research Grade Evaluation. https://
www.usgs.gov/about/organization/science-support/science-quality-and
-integrity/research-grade-evaluation.

Discussion Questions

1. With implicit bias expressed indirectly through the subconscious, people don't often recognize it in themselves. What implicit biases about gender can you identify within your family, peer group, and university? You can learn about your personal biases about women and science by completing online evaluations through Project Implicit, Harvard's research project designed to advance understanding about implicit biases on a variety of topics (see "Resources").
2. Pick two of the implicit biases you articulated in Question 1 and identify steps you can take individually or institutionally to dismantle it.
3. From the perspective of an employee and a supervisor, discuss the pros and cons of the flexible workforce in academia and federal government for women.

Activity: Searching and Applying for a Federal Job

Learning Objectives: Participants will navigate the USAJOBS.gov website to search for a federal job and develop and review resumes specific for it.

Time: 3–5 hrs.

Description of Activity

Materials needed: Access to the internet and computer

Process:

Part 1: USAJOBS (30 min.)

Using USAJOBS.gov to search for a federal position can be intimidating and confusing. However most federal jobs are *only* advertised here. Job seekers interested in positions in the wildlife and natural resource sciences fields need to learn to navigate this system. Within the website's search field, type a search term (wildlife, biologist, or ecologist, for example). The website will then upload a list of jobs. Pick one, click on the job title, and review the job description. Things to note:

- Open and closing dates to apply for the position.
- Pay scale and grade (GS) for position. Usually jobs that are GS-3 or GS-4 are internship positions or student jobs. Most entry-level posi-

tions are GS-5 to GS-7. Mid-level positions are GS-9 to GS-12. GS-13 to GS-15 are the highest and usually require previous experience at lower grade levels. Appointment type identifies whether this position is permanent or temporary.

- Salary range.
- Location.
- Some jobs are open only to existing federal employees, while others are open to the public. This information will be available toward the beginning of the job announcement.
- The "Duties" section describes the job expectations, responsibilities, and tasks that the government expects of the employee. The "Requirements" section describes the conditions of employment, including whether a uniform is required, driver's license needs, and more, depending on the job. This section also includes the qualifications and education necessary for hiring

Part 2: Resume Builder (1–2 hr.)

All federal jobs require several components, including a resume. USAJOBS .gov has a resume builder available; however, you have more control over appearance and format if you create one yourself. Participants will create their resume with a focus on matching their skills to the specific job duties and requirements in the announcement they selected. Do not make formatting too complex.

Specific information to include:

- Full name, mailing address, phone number, email address, and if applicable, eligibility for veteran's preference.
- Education history and degrees earned, with dates of graduation.
- All employment or experience including the job title, agency/company name, supervisor's name, address, contact telephone numbers, and dates of employment. For each job, include a list of your responsibilities using action or achievement verbs. List current or most recent jobs first.
- Research publications and presentations.
- Leadership experience.
- Specialized training and dates completed, including computer software, equipment certificates, awards or honors, wilderness first aid, and membership in professional societies.
- Special skills such as experience with mechanics, boating, carpentry, backpacking, birding, orienteering, rock climbing, etc., but make them relevant to the job.

- A list of three professional references with contact information (even if already included as a supervisor contact). Do not omit references with a message to "please contact me for a list of references."

Part 3: Cover Letter (1 hr.)

Most wildlife job applications also require a cover letter, which is the first thing a potential employer will read and use to develop their first impressions. Cover letters are an opportunity for the applicant to describe how their experience, skills, and education together make them the best applicant for the position. Focus on how you will help your future employer rather than how the job will help advance your career. Before writing the cover letter, read through the job announcement and physically highlight the key words. Cover letters should include:

- Your return address at the top.
- Date.
- Name, title, and complete physical mailing address for the recipient.
- Salutation.
- Paragraph 1 should identify the position you are applying for, where you found the announcement, why you are interested in this position, and why they should hire you.
- Paragraph 2 includes highlights of your specific qualifications and aligns them with the qualifications included in the job announcement. Use the same key words from the job announcement in this paragraph.
- Paragraph 3 concludes the cover letter repeating your interest and enthusiasm for the position, summarizing in a single paragraph why you are an ideal and qualified candidate, and thanking the reader for their time and consideration.
- Ends with a closing and your name.

Part 4: Peer Review (30–45 min.)

Participants will bring two paper copies of their resume, cover letter, and the job announcement from USAJOBS.gov. In small groups, participants will exchange them with two people for peer review.

Reviewer Questions
- Does the resume and cover letter highlight or match the skills and qualifications listed in the position? Often human resource personnel reviewing resumes lack a background in wildlife and will search for exact terms.

- Is the resume organized to aid in locating specific information for the position? Could organization be improved by different formatting?
- Is the contact information complete for the applicant and references?
- Are there any typos or grammatical errors in the resume or cover letter?

Part 5: Large Group Discussion and Reflection (15–20 min.)

Questions

1. What did you learn from reviewing your peers' resumes and cover letters?
2. Are there skills or abilities that you plan to work on to make yourself more competitive for a wildlife job?

References and Resources

Henke, Scott E., and Paul R. Krausman, eds. *Becoming a Wildlife Professional.* Baltimore: Johns Hopkins University Press, 2017.

Kursmark, L. M. *Best Resumes for College Students and New Grads.* 3rd ed. St. Paul, MN: JIST Works, 2012.

Further Reading

Aragon-Long, S. C., V. R. Burkett, H. S. Weyers, S. M. Haig, M. S. Davenport, and K. L. Warner. 2018. *A Snapshot of Women of the US Geological Survey in STEM and Related Careers.* USGS circular 1443. doi: 10.3133/cir1443.

Britton, D. M. 2017. "Beyond the Chilly Climate: The Salience of Gender in Women's Academic Careers." *Gender and Society* 31(1): 5–27.

Ignotofsky, R. 2016. *Women in Science: 50 Fearless Pioneers Who Changed the World.* New York: Ten Speed Press.

Schnall, M. 2013. *What Will It Take to Make a Woman President? Conversations about Women, Leadership and Power.* New York: Seal Press.

Literature Cited

Addessi, E., M. Borgi, and E. Palagi. 2012. "Is Primatology an Equal-Opportunity Discipline?" *PLoS ONE* 7(1):e30458.

American Association of University Professors (AAUP). 2019. *Annual Report on the Economic Status of the Profession.* www.aaup.org/our-work/research/FCS.

Amon, M. J. 2017. "Looking through the Glass Ceiling: A Qualitative Study of STEM Women's Career Narratives." *Frontiers in Psychology* 8:236.

Angus, S. 1995. "Women in Natural Resources: Stimulating Thinking about Motivations and Needs." *Wildlife Society Bulletin* 23(4): 579–582.

Aragon-Long, S. C., V. R. Burkett, H. S. Weyers, S. M. Haig, M. S. Davenport, and K. L. Warner. 2018. *A Snapshot of Women of the US Geological Survey in STEM and Related Careers.* US Geological Survey circular 1443. doi: 10.3133/cir1443.

Bakker, M. M., and M. H. Jacobs. 2016. "Tenure Track Policy Increases Representation of Women in Senior Academic Positions but Is Insufficient to Achieve Gender Balance." *PLoS ONE* 11(9): e0163376.

Bian, L., S. J. Leslie, and A. Cimpian. 2017. "Gender Stereotypes about Intellectual Ability Emerge Early and Influence Children's Interests." *Science* 355(6323): 389–91.

Bishop C. J., L. P. Waits, and J. R. Mawdsley. 2021. "A Place for Universities on the Roadmap." *Wildlife Professional* 15(6): 43–47.

Bornmann, L., R. Mutz, and H. D. Daniel. 2007. "Gender Differences in Grant Peer Review: A Meta-analysis." *Journal of Informetrics* 1(3): 226–238.

Cameron, E. Z., A. M. White, and M. E. Gray. 2016. "Solving the Productivity and Impact Puzzle: Do Men Outperform Women, or Are Metrics Biased?" *BioScience* 66(3): 245–52.

Carr, P. L., A. S. Ash, R. H. Friedman, A. Scaramucci, R. C. Barnett, L. Szalacha, A. Palepu, and M. A. Moskowitz. 1998. "Relation of Family Responsibilities and Gender to the Productivity and Career Satisfaction of Medical Faculty." *Annals of Internal Medicine* 129(7): 532–38.

Ceci, S. J., and W. M. Williams. 2011. "Understanding Current Causes of Women's Underrepresentation in Science." *Proceedings of the National Academy of Sciences* 108(8): 3157–62.

Costa, D. L., and M. E. Kahn. 2000. "Power Couples: Changes in the Locational Choice of the College Educated, 1940–1990." *Quarterly Journal of Economics* 115(4): 1287–315.

Cundiff, J. L., T. K. Vescio, E. Loken, and L. Lo. 2013. "Do Gender-Science Stereotypes Predict Science Identification and Science Career Aspirations among Undergraduate Science Majors?" *Social Psychology of Education* 16(4): 541–54.

D'Agostino, M J. 2015. "The Difference That Women Make: Government Performance and Women-Led Agencies." *Administration & Society* 47(5): 532–48.

Diekman, A. B., E. R. Brown, A. M. Johnston, and E. K. Clark. 2010. "Seeking Congruity between Goals and Roles: A New Look at Why Women Opt Out of Science, Technology, Engineering, and Mathematics Careers." *Psychological Science* 21(8): 1051–57.

Duch, J., X. H. T. Zeng, M. Sales-Pardo, F. Radicchi, S. Otis, T. K. Woodruff, and L. A. Nunes Amaral. 2012. "The Possible Role of Resource Requirements and Academic Career-Choice Risk on Gender Differences in Publication Rate and Impact." *PLoS ONE* 7(12): e51332.

Eagly, A. H., and L. L. Carli. 2007. *Through the Labyrinth: The Truth about How Women Become Leaders.* Boston: Harvard Business Press.

Eddy, S. L., and S. E. Brownell. 2016. "Beneath the Numbers: A Review of Gender Disparities in Undergraduate Education across Science, Technology, Engineering, and Math Disciplines." *Physical Review Physics Education Research* 12(2): 020106.

Eriksen, C., G. Waitt, and C. Wilkinson. 2016. "Gendered Dynamics of Wildland Firefighting in Australia." *Society & Natural Resources* 29(11): 1296–310.

Esarey, J., and K. Bryant. 2018. "Are Papers Written by Women Authors Cited Less Frequently?" Political Analysis 26(3): 331–34.

FederalPay. N.d. "Pay Rates for 'Wildlife Biologist.'" Accessed February 10, 2022. www .federalpay.org/employees/occupations/wildlife-biology.

Foley, D. J., L. A. Selfa, and K. H. Grigorian. 2019. *Number of Women with U.S. Doctorates in Science, Engineering, or Health Employed in the United States More Than Doubles since 1997.* InfoBrief, National Center for Science and Engineering Statistics (NCSES). NSF 19-307. Washington, DC: National Science Foundation, Washington.

Fox, C. W., C. S. Burns, A. D. Muncy, and J. A. Meyer. 2016. "Gender Differences in Patterns of Authorship Do Not Affect Peer Review Outcomes at an Ecology Journal." *Functional Ecology* 30(1): 126–39.

Fox, M. F. 2005. "Women in Science: Career Processes and Outcomes." *Contemporary Sociology* 34(4): 361.

Fox, M. F., and C. Colatrella. 2006. "Participation, Performance, and Advancement of Women in Academic Science and Engineering: What Is at Issue and Why." *Journal of Technology Transfer* 31(3): 377–86.

Fox, M. F., and S. Mohapatra. 2007. "Social-Organizational Characteristics of Work and Publication Productivity among Academic Scientists in Doctoral-Granting Departments." *Journal of Higher Education* 78(5): 542–71.

Fuchs, S., J. Von Stebut, and J. Allmendinger. 2001. "Gender, Science, and Scientific Organizations in Germany." *Minerva* 39(2): 175–201.

Gardiner, M., M. Tiggemann, H. Kearns, and K. Marshall. 2007. "Show Me the Money! An Empirical Analysis of Mentoring Outcomes for Women in Academia." *Higher Education Research & Development* 26(4): 425–42.

Gardner, S. K., and A. Blackstone. 2013. "'Putting in Your Time': Faculty Experiences in the Process of Promotion to Professor." *Innovative Higher Education* 38(5): 411–25.

Girod, S., S. K. Gilmartin, H. Valantine, and L. Schiebinger. 2011. "Academic Couples: Implications for Medical School Faculty Recruitment and Retention." *Journal of the American College of Surgeons* 212(3): 310–19.

Glass, C., and K. L. Minnotte. 2010. "Recruiting and Hiring Women in STEM Fields." *Journal of Diversity in Higher Education* 3(4): 218.

Goforth, W. R. 2006. *The Cooperative Fish and Wildlife Research Units Program*. Special Publication of the US Geological Survey. https://usgs-cru-department-data.s3 .amazonaws.com/headquarters/unit_docs/CRU_Program_BookletS-1.pdf.

Goulden, M., M. A. Kroll, and K. Frasch. 2011. "Keeping Women in the Science Pipeline." *Annals of the American Academy of Political and Social Science* 638(1): 141–62.

Gumpertz, M., R. Durodoye, E. Griffith, and A. Wilson. 2017. "Retention and Promotion of Women and Underrepresented Minority Faculty in Science and Engineering at Four Large Land Grant Institutions." *PLoS ONE* 12(11): e0187285.

Harris, C., B. Myers, and K. Ravenswood. 2019. "Academic Careers and Parenting: Identity, Performance and Surveillance." *Studies in Higher Education* 44(4): 708–18.

Haubold, E. M., and D. Ross-Winslow. 2021. "It's about the People: The USFWS Is Charting a Course to Embrace Broader Constituencies." *Wildlife Professional* 15(6): 38–42.

Hewlett, S. A., and C. B. Luce. 2005. "Off-Ramps and On-Ramps: Keeping Talented Women on the Road to Success." *Harvard Business Press* 83(3): 43–46.

Higgins, M. C., and K. E. Kram. 2001. "Reconceptualizing Mentoring at Work: A Developmental Network Perspective." *Academy of Management Review* 26(2): 264–88.

Holmes, M. A. 2015. "Best Practices to Achieve Gender Parity: Lessons Learned from NSF's Advance and Similar Programs." *Women in the Geosciences: Practical, Positive Practices Toward Parity* 70: 33.

Homyack, J. A., S. H. Schweitzer, and T. Graves. 2014. "Glass Ceilings and Institutional Biases: A Closer Look at Barriers Facing Women in Science and Technical Fields." *Wildlife Professional*, Fall, 48–52.

Hunter, L. A., and E. Leahey. 2010. "Parenting and Research Productivity: New Evidence and Methods." *Social Studies of Science* 40(3): 433–51.

Joecks, J., K. Pull, and U. Backes-Gellner. 2014. "Childbearing and (Female) Research Productivity: A Personnel Economics Perspective on the Leaky Pipeline." *Journal of Business Economics* 84(4): 517–30.

Kaminski, D., and C. Geisler. 2012. "Survival Analysis of Faculty Retention in Science and Engineering by Gender." *Science* 335(6070): 864–66.

Kaplan, K. 2010. "Negotiating for Two." *Nature* 466(7310): 1145–46.

Kern, C. C., L. S. Kenefic, and S. L. Stout. 2015. "Bridging the Gender Gap: The Demographics of Scientists in the USDA Forest Service and Academia." *BioScience* 65(12): 1165–72.

Kulp, A. M., L. E. Wolf-Wendel, and D. G. Smith. 2019. "The Possibility of Promotion: How Race and Gender Predict Promotion Clarity for Associate Professors." *Teachers College Record* 121(5).

Kyvik, S., and M. Teigen. 1996. "Child Care, Research Collaboration, and Gender Differences in Scientific Productivity." *Science, Technology, & Human Values* 21(1): 54–71.

Langin, K. 2019. "Private Sector Nears Rank of Top Ph.D. Employer." *Science* 363: 1135.

Laver, K. E., I. J. Prichard, M. Cations, I. Osenk, K. Govin, and J. D. Coveney. 2018. "A Systematic Review of Interventions to Support the Careers of Women in Academic Medicine and Other Disciplines." *BMJ Open* 8(3): e020380.

Lee, C. J, C. R. Sugimoto, G. Zhang. 2013. "Bias in Peer Review." *Journal of the American Society for Information Science and Technology* 64(1): 2–17.

Lee, S., and B. Bozeman. 2005. "The Impact of Research Collaboration on Scientific Productivity." *Social Studies of Science* 35(5): 673–702.

Levinson, W., K. Kaufman, B. Clark, and S. W. Tolle. 1991. "Mentors and Role Models for Women in Academic Medicine." *Western Journal of Medicine* 154(4): 423–26.

Macfarlane, B. and D. Burg. 2019. "Women Professors and the Academic Housework Trap." *Journal of Higher Education Policy and Management* 41(3): 262–74.

Martin, L. J. 2012. "Where Are the Women in Ecology?" *Frontiers in Ecology and the Environment* 10(4): 177–78.

McCampbell, A. S. 1996. "Benefits Achieved through Alternative Work Schedules." *People and Strategy* 19(3): 30–37.

McGuire, K. L., R. B. Primack, and E. C. Losos. 2012. "Dramatic Improvements and Persistent Challenges for Women Ecologists." *BioScience* 62(2): 189–96.

Mitchell, M., and M. McKinnon. 2019. "'Human' or 'Objective' Faces of Science? Gender Stereotypes and the Representation of Scientists in the Media." *Public Understanding of Science* 28(2): 177–90.

Montgomery, B. L. 2017. "Mapping a Mentoring Roadmap and Developing a Supportive Network for Strategic Career Advancement." *Sage Open* 7(2): 2158244017710288.

Morimoto, S. A., A. M. Zajicek, V. H. Hunt and R. Lisnic. 2013. "Beyond Binders Full of Women: NSF ADVANCE and Initiatives for Institutional Transformation." *Sociological Spectrum* 33(5): 397–415.

Morton, S., and J. A. Kmec. 2018. Risk-Taking in the Academic Dual-Hiring Process: How Risk Shapes Later Work Experiences." *Journal of Risk Research* 21(12): 1517–32.

Moss-Racusin, C. A., J. F. Dovidio, V. L. Brescoll, M. J. Graham, and J. Handelsman. 2012. "Science Faculty's Subtle Gender Biases Favor Male Students." *Proceedings of the National Academy of Sciences* 109(41): 16474–79.

National Academies of Sciences, Engineering, and Medicine (NASEM). 2018. *Sexual Harassment of Women: Climate, Culture, and Consequences in Academic Sciences, Engineering, And Medicine*. Washington, DC: National Academies Press.

National Center for Science and Engineering Statistics (NCSES). 2001. "Characteristics of Doctoral Scientists and Engineers in the United States: 2001." https://wayback.archive-it.org/5902/20150629124530/http:/www.nsf.gov/statistics/nsf03310/sect3.htm.

National Center for Science and Engineering Statistics (NCSES). 2018. *Survey of Doctorate Recipients 2017*. Alexandria, VA: National Science Foundation. http://ncsesdata.nsf.gov/doctoratework/2017/.

National Research Council (NRC). 2010. *Gender Differences at Critical Transitions in the Careers of Science, Engineering, and Mathematics Faculty*. Washington, DC: National Academies Press.

Natural Sciences and Engineering Research Council (NSERC). 2019. Competition Statistics Discovery Grants and Research Tools and Instruments 2013–2019. http://publications.gc.ca/site/eng/9.840912/publication.html.

National Science Foundation, Division of Science Resources Statistics (NSF). 2011. *Women, Minorities, and Persons with Disabilities in Science and Engineering: 2011*. Special report NSF11-309. www.nsf.gov/statistics/wmpd/.

Nica, E. 2018. "Has the Shift to Overworked and Underpaid Adjunct Faculty Helped Education Outcomes?" *Educational Philosophy and Theory* 50: 213–16.

Nicholson, K., P. R. Krausman, and J. A. Merkle. 2008. "Hypatia and the Leopold Standard: Women in the Wildlife Profession 1937–2006." *Wildlife Biology in Practice* 4(2): 57–72.

O'Brien, K. R., and K. P. Hapgood. 2012. "The Academic Jungle: Ecosystem Modelling Reveals Why Women Are Driven Out of Research." *Oikos* 121(7): 999–1004.

Pautasso, M. 2013. "Focusing on Publication Quality Would Benefit All Researchers." *Trends in Ecology and Evolution* 6(28): 318–20.

Peñas, C. S., and P. Willett. 2006. "Brief Communication: Gender Differences in Publication and Citation Counts in Librarianship and Information Science Research." *Journal of Information Science* 32(5): 480–85.

Pitts, D. 2009. "Diversity Management, Job Satisfaction, and Performance: Evidence from US Federal Agencies." *Public Administration Review* 69(2): 328–38.

Powell, A., T. M. Hassan, A. R. J. Dainty, and C. Carter. 2009. "Note: Exploring Gender Differences in Construction Research; A European Perspective." *Construction Management and Economics* 27(9): 803–7.

Rinkus, M. A., J. R. Kelly, W. Wright, L. Medina, and T. Dobson. 2018. "Gendered Considerations for Safety in Conservation Fieldwork." *Society & Natural Resources* 31(12): 1419–26.

Risman, B. J., and G. Davis. 2013. "From Sex Roles to Gender Structure." *Current Sociology* 61: 733–55.

Sabharwal, M. 2014. "Is Diversity Management Sufficient? Organizational Inclusion to Further Performance." *Public Personnel Management* 43(2): 197–217.

Sandberg, S., and N. Scovell. 2013. *Lean In: Women, Work, and the Will to Lead.* New York: Alfred Knopf.

Sax, L. J., L. S. Hagedorn, M. Arredondo, and F. A. DiCrisi. 2002. "Faculty Research Productivity: Exploring the Role of Gender and Family-Related Factors." *Research in Higher Education* 43(4): 423–46.

Scandura, T. A., and M. J. Lankau. 1997. "Relationships of Gender, Family Responsibility and Flexible Work Hours to Organizational Commitment and Job Satisfaction." *Journal of Organizational Behavior: The International Journal of Industrial, Occupational and Organizational Psychology and Behavior* 18(4): 377–91.

Schiebinger, L. L., A. Davies Henderson, and S. K. Gilmartin. 2008. *Dual-Career Academic Couples: What Universities Need to Know.* Stanford, CA: Michelle R. Clayman Institute for Gender Research, Stanford University.

Schiebinger, L., I. Klinge, I. Sánchez de Madariaga, H. Y. Paik, M. Schraudner, and M. Stefanick, eds. 2011–2018. *Gendered Innovations in Science, Health & Medicine, Engineering and Environment. Part 2: Subtle Gender Bias and Institutional Barriers.* https://genderedinnovations.stanford.edu/institutions/bias.html.

Shauman, K. 2017. "Gender Differences in the Early Employment Outcomes of STEM Doctorates." *Social Sciences* 6(1): 24.

Shaw, A. K., and D. E. Stanton. 2012. "Leaks in the Pipeline: Separating Demographic Inertia from Ongoing Gender Differences in Academia." *Proceedings of the Royal Society B: Biological Sciences* 279(1743): 3736–41.

Sheridan, J., J. N. Savoy, A. Kaatz, Y. G. Lee, A. Filut, and M. Carnes. 2017. "Write More Articles, Get More Grants: The Impact of Department Climate on Faculty Research Productivity." *Journal of Women's Health* 26(5): 587–96.

Sieverding, M., C. Eib, A. B. Neubauer, and T. Stahl. 2018. "Can Lifestyle Preferences Help Explain the Persistent Gender Gap in Academia? The 'Mothers Work Less' Hypothesis Supported for German but Not for US Early Career Researchers." *PLoS ONE* 13(8): e0202728.

Smith, F. A., and D. M. Kaufman. 1996. "A Quantitative Analysis of the Contributions of Female Mammalogists from 1919 to 1994." *Journal of Mammalogy* 77(3): 613–28.

Stack, M. 2018. "Why I'm Not Surprised Nobel Laureate Donna Strickland Isn't a Full Professor." *Conversation.* https://theconversation.com/why-im-not-surprised-nobel-laureate-donna-strickland-isnt-a-full-professor-104459.

Stepan-Norris, J., and J. Kerrissey. 2016. "Enhancing Gender Equity in Academia: Lessons from the ADVANCE Program." *Sociological Perspectives* 59(2): 225–45.

Stephens, T. 2020. "Allyship in Academia: An Action Call For." August. https://societyforpsychotherapy.org/allyship-in-academia-an-action-call-for.

Sugimoto, C. R., and C. Blaise. 2013. "Citation Gamesmanship: Testing for Evidence of Ego Bias in Peer Review." *Scientometrics* 95(3): 851–62.

Symonds, M. R. E., N. J. Gemmell, T. L. Braisher, K. L. Gorringe, and M. A. Elgar. 2006. "Gender Differences in Publication Output: Towards an Unbiased Metric of Research Performance." *PLoS ONE* 1(1): e127.

Thomas, N. R., D. J. Poole, and J. M. Herbers. 2015. "Gender in Science and Engineering Faculties: Demographic Inertia Revisited." *PLoS ONE* 10(10): e0139767.

Tower, G., J. Plummer, and B. Ridgewell. 2007. "A Multidisciplinary Study of Gender-Based Research Productivity in the World's Best Journals." *Journal of Diversity Management* 2(4): 23–32.

Trube, M. B., and B. vanDerveer. 2015. "Support for Engaged Scholars: The Role of Mentoring Networks with Diverse Faculty." *Mentoring & Tutoring: Partnership in Learning* 23(4): 311–27.

Uriarte, M., H. A. Ewing, V. T. Eviner, and K. C. Weathers. 2007. "Constructing a Broader and More Inclusive Value System in Science." *BioScience* 57(1): 71–78.

USAJOBS. N.d. "Military Spouses." www.usajobs.gov/Help/working-in-government/unique -hiring-paths/military-spouses/.

US Department of the Interior (DOI). 2017. "Women's History Month: Breaking the Glass Ceiling at Interior." March 1. www.doi.gov/blog/womens-history-month-breaking-glass -ceiling-interior.

US Forest Service (FS). 2015. *Forest Service Research and Development Performance and Accountability Report.* USFS 1076. Washington, DC: Department of Agriculture.

US Geological Survey (USGS). N.d. (a). *USGS Workforce Plan 2015–2020.* Cooperative Fish and Wildlife Research Units Program. www1.usgs.gov/coopunits/.

US Geological Survey (USGS). N.d. (b). www.usgs.gov/human-capital/usgs-workforce-plan -2015-2020.

US Office of Personnel Management (OPM). N.d. Policy, Data, Oversight: Work-Life. http://www.OPM.gov/policy-data-oversight/worklife/.

US Office of Personnel Management (OPM). 2006. *Research Grade Evaluation Guide 2006.* OPM. www.opm.gov/policy-data-oversight/classification-qualifications/classifying -general-schedule-positions/functional-guides/gsresch.pdf.

van Arensbergen, P., I. van der Weijden, and P. van den Besselaar. 2012. "Gender Differences in Scientific Productivity: A Persisting Phenomenon?" *Scientometrics* 93(3): 857–68.

Wang, M. T., and J. L. Degol. 2017. "Gender Gap in Science, Technology, Engineering, and Mathematics (STEM): Current Knowledge, Implications for Practice, Policy, and Future Directions." *Educational Psychology Review* 29(1): 119–40.

Ward, K., and L. Wolf-Wendel. 2016. "Academic Motherhood: Mid-career Perspectives and the Ideal Worker Norm." *New Directions for Higher Education* 1(176): 11–23.

West, J. D., J. Jacquet, M. M. King, S. J. Correll, and C. T. Bergstrom. 2013. "The Role of Gender in Scholarly Authorship." *PLoS ONE* 8(7): e66212. doi: 10.1371/journal. pone.0066212.

White House. 2014. "Presidential Memorandum—Enhancing Workplace Flexibilities and Work-Life Program." June 23. https://obamawhitehouse.archives.gov/the-press-office/2014 /06/23/presidential-memorandum-enhancing-workplace-flexibilities-and-work-life-.

Whittington, K. B. 2011. "Mothers of Invention? Gender, Motherhood, and New Dimensions of Productivity in the Science Profession." *Work and Occupations* 38(3): 417–56.

The Wildlife Society (TWS). 2020. "A message from The Wildlife Society." June 3. https:// wildlife.org/a-message-from-the-wildlife-society/.

Williams, W. M., and S. J. Ceci. 2015. "National Hiring Experiments Reveal 2:1 Faculty Preference for Women on STEM Tenure Track." *Proceedings of the National Academy of Sciences* 112(17): 5360–65.

Zeldin, A. L., and F. Pajares. 2000. "Against the Odds: Self-Efficacy Beliefs of Women in Mathematical, Scientific, and Technological Careers." *American Educational Research Journal* 37(1): 215–46.

Zhang, H., and J. A. Kmec. 2018. "Non-normative Connections between Work and Family: The Gendered Career Consequences of Being a Dual-Career Academic." *Sociological Perspectives* 61(5): 766–86.

Diana Doan-Crider, Jamila Blake, and
Ruth Plenty Sweetgrass-She Kills

Creating an Equitable Environment

Learning from a River

It is still a lonely place in academia and as an indigenous person and a sometimes
volatile space . . . [T]here's a significant lack of representation of indigenous
peoples as doctoral holders. For me, the reason why I got the degree was to have the
degree. To get the seat at the table.

—Dr. Stephanie Carroll, "Run to be Visible"

Exclusion and Lack of Diversity within the Wildlife Profession: A River of No Return?

A State of Hysteresis

Ruth Cordova of the Pomo People said, "*My grandmother was born on a gravel*
bed along the creek many moons ago, before this land was taken from us, the salmon
were so thick in the creek you could walk across them" (Cordova n.d.).

The Columbia River drainage basin (Figure 5.1) feeds more water into the
ocean than any other river in North America and covers more than 258,000
square miles of land. Prior to European colonization, the Columbia and its trib-
utaries flowed freely, providing habitat for wildlife and humans alike. The
crowds of chinook (*Oncorhynchus tshawytscha*), coho (*O. kisutch*), sockeye
salmon (*O. nerka*), and steelhead trout (*O. mykiss irideus*) were so numerable that
William Clark noted that "*the multitudes of this fish are almost inconceivable*" during the Corps of Discovery expedition (Coues 1987). These fish depended on
certain physical and environmental conditions to lay their eggs and were
adapted to jumping over the numerous natural barriers along the river. While

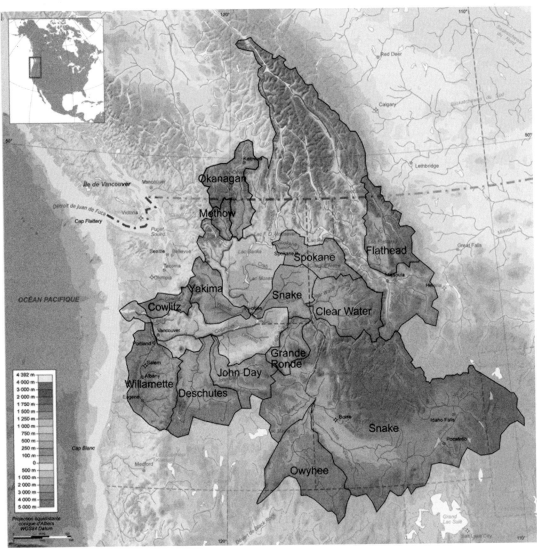

FIGURE 5.1. Post-colonization names of the Columbia River Basin.
Wikimedia Commons, courtesy of Bourrichon.

some didn't survive localized events, the system's resilience compensated for any losses (Oliver et al. 2015). The areas along the rivers were also essential for native microbes and vegetation that helped to stabilize the soil and regulate the water's flow. These intertwined relationships helped to feed the "Salmon People," or the Wy-Kan-Ush-Pum (in the Sahaptin language), who lived along the river. Their own relationship with and dependence upon the fish contributed to their knowledge, and their knowledge contributed to the river system's health over centuries (Columbia River Inter-Tribal Fish Commission n.d.[a]).

Today, there are more than 60 dams along the Columbia River drainage basin. The immense data set of long-term ecological knowledge stored up in the stories and experiences of the Salmon People was confined to reservations and largely ignored. Anadromous fish stocks have been reduced dramatically since European arrival (Johnson et al. 2019) and are now further endangered by rising water temperatures (Keefer et al. 2018). Much like many other rivers across the planet, overuse of the Columbia's resources has resulted in the crossing of major ecological thresholds that will be very difficult to reverse (Brooks et al. 2003).

Ecologists and engineers alike are now contending with complex systems that have reached a state of **hysteresis**, which means that they cannot be easily returned to former states once they have degraded (Beisner et al. 2003). Furthermore, it is where the observed equilibrium cannot be predicted solely by environmental variables but also requires knowledge of the system's history (Cornell University Laboratory of Atomic and Solid State Physics 1994).

If you are an ecologist or engineer, you know what it takes to restore a river to a healthy state. You can't just toss a few fish in and hope that they make it. If you are three women of color being asked what can we do to fix the diversity problem in the wildlife profession, you can't just open the doors and say, "okay, you can come back in now." Our profession is about our experiences of trying to "fit in" to a system that was not created with our cultural communities, belief systems, or limitations in mind, and having to eat the breadcrumbs of what everyone else ordered. The current state of our profession is about rethinking a dysfunctional system rather than just "managing diversity" so that everyone looks good . . . on the surface. It is about *listening* and *cultivating* an environment where all participants have equitable footing and a place at the table.

In this chapter, we shed some light on the **history** that got us here. We highlight the **individual, cultural, and traditional values** that must be included in the solutions we develop to address the issues with our system. We emphasize the right to **emotional well-being** and **self-determination** in the pursuit of careers in a country, and profession, where the "majority" has historically made decisions for everyone else. Finally, we cover the changes that will be necessary in the **academic system and workforce** to help us tailor this profession for individuals who want to apply the profession in different places and ways. We do not propose a one-size-fits-all model. More importantly, we demonstrate that while repairing this system is no simple matter, it is not beyond our reach, and much like a healthy river system, it will strengthen individual success and resilience for the community as a whole.

We chose to use the analogy of a river system to speak to the two currents of objectivity and subjectivity for three reasons: (1) because while scientists try to separate themselves, compartmentalize, study, and understand ecological systems using quantitative and objective methods, the numbers don't always explain everything needed to ensure a system's health; (2) because the human and cultural relationships that were once part of this river were woven into a complex ecological system that could not be separated out without damaging the system; and (3) because when functioning properly, a river and all of its complexity provides enormous benefits to all living things, but when damaged, it is not repaired by simply fixing the mechanical actions that got it there. Our hope is to shed light on the penetrating changes that must take place to create an equitable environment in our workforce that not only reflects the face of this nation but the spirit of freedom as well.

While we speak as women of color, we cannot speak for all marginalized or underrepresented groups. There are many others who strive for what we advocate in this chapter. Our commonality is that we speak as individuals who do not seem like they "fit in" with the majority and whose views are rarely considered in decision-making processes. For describing the diverse sectors of the population in this chapter, we have chosen not to use the term "minority," in favor of "marginalized" and "underrepresented," given that some groups consider themselves sovereign nations and not as a subcategory of the overall population. Underrepresented groups occupy a spectrum of identities influenced by socioeconomic levels, sexual orientation, mental and physical ability, gender, race, or veteran status, although we may not address each one. Although this chapter is directed to the wildlife profession, we sometimes refer to "natural resource or environmental professions" given the similarities and coverage in the literature. Some of the barriers we address are documented by quantitative research, but others are qualitative in nature or have yet to be studied and addressed sufficiently to provide a real understanding of the problem.

Understanding these complex variables and their interactions will be essential if we are to identify what thresholds—or goals—need to be reached so we can push beyond our own current state of hysteresis. But although this might sound intimidating, we should not let it be. Much like the restoration of a river, it can be exciting to envision a beautiful and healthy system where we all belong. Like any other arduous problem-solving effort in conservation, the most challenging experiences can be the most rewarding.

Some Historical Perspective on Exclusion

Understanding the pathways and identifying the variables that brought us to this point of hysteresis is important. Post-colonization in North America, the human composition across these landscapes changed dramatically. The use of those landscapes all favored the power and gain of the new privileged population, whereas the people who already lived in these places—or were forcefully brought here—experienced loss and powerlessness. Furthermore, decision making was now in the hands of a culture that was unfamiliar with those new landscapes.

Academic and science systems in North America were largely shaped by the Royal Society of England in 1660 (Brasch 1931). Its mission was to *"recognise, promote, and support excellence in science and to encourage the development and use of science for the benefit of humanity"* (Royal Society 2012). Its journal, *Philosophical Transactions*, set the standard for scientific peer review and priority as early as 1665. Because the organization was dominated by white men and women were excluded, and most certainly those of color, today's academic systems and scientific philosophies are heavily weighted toward those perspectives. Most land use decisions were made for the growing, largely white society. Unfortunately, those outcomes didn't bode well for most underrepresented groups (or for the land), leading to the deeply embedded sense of distrust and resentment that we see today (Koch 2019). This part of history has been "diplomatically ignored" in a conservation movement that was influenced, and still is, by the founding Anglo science and academic system (Kashwan 2020).

In 1492 when Christopher Columbus "discovered" America, scores of people were already living on the continent and had intimate knowledge about their home, which was to be renamed "America" (Dunbar-Ortiz and Gilio-Whitaker 2016). For centuries, Columbus, and others, were revered as the continent's founders while millions of Indigenous citizens were, and continue to be, silenced. "Columbusing" is a term that was coined to name the expropriation of something that already existed, giving credit to the "discoverer" (Salinas 2014). The highly esteemed Aldo Leopold, the father of contemporary wildlife management in North America, is noted for his views on the interconnectivity of the natural world expressed in "The Land Ethic" (Leopold 1949). However, long before Leopold's publication, Indigenous people being removed from their lands already knew and lived by these precepts (Cornell 1985). It does not diminish Leopold's important role in conservation to remember that the underrepresented have always been part of the environmental movement. We have engaged with natural resources in ways that were ignored prior to and at the early stages of this country's establishment. Part of this is due to the influence of publishing in Westernized academia, as the official badge of recognition. However, many groups collect information and transfer knowledge within their communities through storytelling and experiential learning. These practices fit into the broad activity of "science," which is observing the world, watching and listening, and observing and recording (NASA 2019). Yet Indigenous agro-ecological groups (on any colonized continent) are still overlooked for their traditional knowledge and effectiveness in managing landscapes, because they are not considered "scientists" (Fang and Casadevall 2011; Chigonda 2018). For example, fire (Anderson and Lake 2013), livestock and wildlife management, land clearing, and farming were in use for thousands of years before Leopold wrote about the tools of the ax, plow, fire, and cow (Leopold 1986). One of the authors of this chapter, Diana Doan-Crider, overheard the knowledge of Indigenous presenters at an international scientific wildlife meeting dismissed as just "folklore," despite their living with and hunting the species in question for thousands of years and documenting long-term weather patterns and sea ice changes (Johnson et al. 2020). Black, Indigenous, and people of color (BIPOC) have historically gone unrecognized, despite significant contributions to our understanding of the natural world.

If you look closely, academic and scientific books are often written about other cultures and countries by Anglo scientists, usually with an objectivity that separates them from their specimens. Ota Benga (Box 5.1), a young Mbuti man who was captured and brought to the United States and exhibited in a zoo in 1904, is just one extreme case of how Anglo societies dehumanize other cultures. Most Latin plant and animal names are tied to Anglo scientists and their naming systems (IAPT, n.d.) or "renamed" (Littlefield and Underhill 1971) without consideration of other cultures' views or rights. For example, the saguaro cactus (*Carnegiea gigantea*) is an integral component in the ecological, sociological, spiritual relationship spanning thousands of years with the Tohono O'odham people of the Southwest and is called *hashan* in their language. Yet the plant was renamed after a wealthy Scottish businessman because his institution founded a botanical laboratory in the 1900s. Even recently, some countries and Tribes have had to develop policies to discourage a trend of visiting scientists who were documenting cultural knowledge without permission, or collecting and removing specimens from their lands and then publishing "discoveries" that were already known by local inhabitants (Roht-Arriaza 1996). Some scientists see this as a type of legitimized "theft" in their search for knowledge and perceived benefit for humanity. Several important publications for working with Indigenous communities point out how Eurocentric researchers detach themselves from and objectify populations who have limited power, creating a wariness of scientists and researchers in general (Baskin 2005; Cram et al. 2006, TeApatu et al. 2014; Snow et al. 2016).

Many in our profession fail to see people who have been historically overlooked and excluded because they are simply carrying on with generations of

BOX 5.1 The Dehumanization of Ota Benga

The 19th and 20th centuries were an exciting time of exploration of the natural world for Anglo cultures, bringing to mind expeditions like Darwin on the HMS *Beagle* or Shackleton "conquering" Antarctica. These explorers fanned out in waves across the ever-expanding world, often making their names by bringing back new and exciting organisms. But in and of itself, it was an opportunity for Anglo society to lay "scientific claim" on the world. These explorers not only collected and renamed thousands of wildlife and plant species but also came into contact with peoples they had never encountered before—sometimes bringing human "samples" home as "discoveries" to be put on display.

Organizers of the 1904 St. Louis World's Fair commissioned explorers to bring back thousands of Indigenous peoples, including "pygmies" and neighboring tribespeople from Africa, Native Americans, Filipinos, and Ainu (Newkirk 2015). For seven months, these people were put on display for the public as part of the exhibit of exotic races—sometimes referred to as the "human zoo." A man from the Congo named Ota Benga was part of the World's Fair exhibit and was also put on display in the Primate House of the Bronx Zoo, which was sanctioned by the zoo's superintendent William Hornaday, one of Anglo society's leading zoologists of the day (Figure 5.2).

Ota Benga was released from the Bronx Zoo after a few weeks as a result of public backlash—largely spear-headed by Reverend Dr. Robert Stuart MacArthur and Reverend James H. Gordon (Newkirk 2015). After his release, Ota Benga spent time at an orphanage and then worked on farms to earn money for passage home.

This dehumanization gives us a glimpse of the origin of some of the deep-seated cultural biases that are camouflaged in our natural resource and wildlife professions, and it reflects how many groups are still viewed and treated today. When we think of the many conservation heroes who have contributed to this profession, it's easy to allow the Ota Benga stories to remain hidden in the back of some dusty museum storeroom. However, justice requires that we face our history and the current dysfunctional system that is a direct result of how marginalized groups have endured a history of violence and dehumanization, especially those that were sanctioned by scientific or professional institutions. After years of psychological trauma and an unfulfilled yearning to return home, Ota Benga took his own life in 1916 (Newkirk 2015). We cannot let this go unnoticed.

FIGURE 5.2. Ota Benga, labeled as "pygmy." AMNH Anthropology catalog #99/4404 B, courtesy of the Division of Anthropology, American Museum of Natural History.

professional training. Publishing scientific material can certainly benefit humanity, and in fact, many individuals from underrepresented groups are quite successful in today's scientific and academic atmosphere. However, to ask people to ignore their own belief systems and use only current academic and scientific standards is awkward and unfair, to say the least. While Westernized academia and science separate the natural sciences into different disciplines through a rather "reductionist" approach (Fang and Casadevall 2011), agroecological cultures view themselves as part of the Earth's systems. Already, some people believe that they have to "choose" between their cultural-spiritual beliefs and science, which has resulted in a gap of trust and understanding between scientists and the public when we need to join forces to address issues such as climate change (McPhetres and Zuckerman 2018; Scheitle and Ecklund 2018).

Different views have value, and they strengthen our ability to solve our problems at much faster rates and with better outcomes (Østergaard et al. 2011). For example, the Salmon People have ties to the land that begin with their own genesis stories dating back thousands of years. Their relationship and role in that system should be part of the solution to the Columbia's hysteresis problem. Excluding the millions of people who have made significant contributions to science just because they don't "fit" the model of Westernized science hinders our ability to solve some of our complex socioecological problems. This brings us to where we are today . . . in an unstable state.

We're Right Here—Underrepresented Women in the Wildlife Profession

If diversity in our profession is just an issue of gender, then we should be satisfied with just two kinds of people, which is a goal that is within reach, based on recent increases in the number of women in our classrooms and offices. However, is that all we want? To continue with the Columbia River analogy, any ecologist knows that a healthy ecosystem is one that supports species diversity. Some species are more numerous and visible, but just because a species is less so doesn't diminish its value or importance in the system. Similarly, a diverse workplace and profession is one that includes everyone in the conversation, despite their numbers or proximity to the mainstream. One principle of hysteresis is that you can't get out of a situation by simply backtracking the way that you got there. Hence, you can't just tweak a few barriers and expect the system to begin flowing smoothly. Creating an equitable environment will produce complex challenges with no clearly visible pathway. While "women in wildlife" has been a recent discussion point in our profession, the topic of this chapter—underrepresented women in wildlife—has had us stumped perhaps for that reason. How *do* we find our way out of this, and is our profession willing to commit? Underrepresented women are waiting to be part of this conversation.

Despite some increases of underrepresented groups in the environmental workforce, the wildlife profession remains largely homogenous (Bowser et al. 2021). Underrepresented ethno-racial groups make up only 16% of employees found in 191 conservation and preservation organizations, 74 government environmental agencies, and 28 environmental grant-making foundations, and do not come close to representing the composition of our nation (Taylor 2014). We can assume that the actual numbers of wildlife professionals are much less. The Census Bureau projects that the United States will become a majority-minority (Alba 2018) country by 2044 (Colby and Ortman 2015). These changing demographics mean that institutions will be facing a potential workforce diversity crisis if they can't solve these problems. However, in doing so, we will only enhance our ability to achieve new and greater conservation successes.

The feminist movement of the 1970s created a momentum for women's rights that has slowly carried over into natural resources professions. However, underrepresented women struggled to gain attention and equitable footing with white women while working to address sexist and racist oppression in their own communities (Garcia 1989). For underrepresented women (e.g., women of color), gender equality is a multidimensional and intersectional issue (Garcia 1989) of social and political discrimination overlapping with gender discrimination (International Women's Development Agency 2018). Many women of color perceived the "white feminist movement" ignoring intersectionality and saw it as an elitist group where they didn't feel welcome or comfortable (Garcia 1989). The fight for equality is indeed intersectional, and we must learn how these camouflaged undercurrents can marginalize an individual or group of people.

While representation of women in the environmental-related workforce has increased to 55%, those successes have not been observed for underrepresented women, who are lagging significantly (Taylor 2014; Taylor et al. 2018). While we acknowledge that white women have not had it easy entering a world dominated by white men, they have made significant headway over the past 30 years (Taylor 2014). White women may have simply had some advantage in their higher numbers, and while the exclusion of underrepresented women was not done consciously, we draw attention to this discrepancy to ensure that we don't remain at the status quo.

One of the authors of this chapter, Ruth Plenty Sweetgrass-She Kills, interviewed a number of Native American women in the natural resources profession to detect the causes of low graduate program retention (Gervais et al. 2017). One interviewee stated that her mentor, who was a Native American woman, had to "deform" herself to get her PhD in botany by leaving behind everything that comprised her cultural identity, "becoming like a white man." The mentor was angry and felt very betrayed by white women around her because they were being accepted and celebrated by their white male peers, conditioned, in part, by not rocking the boat or challenging the assumptions and beliefs of the dominant group. It's possible that because white women share in white privilege, they are more easily accepted by white men and less motivated to risk their membership in the group by fighting for others' equality. These sometimes-hidden biases require serious scrutiny and reflection to see just how deeply ingrained they are in our profession.

Exploring Barrier Myths and Complexities

What's Love Got to Do with It?

Any ecological assessment requires a well-coordinated strategy. In the case of professional diversity, many of our colleagues have undertaken the effort to investigate as many individual barriers as possible (Taylor 2014; Balcarczyk et al. 2015; Haynes et al. 2015; Gervais et al. 2017; Taylor 2017; Taylor et al. 2018). However, those that are yet camouflaged warrant serious attention because of their different root causes and potential to cause logjams in the overall system.

Looking at the research that has been conducted, some barriers have been misunderstood. Assumptions related to recruitment of underrepresented groups require a new review to understand which ones are contributing to the real bottlenecks (Taylor 2014). If so, we should constantly expand our discussions and look at this dilemma from varying angles. For example, the assumption that underrepresented groups—especially people of color—were simply not interested in environmental sciences (O'Connell and Holmes 2011) was borne out by surveyed students (Taylor 2017, 2018; Taylor et al. 2018). Furthermore, some studies question whether underrepresented students and their career paths are influenced more by salary levels than their previous exposure to nature (FS 2004; Taylor 2018). These studies lead to two important points: (1) assumptions about barriers need to be dissected under the intersectionality microscope, and (2) stakeholders from underrepresented groups need to be directly involved with this research because they raise important questions through their own experiences and perspectives. In addition, new generations (and a changing demography) have different perceptions and experiences, and they now face different barriers, all in the midst of a rapidly changing workforce that is trying to bridge gaps for the previous cohorts. We will constantly have new challenges that need creative solutions, and we should not let our guard down each time we reach a particular threshold that makes us feel good. Just as with the Columbia River, our system will need constant monitoring and tweaking.

For fish in the Columbia River, let's consider the variables that make a system attractive, tolerable, or intolerable for the many species that once lived there. We know that certain characteristics must be present in the habitat, or the species that depend on them will simply not show up. Habitat preference and selection models indicate that there exist *thresholds* that keep certain species at certain levels.

This may be no different for the barriers that limit certain groups of people in our profession. Barriers can interact and accumulate to thresholds that result in a completely impeded and formidably unattractive environment for these groups as a whole. Inclusive diversity management—or good human habitat—is where everyone has the opportunity to lead, be promoted, participate, collaborate in healthy and equitable ways, be compensated fairly, and socialize and be mentored in ways that help them maximize their potential (Taylor 2007; Batavia et al. 2020). One study notes that students specifically look for those characteristics when deciding to work for an organization (Taylor 2007). So, while we know what good habitat *should* be, identifying which barriers play what roles in reaching these thresholds can be confusing.

Dorceta Taylor, one of the most recognized researchers in the field of diversity and inclusion for the environmental workforce, identifies a "*green ceiling*," or a "*green insiders club*" that is overwhelmingly white (Taylor 2014). Today, the renewed mainstream attention on racism reminds us that perhaps not as much progress has been made as we thought, and that this is still a huge logjam for restoration in our workplaces and communities. So, what does this green ceiling look like now, how many barriers are involved, and do we have the knowledge to understand and remove it so that members of underrepresented groups will be able to thrive in those environments?

Psychologists studied the achievement of prominent Black and white women in higher education by using grounded theory (Strauss and Corbin 1994) to assess their success (Sperber et al. 1997). They found three key variables: **persistence**, **connection**, and **passion**. These variables are packaged with a gamut of other intersectional variables for different people and in different scenarios. However, the success of the women in this study hinged on whether they had these three things, most of which were cultivated through various support systems and experiences. Who doesn't want to be passionate about their career path? How can you be passionate about your career path when nobody includes you and your values in the discussions and activities that cultivate passion for everyone else? How can you have connection if nobody has anything in common with you? We ask each of you to reflect on what helped to fuel, or kill, your passion about your relationship with nature and your profession. In the grounded theory study, these women somehow found the support they needed to feel included, they connected with mentors, they felt empowered to help others, and they were able to work on issues that were important to them (Sperber et al. 1997). Although many of these women encountered different barriers at different times, these three overarching variables created a buffer that helped them to navigate better, establish themselves in this environment, and achieve success.

Another study, evaluated more than 1,000 forestry students with strong representation from different marginalized groups and socioeconomic classes (O'Herrin et al. 2018). The authors developed a list of 18 professional support mechanisms for considering a career path in natural science disciplines. Of the 18 support mechanisms, students valued *"the passion or enthusiasm you would have for your work"* as first, and *"meaningfulness of the work you would be doing"* as second. Both ranked above better wages, promotional ladders, and esteemed scientific journals. So at least two studies have demonstrated that passion, connection, and persistence are critical. If that's the case, then why aren't they at the top of the list in our diversity and inclusion discussions? Because these values are intangible, they are hard to quantify, but we should do our best. Let us help: our own passion is just as strong as everyone else's, but again, it may be driven by different values than those found within the walls of our profession. They should also be on the menu.

The Columbia River is an apt analogy for our inclusion dilemma in that barriers can keep people out of the main current and off the pathway to success. For example, it is not enough to hire physically disabled people into the ranks of an organization and build a few ramps. While that person is technically a "part" of the organization or institution and makes it more diverse, are they "included"? It is useful to view "disability" as a relational concept, a function of a person's interaction with the social and physical environments (Jerman 2020). *How* they can interact makes a given particular environment enabling or disabling. The inroads begun with the American Disabilities Act (ADA) of 1990 (Global Greengrants Fund 2019) do not go far enough for meaningful inclusion, and it is up to us to find ways to keep them engaged. These limitations on how we are able to interact with our environments, in a way, can be applied to all marginalized groups, so our strategies should be comprehensive.

In comparison to a healthy ecological river system, small refugia of habitat might allow species to get a foothold and perhaps gain strength to expand to other areas if things change within the system. Most underrepresented groups have the resiliency to dig in their heels and problem-solve from within *if* provided with a decent foothold. However, while these small patches are beneficial, they may not be big enough to push representation above certain diversity thresholds that allow underrepresented groups to prosper as a whole. So in a panic to increase diversity, agencies and institutions implement programmatic approaches that wave underrepresented individuals through the door with well-intentioned promises of scholarships or jobs, diversity-speak, a "sense of community," and marketing ads that include marginalized groups. Sadly, these recruits find themselves in an unsuitable habitat that does not support their needs. By now, it is clear that many programmatic approaches are not working the way they were intended (FS 2004; O'Herrin et al. 2018). Among 15 major disciplines recognized by the Census Bureau, agriculture and natural resources has the lowest percent of marginalized individuals with bachelor's degrees in the workforce (FS 2004). Recruitment and first-year retention of newly hired underrepresented students into federal natural resources agencies remains low, similar to college enrollment in those natural resources career paths (Quinn et al. 2016). Because exit interviews are not conducted in most cases, or not conducted thoroughly enough, little is known as to why. But as three women who have walked through those doors, we can guess.

The Anatomy of a Damaged System—Identifying the Barriers

Reconnecting Children through Early and Sustained Education

Barriers to the environmental profession have been identified at various life stages: early education, college, early career, and established professional retention (Taylor 2007, 2014; Balcarczyk et al. 2015; Haynes et al. 2015). However, there is no one-size-fits-all analysis, even for women who would categorize themselves as from the same background. Intersectionality, again, compounds some of these barriers and makes them difficult to tease out (Haynes et al. 2015; International Women's Development Agency 2018). These impacts can begin early in a child's education journey, as we begin to understand and internalize the expectations of our families, communities, and cultures, and even as we enter higher education and, ultimately, the conservation workforce.

Every year, the Salmon People hold a Salmon Camp for their youth along the Columbia River. There, the children learn about their Tribal cultures and fish ecology, and spend time rafting along the river to identify fish habitat and potential barriers (Columbia River Inter-Tribal Fish Commission, n.d.). One of the participants said, "*Salmon Camp has made me more interested in making baskets, how things work, and the First Foods and how to protect them*," illustrating their Tribe's unique relationship to the river. But another student said, "*Salmon Camp gave perspective into Native culture, but also provided thought into future possibilities.*" Tribal members recognize the camp's importance in linking their children's cultures to future opportunities in science and higher education for the ultimate benefit of conserving a vital system. Partners in the effort include federal agencies such as the Bureau of Indian Affairs and the US Fish and Wildlife Service, and it is funded by organizations such as the Bonneville Power Administration, several private foundations, and Tribal governments.

One of the top barriers that has been cited for younger generations in entering the natural resource field is a lack of interest in or desire to be connected with nature (Morgan 2013; Haynes et al. 2015). Even without specific influences from culture or family on what career path to follow, people's childhood experiences are well known to impact the career fields they choose as adults (Haynes et al. 2015). If underrepresented children are not connected to nature in their daily lives, they likely will grow up without knowing anyone who is connected to a natural resource profession. Furthermore, elementary education curricula are lacking in the natural sciences, but more so in low-income areas (Haynes et al. 2015). This disconnect is reinforced by overt or subtle environmental racism, individually or in their communities. The first studies on diversity in environmental activities in the 1960s found that people who identified as nonwhite were far less likely to participate in outdoor recreation than those who identified as white, although this has not been dissected thoroughly enough (Taylor 2014). Regardless, recent reports indicate that these numbers may be changing for the better, at least for some groups (Outdoor Foundation 2016). It can still be difficult to tease out differences in types of recreational activities such as camping, hunting, fishing, or hiking that are offered through various traditional clubs, such as Girl Scouts or 4H, and how they might influence decisions to pursue state and federal career paths in wildlife, forestry, soils, fisheries, or range management. Although some children from some groups, like Hispano-agricultural and Native American communities, may be more connected to the environment through their participation in traditional activities such as food gathering, horseback riding, and fishing, they are not as likely to join clubs with stronger connections to academic institutions that include natural resources career paths. Therefore, they do not become familiar with those career paths and their requirements. Over time, the positive connection between nonwhite children and the outdoors may weaken without programs to strengthen their interest. Most agencies tend to target high school and college stages for educational outreach, even though most long-lasting impacts are made at much younger ages (Haynes et al. 2015).

Creativity and partnerships are vital pieces in addressing barriers and connecting with any group. Homogenous programmatic outreach efforts will simply not be enough to engage young people in the profession without addressing the values of their cultures, their different views, or even the limitations that keep them from engaging in the natural resources profession in the same ways of mainstream society. Not everyone can leave their communities and families to meet job mobility requirements. Not everyone has the same physical abilities to fulfill the field requirements. Too often, schools and agencies often target the pools where they can get the quickest return on applicants to fill positions, but unfortunately, that approach obviously misses people who are not swimming in the main pond.

Lack of Knowledge about Career Paths

If the interest of a young girl can somehow be cultivated enough to keep her engaged in the idea of a career in natural resources past early childhood barriers, she will need careful navigation and a strong support network to keep her moving along the right current. Accessing a higher-level education already poses challenges for underrepresented women, such as logistics, stress, finances, and lack of cultural support (Gervais et al. 2017). Clear information about career paths and directions on how to get there aren't often readily available to these groups. If a young woman hasn't learned much about wildlife and natural resource professions, she's unlikely to take the relevant coursework and go into the field (Nagappan 2018). The lack of a relevant degree is now surfacing as a major barrier for entry into natural resources professions, particularly for those students who attend minority-serving institutions (MSIs). "Lack of knowledge about careers" was the most-mentioned barrier in a study assessing culturally diverse hires in the US Fish and Wildlife Service (Balcarczyk et al. 2015). This was also a challenge with high school and college career counselors surveyed in Texas and New Mexico, who sometimes unknowingly divert students into nonspecific academic programs (Shippy et al. 2021). Furthermore, most accredited natural resource programs—at least those that meet the criteria for federal "mission critical occupations" as identified by the Office of

Personnel Management (OPM n.d.)—are generally found only at larger universities that can afford to support additional faculty for those specializations (DOI, n.d.). While many MSIs do include broad degree programs such as biology, conservation, or environmental science, their curricula tend to fall short of the courses required by the OPM, thus leaving students unqualified for federal specialist level series (i.e., wildlife biologist—0486 series, forester—0460 series, or range management—0454 series) and ineligible for upper-level professional positions. Furthermore, students cannot learn about natural resource careers if they attend schools that don't have accredited programs in those areas. Schools will not support those disciplines if there is insufficient demand. Most outreach efforts conducted by professional organizations target schools with those specialized programs, and agencies tend not to recruit at MSIs for mission critical positions, so again, underrepresented students end up missing the memo (Taylor 2014).

Some of the cyclic issues encountered by underrepresented women may carry over to activities such as internships, research opportunities, and technician positions. In the wildlife profession, volunteer work and internships are still seen as a vital part of establishing yourself in the field. However, many underrepresented students and early-career professionals, particularly those from low-income households or who are caregivers, cannot afford to take an unpaid position, which significantly lowers their ability to persist in a science, technology, engineering, or math (STEM)–related profession (Fournier et al. 2019). Additionally, because many internship programs still lack diversity, already marginalized students again encounter the barrier of social isolation or exclusion. However, mobility appears to be a more significant barrier for recruitment and retention in new research (Quinn et al. 2016). One example from coauthor Ruth Plenty Sweetgrass-She Kills involves a federal agency that consistently requested interns from a particular group of underrepresented students in her office and yet was unable to assist with the mobility and financial needs of students who had families. Agency restrictions did not allow for this type of accommodation, and the internships were consistently unfilled with no creative attempts to make adjustments. Another author, Diana Doan-Crider, had three students who were women of color with children, and they needed family housing to accept internship positions. Because there was some flexibility in a collaborative grant, funds for housing were made available—but only through some creative maneuvering that was not the norm. Although salary may not be a primary barrier (Taylor 2018), financial stress is a factor for retention of Native American women throughout the sometimes-tricky pathways of becoming a wildlife professional (Gervais et al. 2017).

Good news—these barriers are surmountable. Professional organizations need to begin to reach out to a wider range of students. Career counselors and teachers should be provided with sufficient information to point their students in the right direction. Larger land grant universities have to build or strengthen bridge programs with smaller colleges. However, we caution that these efforts must be accompanied by a good dose of consideration and sincerity. Targeted recruitment of students from underrepresented backgrounds raises skepticism of institutional motivations to increase its own diversity numbers or funding opportunities (Quinn et al. 2016). To most underrepresented individuals, these are tell-tale signs of shortcomings in the type of organic support needed for those individuals to succeed and become passionate about their jobs. Larger universities need to explore their roles in strengthening place- and culture-based academic institutions that help to empower communities. This could be done through community-relevant research or collaborating to coteach essential courses that are place and culture based but without the high tuition costs of larger universities. Colearning between institutions would also help less marginalized students learn about the societal challenges and cultural values of underrepresented communities, which will only make them better professionals and problem solvers at a more inclusive scale. These improved relationships will eventually cultivate more equitable and collaborative efforts down the road that will benefit everyone involved.

Cultural and Societal Pressures

The potential impact of family pressure, preference, and overall influence is widely published and should not be ignored for young women of certain cultures as we recruit or help them stay in our profession (Balcarczyk et al. 2015; Demby and Meraji 2016). The numerous factors that contribute to a general lack of familiarity of some cultures with natural resource professions can sometimes translate into, at best, a lack of support for women wishing to enter the field, and at worst, outright opposition and pressure to enter a more prestigious professional career. Coauthor Doan-Crider notes that while one of her colleagues was recruiting students from a largely Hispanic community, they were approached by a parent who angrily stated, "While we've spent our lives trying to keep our kids out of the fields, you all are trying to take them back there. We want our children to be lawyers or doctors!" Public misconceptions of natural resources careers can often be influenced by what people see in the media. However, while we know these cultural and social expectations may still influence some young women's career choices, things are slowly changing for the better.

Because natural resource careers do often involve field work, any negative historical perceptions about the outdoors may influence how these jobs are accepted in some underrepresented communities, especially for women. Yet while the wildlife profession was once dominated by men, student enrollment ratios in wildlife biology programs today favor women (Lopez and Brown 2011). However, certain patriarchal cultures hold distinct norms about interacting with men. For example, for the Apsáalooke, it may be culturally inappropriate for young women to speak to strange men or to be left alone with men who are not their husbands (Little Big Horn College n.d.). In addition, traditional families of some cultures may not approve of women conducting what would be considered masculine activities (Schmitz et al. 2004). The remote nature of many wildlife-related jobs and projects can also cause concern for women of color because of a history of violence against them, which continues today, as with those highlighted through the Missing and Murdered Indigenous Women campaign (Native Women's Wilderness n.d.). This can be a barrier to working mostly with men and having limited communication with their families.

Social and cultural pressures as barriers can sometimes be camouflaged. Things like discrimination or exclusion can be easily tagged as the primary barriers, when in reality, other influences may amplify their impact. For example, a woman who considers a profession culturally inappropriate for her may also feel less empowered to speak out against discrimination and exclusion for fear of "rocking the boat" or because she does not feel worthy or deserving. The intersection of these barriers is difficult to quantify, but it certainly requires scrutiny. This information could be very helpful in communicating and reinforcing healthy perspectives among young women who have an interest in the profession.

Recruitment and Retention in the Workplace

Racial and ethnic discrimination in recruitment, hiring, and promotion has been cited as one of the top five underlying reasons for low diversity in the STEM fields (Funk and Parker 2018). However, there are many strong undercurrents that keep underrepresented women from getting a competitive start in the workforce, which are often difficult to see. Obtaining quantitative data on the impacts of variables like self-confidence to apply for certain jobs, application readiness, academic qualifications, mobility challenges, and discomfort in the workplace is difficult. It's hard to know which women we've lost because many of them never even show up.

If, as students, women make it past the many barriers and gain mid- to upper-level employment in the natural resources field, then they are confronted with leaping over the tall waterfalls of the gender pay gap, where Black, Native American, and Hispanic women earn lower wages compared to white and Asian American women (Gould and Scheider 2017; Khan et al. 2019). It's no wonder we see so few in high-level executive or leadership positions (Taylor 2014), which is discouraging and contributes to low retention rates.

In higher education, particularly in the science fields, Latinx, Black, and Native American women perceive a need to work harder than their colleagues to be seen as a legitimate scholar (Flaherty 2019). Furthermore, while colleges and universities have increased their diversity hiring in the past 20 years, most have been for non-tenure-track positions (Finkelstein et al. 2016), which leaves underrepresented faculty members vulnerable and with little influence in academia. Again, discrimination can be camouflaged in many ways. In "11 Sneaky Ways Companies Get Rid of Older Workers," the author states that some companies, agencies, and institutions are getting more creative in disguising overt discrimination and covering their tracks as more women pursue litigation (Jacobs 2013). Underrepresented women are more likely to experience harmful stereotypes in the workplace resulting in "imposter syndrome"—feeling like you don't truly belong or deserve what you have and that you'll be exposed as a fraud when there is no basis for this perception (Beilock 2019). Women also temper their behavior to avoid playing into any existing stereotypes of their marginalized group (Corbin et al. 2018), which further contributes to pay discrepancies by creating fear in salary negotiations. For those underrepresented women who do make the leap over the top of the employment waterfall, most have stories of being inadvertently tokenized to highlight an organization's diversity goals. These women are pressured by default to represent their entire identity group—thus making them feel much more visible and more heavily scrutinized, and in many cases, overworked. One of the authors of this chapter, Jamila Blake, was asked on the first day of her first full-time position whether she was only hired because she added diversity to the organization. While anyone would feel stressed and potentially a little lost on their first day, she felt a level of added scrutiny to not only demonstrate that she was qualified but also to avoid confirming negative stereotypes associated with her identity groups. Ultimately, all these factors contribute to vicious cycles of "not fitting in," especially where no support system exists, which makes them look elsewhere. The issue is particularly compounded in the natural resource field, where the traditional stakeholders are typically white and are not used to dealing with people who are different from them. As a whole, the natural resources field does not have a reputation for inclusiveness or diversity, as seen in advertisements, recruiting campaigns, book publishing, and news reports dominated by white men and some white women. But thankfully this appears to be changing.

Even if some underrepresented women find enough of a microhabitat for their persistence, connection, and passion, they don't find the same kind of refuge as do their white colleagues, where "they look like me / I look like them," nor do they have the advantage of the same social and historic stepping-stones (Sperber et al. 1997). One key component for the success of women in the workplace has always been the presence of mentors (Thomas 2001; Young et al. 2017; Taylor et al. 2018). Yet only about 17% of programs surveyed in 2018 included mentoring programs, even though students indicated that it was highly desirable.

Coauthor Doan-Crider experienced discrimination in her workplace but found no support in navigating some objectionable practices to manipulate her non-tenured faculty position into a staff position so that monies could be used for another tenured-track faculty position. She was already the lowest-paid faculty member in her department but was still expected to perform teaching and advising duties with no increase in salary and no faculty voting privileges. She was unable to find some sort of advocacy through her department and college, and she was asked to leave after challenging some of those decisions on her own and not clearly documenting what had happened to her. Had it come to that, a lawsuit would have resulted in negative press and high legal expenses for the institution and undue stress for the woman, becoming another trigger for the diversity blame game.

While underrepresented women are gaining momentum in amending gender pay gaps in court (Flaherty 2020), most women would rather avoid taking these extreme measures. Underrepresented women would prefer fair institutions and organizations, thorough and systematic reviews to ensure inclusion, happier workplaces, and a decent compass (such as mentoring) to help us navigate toward equitable access to resources. Prioritizing and developing organic and consistent mentoring programs will not only strengthen women but improve conditions for everyone. These groups, including those at professional societies, working groups, breakout sessions, and niche organizations, can support women through sharing their own unique experiences and connecting them to resources but also help them mitigate conflicts and navigate into leadership positions.

We know that these efforts work. Federal agencies have taken immediate action to repair identified barriers to recruitment and are making strides to remove barriers through partnerships (Quinn et al. 2016). Academic institutions and professional societies, across the board, are busily addressing diversity and inclusion issues, and doing so with sincere motives, despite the lack of improvement in crossing meaningful thresholds (Ganguli and Launchbaugh

2013; ED 2016). This lends hope that a greater effort can follow through with a comprehensive strategy to build the momentum that we need.

Where Do We Go from Here?

For our hysteresis equation, we hope we have provided some insight on the history that *got* us here, the variables that are *keeping* us here, and now, what it will take to push us over the threshold to get us *out* of here. However, we also hope that our readers can see the potential for a restored system that is fairer and more equitable. While finding effective and sustainable solutions to barriers that exclude people from our profession will require the deconstruction of a few dams, we think the resulting system will be healthier for *all* stakeholders, not just one group. It depends on whether governments, environmental organizations, and agencies will prop the doors open with sustainable solutions to make a real impact on conservation workforce diversity. To increase diversity does not mean we ask for pity or charity to placate our hurt feelings or to simply bring more chairs into the room. What we ask is for space front and center at the head table and the chance to direct our own path. Healthy, productive, and inclusive diversity management practices give full and equitable participation to *all* members, not just a few (Batavia et al. 2020). We want equitable access to the tools that allow us to help ourselves. We want knowledge about career paths. We want professions that allow us to address our own community needs. We want an education system that is inclusive of other cultures and their histories. We want partnerships that strengthen our efforts and help us get jobs.

As this chapter is being written, a proposal to remove four dams along the Snake River in the Columbia River basin that are part of critical salmon habitat is under consideration (Leslie 2021). If this happens, and if the authorities that be, and the public, deem the benefit worth the effort, this could be a game changer for the river system and its inhabitants. This is an ironic and timely analogy to our own profession's dilemma and helps us see that change is not a bad thing, and that the benefit of the whole system is better than the needs of "almost" everyone else. Ultimately, it highlights the urgency of engaging with everyone so they can envision what that new system should look like and what they will gain . . . not what they will lose. We believe it will be good for everyone.

How to Be an Ally

- Create a welcoming and inclusive environment that everyone would *want* to inhabit.

- Understand community needs of marginalized groups, and work with underrepresented colleagues to create culturally and community appropriate materials or outreach programs.
- Rethink current academic models and look for ways to develop culture- and place-based education approaches that provide youth access to information and programs about natural resources that are relevant to their own communities and worldviews.
- Empower marginalized and underrepresented groups so they are encouraged and supported to integrate their own cultural beliefs and worldviews in learning processes and in professional mainstream decision making.
- Build relationships that strengthen trust with marginalized groups by investing time and effort into activities that enhance interaction, dialogue, understanding, collaboration, and equitable engagement.
- Partner with marginalized and underrepresented groups by committing short- and long-term support, such as financial, logistic, academic, or technical service that will ensure their independence to conceptually design and implement projects that address their needs.
- Invest in community participatory research and engage students and professionals in projects that matter to them.
- Educate yourself and others about the complicated history of Black, Indigenous, and people of color in conservation and the outdoors, and help make their influence and perspectives more visible in our profession.
- Share credit where it is due with past generations of minoritized people and their historical information, and acknowledge those who contribute with their local knowledge in publications and presentations.
- Larger 1862 land grant universities can help empower communities by bridging programs with smaller colleges (e.g., 1994 Tribal land grant colleges and universities and community colleges) to share learning opportunities and access to course material.
- Engage and educate children and young adults through their cultures, which will strengthen communities and conservation as a whole.
- Professional organizations should reach out to MSIs and smaller colleges to engage directly with underrepresented groups that are not exposed to natural resources career paths.
- Establish partnerships to provide affordable or no-cost support for family mobility and caregiving for interns and new employees.
- Invest in organic and consistent mentoring programs with histories of success supporting minoritized and underrepresented people.

- See underrepresented women as qualified and highly valued in their profession and help them and others to view them as so.
- Teach women how to identify exclusionary and discriminatory actions, while also providing support in addressing the issues while protecting their professional reputation and job security.
- Call out and deal with discrimination or disrespect between colleagues and students immediately and assertively.
- Don't pity marginalized groups—empower them.
- Ensure that academic and work activities are accessible to all individuals with thoughtful consideration to possible barriers for those with varying forms of disability (visible and invisible), making an inclusive and productive environment for everyone.
- To cultivate an environment of accessibility, standardize the use of assistive technologies (e.g., close captioning) and offer other accommodations to facilitate access.

Summary

We compared the diversity challenges in our profession to the restoration of a large river system that has reached a state of hysteresis, where the pathways that caused the imbalance will not be the same to get out. We discussed what it will take to get us over the threshold given the complexities and intersectionality of barriers, and how deconstructing these barriers and developing solutions will not be fixed through programmatic solutions. The goal should be to create an environment where everyone can feel relevant and be passionate about their work.

Resources: Professional Associations and Their State or Section Chapters
- The Wildlife Society: Early Career Professionals Working Group; Inclusion, Diversity, Equity and Awareness Working Group; International Wildlife Management Working Group; Native Peoples' Wildlife Management Working Group; Student Development Working Group; Out in the Field; Women of Wildlife
- American Fisheries Society
- Ecological Society of America: Black Ecologist Section; Early Career Ecologists; Inclusive Ecology Section; Strategies for Education in Ecology, Diversity and Sustainability (SEEDS); Students Section
- Society of American Foresters: Diversity and Inclusion Working Group

- Native American Fish and Wildlife Society
- National Wildlife Federation—Artemis
- Minorities in Agriculture, Natural Resources and Related Sciences

Discussion Questions

1. Underrepresented women continue to experience feelings of isolation, exclusion, and distrust. Can you think of a particular situation where you can help create a better sense of community, safety, and trust?

2. How can you make your environment more culturally relevant to everyone in your school or workplace? Think of individuals working on particular issues and how you can highlight their efforts.

3. Strong mentors support the recruitment and retention of underrepresented women. What hinders good mentor relationships, and how can more women be supported to act as mentors?

4. Throughout the history of the women's rights and feminist movements, there has been a lack of connection, empathy, and unity between women of different race, class, and sexual orientation. Can you identify particular examples where the lens of intersectionality might cause these divergent factors to persist, even in today's feminist movements?

Activity: Creating the Right Environment

Learning Objectives: Participants will explore, identify, and discuss the barriers and pathways to careers in the wildlife profession. Participants will identify inhospitable components of the existing system and what needs to happen to restore it to good habitat for underrepresented groups.

Time: 35 min.

Description of Activity

Materials needed: Copies of the worksheet for anyone without a book and writing utensils.

Process:

Part 1 (5 min.): Sometimes environments have to be modified to make them hospitable to certain inhabitants. Logjams and dams must be removed to create pathways for salmon, or the ambient water temperature needs to be

Did these things hinder or help you?	Barrier	Neutral	Pathway
Lower pay	☐	☐	☐
Microaggressions	☐	☐	☐
Feeling like you "fit in"	☐	☐	☐
Inclusion of your culture	☐	☐	☐
Strong support from your family/friends	☐	☐	☐
Feeling like an imposter	☐	☐	☐
Early life experience with natural resources	☐	☐	☐
Cultural differences from coworkers	☐	☐	☐
Ability to balance work with personal life	☐	☐	☐
Mobility/job and locations were challenging	☐	☐	☐
Feeling excluded from groups or activities	☐	☐	☐
Access to natural resource courses/ programs	☐	☐	☐
Lack of role models that looked like you	☐	☐	☐
Unpaid internships or volunteer work	☐	☐	☐
Access to mentoring	☐	☐	☐
Understanding the employment application process	☐	☐	☐
Including community in your work	☐	☐	☐
Environment fostered passion for your work	☐	☐	☐
Lack of training in professional skills	☐	☐	☐
Early knowledge of wildlife profession and opportunities	☐	☐	☐
Understanding graduate school application process	☐	☐	☐

changed so species can thrive. Read through the list of potential "barriers" and "pathways" for pursuing a career in the wildlife and natural resources profession, and check the box in terms of how you view each one.

Part 2 (10 min.): Form pairs or small groups to discuss the items in the table identified as most influential to their education and career experience. After discussing these different barriers and pathways, ask groups what a healthy and diverse academic and professional environment should be like for everyone (including non-underrepresented groups) where these barriers did not exist.

Part 3: Debriefing and Discussion Questions (20 min.)

1. Did group members rank top barriers differently? What does this suggest about people's perceptions of barriers and challenges and impacts on their wildlife careers?

2. What pathways cultivated passion and helped people thrive in their positions? What was the most surprising thing you learned during this discussion?

3. What were some of the visions of barrier free systems? What kinds of structural or institutional changes were necessary? Who or what have the power or authority to fulfill your visions?

4. How would you use the information you learned or your experience from this activity in your life as a wildlife professional, student, or citizen?

References and Resources

This activity was modified from Jonamay Lambert and Selma Myers, "Dealing with Change: Barriers and Strategies," in *The Diversity Training Activity Book: 50 Activities for Promoting Communication and Understanding at Work* (New York: AMACOM American Management Association, 2009).

Literature Cited

Alba, R. 2018. "What Majority-Minority Society? A Critical Analysis of the Census Bureau's Projections of America's Demographic Future." *Socius* 4:1–10. doi: 10.1177/2378023118796932.

Anderson, M. Kat, and Frank K. Lake. 2013. "California Indian Ethnomycology and Associated Forest Management." *Journal of Ethnobiology* 33(1): 33–85, doi: 10.2993/0278-0771-33.1.33.

Balcarczyk, K. L., D. Smaldone, S. W. Selin, C. D. Pierskalla, and K. Maumbe. 2015. "Barriers and Supports to Entering a Natural Resource Career: Perspectives of Culturally Diverse Recent Hires." *Journal of Forestry* 113(2): 231–39.

Baskin, C. 2005. "Storytelling Circles: Reflections of Aboriginal Protocols in Research." *Canadian Social Work Review* 22(2): 171–87.

Batavia, Chelsea, Brooke E. Penaluna, Thea Rose Lemberger, and Michael Paul Nelson. 2020. "Consider the Case for Diversity in Natural Resources." *Bioscience* 70(8): 708–18.

Beilock S. 2019. "Research-Based Advice for Women Working in Male-Dominated Fields." *Harvard Business Review.* Last modified March 3. https://hbr.org/2019/02/research -based-advice-for-women-working-in-male-dominated-fields.

Beisner B. E., D. T. Haydon, and K. Cuddington. 2003. "Alternative Stable States in Ecology." *Frontiers in Ecology and the Environment* 1(7): 376–82.

Bowser, G., N. Roberts, D. Simmons, and M. Perales. 2012. "The Color of Climate: Ecology, Environment, Climate Change, and Women of Color—Exploring Environmental Leadership from the Perspective of Women of Color in Science." In *Environmental Leadership: A Reference Handbook,* ed. D. R. Gallagher. Thousand Oaks, CA: Sage, 2:60–67. https://dx.doi.org/10.4135/9781452218601.n7.

Brasch, Frederick E. 1931. "The Royal Society of London and Its Influence upon Scientific Thought in the American Colonies." *Scientific Monthly* 33(5): 448–69. www.jstor.org /stable/15250.

Brooks, A. P., G. J. Brierley, and R. G. Millar. 2003. "The Long-Term Control of Vegetation and Woody Debris on Channel and Flood-Plain Evolution: Insights from a Paired Catchment Study in Southeastern Australia." *Geomorphology* 51: 7–30.

Chigonda, Tanyaradzwa. 2018. "More Than Just Story Telling: A Review of Biodiversity Conservation and Utilisation from Precolonial to Postcolonial Zimbabwe." *Scientifica* (August): article ID 6214318. doi: 10.1155/2018/6214318.

Colby, Sandra L., and J. M. Ortman. 2015. "Projections of the Size and Composition of the U.S. Population: 2014 to 2060." US Census Bureau, March 3. www.census.gov/library /publications/2015/demo/p25-1143.html.

Columbia River Inter-Tribal Fish Commission. N.d. (a). CRITFC Salmon Camp. Accessed October 5, 2020. www.critfc.org/for-kids-home/salmon-camp/.

Columbia River Inter-Tribal Fish Commission. N.d. (b). "We Are All Salmon People." Accessed October 10, 2020. www.critfc.org/salmon-culture/we-are-all-salmon -people/.

Corbin, N. A., W. A. Smith, and J. R. Garcia. 2018. "Trapped between Justified Anger and Being the Strong Black Woman: Black College Women Coping with Racial Battle Fatigue at Historically and Predominantly White Institutions." *International Journal of Qualitative Studies in Education* 31(7): 626–43, doi: 10.1080/09518398.2018.1468045.

Cordova, Ruth. N.d. Pomo Elder Quote. Dry Creek Rancheria Band of Pomo Indians. Accessed October 10, 2020. www.drycreekrancheria.com/pomo-elder-quote/.

Cornell, George L. 1985. "The Influence of Native Americans on Modern Conservationists." *Environmental Review* 9(2): 105–17. doi: 10.2307/3984336.

Cornell University Laboratory of Atomic and Solid State Physics. 1994. "What's Hysteresis?" www.lassp.cornell.edu/sethna/hysteresis/WhatIsHysteresis.html.

Coues, Elliot, ed. *The History of the Lewis and Clark Expedition.* New York: Dover, 1987; reprint of the 1893 Francis P. Harper 4-vol. edition.

Cram, F.A., A. Ormond, and L. Carter. 2006. "Researching Our Relations: Reflections on Ethics and Marginalization." *AlterNative: An International Journal of Indigenous People* 2:174–92.

Demby, Gene, and Shereen Marisol Meraji. 2016. "Made for You and Me." *Code Switch Podcast.* National Public Radio. June 8. https://one.npr.org/?sharedMediaId =480942626:481181970.

Dunbar-Ortiz, Roxanne, and Dina Gilio-Whitaker. 2016. *All the Real Indians Died Off: And 20 Other Myths about Native Americans.* Boston: Beacon Press.

Fang, F. C., and A. Casadevall. 2011. "Reductionistic and Holistic Science." *Infection and Immunity* 79(4): 1401–4. doi: 10.1128/IAI.01343-10.

Finkelstein, Martin, Valerie Martin Conley, and Jack H. Schuster. 2016. "Taking the Measure of Faculty Diversity." TIAA Institute. April. www.tiaainstitute.org/publication /taking-measure-faculty-diversity.

Flaherty, Colleen. 2019. "What Faculty Members Think." *Inside Higher Ed*, March 5. www .insidehighered.com/news/2019/03/05/major-survey-shows-professors-worry-about -discrimination-arent-prepared-deal.

Flaherty, Colleen. 2020. "Paying for Inequity." *Inside Higher Ed*, October 8. www.insidehighered .com/news/2020/10/08/princeton-settles-federal-government-gender-based-faculty-pay -gap-case.

Fournier, Auriel M. V., Angus J. Holford, Alexander L. Bond, Margaret A. Leighton. 2019. "Unpaid Work and Access to Science Professions." *PLOS ONE* 14(6): e0217032. doi: 10.1371/journal.pone.0217032.

Funk, Cary, and Kim Parker. 2018. "Blacks in STEM Jobs Are Especially Concerned about Diversity and Discrimination in the Workplace." Pew Research Center, Social and Demographic Trends. January 9. www.pewresearch.org/social-trends/2018/01/09 /blacks-in-stem-jobs-are-especially-concerned-about-diversity-and-discrimination-in -the-workplace/.

Ganguli, Amy C., and Karen L. Launchbaugh. 2013. "The Evolving Role of Women as Rangeland Educators and Researchers in Colleges and Universities and in the Society for Range Management." *Rangelands* 35: 15–21.

Garcia, Alma M. 1989. "The Development of Chicana Feminist Discourse, 1970–1980," *Gender and Society* 3(2): 217–38. www.jstor.org/stable/189983.

Gervais, B. K., C. R. Voirin, C. Beatty, G. Bulltail, S. Cowherd, S. Defrance, B. Dorame, R. Gutteriez, . . . and S. J. Hoagland. 2017. "Native American Student Perspectives of Challenges in Natural Resource Higher Education." *Journal of Forestry* 115(5): 491–97. doi: 10.5849/jof.2016-065R1.

Global Greengrants Fund. 2019. "Why the Environmental Justice Movement Must Include Persons with Disabilities." March 18. www.greengrants.org/2019/03/18/disability-and -environment/.

Gould, E., and J. Schieder. 2017. "Black and Hispanic Women Are Paid Substantially Less Than White Men." *Economic Policy Institute*. March 7. www.epi.org/publication/black -and-hispanic-women-are-hit-particularly-hard-by-the-gender-wage-gap/.

Haynes, N., S. K. Jacobson, and D. M. Wald. 2015. "A Life-Cycle Analysis of Minority Underrepresentation in Natural Resource Fields." *Wildlife Society Bulletin* 39(2): 228–38.

International Association for Plant Taxonomy (IAPT). N.d. Accessed October 5, 2021. www .iaptglobal.org/.

International Women's Development Agency. 2018. "What Does Intersectional Feminism Actually Mean?" https://iwda.org.au/what-does-intersectional-feminism-actually-mean/.

Jacobs, Deborah L. 2013. "11 Sneaky Ways Companies Get Rid of Older Workers." *Forbes*. November 13. www.forbes.com/sites/deborahljacobs/2013/11/03/11-sneaky-ways -companies-get-rid-of-older-workers.

Jerman, Helen. 2020. "As Americans with Disabilities Act Turns 30, Barriers Remain." *Baptist News Global*. July 28. https://baptistnews.com/article/as-americans-with -disabilities-act-turns-30-barriers-remain/.

Johnson, Bobbi M., McLain S. Johnson, and Gary H. Thorgaard. 2019. "Salmon Genetics and Management in the Columbia River Basin." *Northwest Science* 92(5): 346–63.

Johnson, N., T. Pearce, K. Breton-Honeyman, D. N. Etiendem, and L. L. Loseto. 2020. "Knowledge Co-production and Co-management of Arctic Wildlife." *Arctic Science* 6(3): 124–26. doi: 10.1139/as-2020-0028.

Kashwan, Prakash. 2020. "American Environmentalism's Racist Roots Have Shaped Global Thinking about Conservation." *Conversation*, September 2. https://theconversation .com/american-environmentalisms-racist-roots-have-shaped-global-thinking-about -conservation-143783.

Keefer, Matthew L., Tami S. Clabough, Michael A. Jepson, Eric L. Johnson, Christopher A. Peery, and Christopher C. Caudill. 2018. "Thermal Exposure of Adult Chinook Salmon and Steelhead: Diverse Behavioral Strategies in a Large and Warming River System." *PLoS ONE* 13(9): e0204274.

Khan, Mishal Sameer, Fatim Lakha, Melisa Mei Jin Tan, Shweta Rajkumar Singh, Rina Yu Chin Quek, and Emeline Han. 2019. "More Talk Than Action: Gender and Ethnic Diversity in Leading Public Health Universities." *Health Policy* 393(1017): 594–96. doi: 10.1016/S0140-6736(18)32609-6.

Koch, J. W. 2019. "Racial Minorities' Trust in Government and Government Decisionmakers." *Social Science Quarterly* 100: 19–37. doi:10.1111/ssqu.12548.

Leopold, Aldo. 1949. *A Sand County Almanac*. New York: Oxford University Press.

Leopold, Aldo. 1933. *Game Management*. Madison: University of Wisconsin Press. muse.jhu.edu/book/8765.

Leslie, Jacques. 2021. "Op-Ed: Listen Up: A Republican Says We Have to Breach Four Snake River Dams." *Los Angeles Times*, March 10. https://www.latimes.com/opinion/story/2021-03-10/snake-river-dams-demolition-mike-simpson-idaho-washington.

Little Big Horn College. N.d. Apsáalooke Tribal History Writing Projects. Accessed July 12, 2019. http://lib.lbhc.edu/index.php?q=node/179.

Littlefield, Daniel F., and Lonnie E. Underhill. 1971. "Renaming the American Indian: 1890–1913." *American Studies* 12(2): 33–45. www.jstor.org/stable/40641008.

Lopez, R., and C. H. Brown. 2011. "Why Diversity Matters; Broadening Our Reach Will Sustain Natural Resources." *Wildlife Professional* 5:20–27.

McPhetres J., and M. Zuckerman. 2018. "Religiosity Predicts Negative Attitudes towards Science and Lower Levels of Science Literacy." *PLoS ONE* 13(11): e0207125. doi: 10.1371/journal.pone.0207125.

Morgan, T. C. 2013. "The Major Barriers and Influences to Minority Recruitment and Retention in Natural Resource Occupations." MS thesis, University of Florida, Gainesville.

Nagappan, Padma. 2018. "Now Colleges Can Attract More Minority Students to Environmental Studies and Careers." *Revelator*. February 7. https://therevelator.org/colleges-minority-students-environment/.

National Aeronautics and Space Administration (NASA). 2019. "What Is Science?" Space Place—Explore Earth and Space. Last modified June 27. https://spaceplace.nasa.gov/science/en/.

Native Women's Wilderness. N.d. "Murdered and Missing Indigenous Women." Accessed March 7, 2021. www.nativewomenswilderness.org/mmiw.

Newkirk, Pamela. 2015. *Spectacle: The Astonishing Life of Ota Benga*. New York: Amistad, an imprint of Harper Collins.

O'Connell, Suzanne, and Mary Anne Holmes. 2011. "Obstacles to the Recruitment of Minorities into the Geosciences: A Call to Action." *GSA Today* 21(6): 52–54. doi: 10.1130/G105GW.1.

O'Herrin, K., S. D. Day, P. E. Wiseman, C. R. Friedel, and J. F. Munsell. 2018. "University Student Perceptions of Urban Forestry as a Career Path." *Urban Forestry & Urban Greening* 34:294–304.

Oliver, Tom H., Matthew S. Heard, Nick J. B. Isaac, David B. Roy, Deborah Procter, Felix Eigenbrod, Rob Freckleton, Andy Hector, . . . and James M. Bullock. 2015. "Biodiversity and Resilience of Ecosystem Functions." *Trends in Ecology and Evolution* 30(11): 673–84. doi: 10.1016/j.tree.2015.12.008.

Østergaard, C. R., B. Timmermans B, and K. Kristinsson. 2011. "Does a Different View Create Something New? The Effect of Employee Diversity on Innovation." *Research Policy* 40(3): 500–509.

Outdoor Foundation. 2016. *Outdoor Participation Report*. https://outdoorindustry.org/wp-content/uploads/2017/05/2016-Outdoor-Participation-Report.pdf.

Quinn, William, Mort Kothmann, and Diana Doan-Crider. 2016. *Education, Experience, and Employment for Underrepresented Students in Natural Resources.* US Forest Service Final Report: 2010–2015, grant number 2010–02015.

Roht-Arriaza, Naomi. 1996. "Of Seeds and Shamans: The Appropriation of the Scientific and Technical Knowledge of Indigenous and Local Communities." *Michigan Journal of International Law* 17:4.

Royal Society. 2012. Accessed March 24, 2019. https://royalsociety.org/.

Salinas, Brenda. 2014. "'Columbusing': The Art of Discovering Something That Is Not New." *NPR Code Switch*, Remixed, July 6. www.npr.org/sections/codeswitch/2014/07/06/328466757/columbusing-the-art-of-discovering-something-that-is-not-new.

Scheitle, C. P., and E. H. Ecklund. 2018. "Perceptions of Religious Discrimination among U.S. Scientists." *Journal for the Scientific Study of Religion* 57: 139–55. doi:10.1111/jssr.12503.

Schmitz, K., S. Diefenthaler, C. K. Oyster. 2004. "An Examination of Traditional Gender Roles among Men and Women in Mexico and the United States." www.semanticscholar.org/paper/An-Examination-of-Traditional-Gender-Roles-Among-in-Schmitz-Diefenthaler/ebac7563759804e759b3faf5e2aede75e698b104.

Shippy, Tina, Diana Doan-Crider, and William Quinn. 2021. "Highschool and College Counselor Knowledge of Distinct Career Paths in Natural Resources." Unpublished manuscript. March 8.

Snow, K. C., D. G. Hays, G. Caliwagan, D. J. Ford Jr., D. Mariotti, J. Maweu Mwendwa, and W. E. Scott. 2016. "Guiding Principles for Indigenous Research Practices." *Action Research* 14(4): 357–75.

Sperber, Richie B., R. Fassinger, Linn S. Geschmay, and J. Johnson. 1997. "Persistence, Connection, and Passion: A Qualitative Study of the Career Development of Highly Achieving African American-Black and White Women." *Journal of Counseling Psychology* 44: 133–48. doi: 10.1037/0022-0167.44.2.133.

Strauss, A., and J. Corbin. 1994. "Grounded Theory Methodology: An Overview." In Handbook of Qualitative Research," edited by N. K. Denzin and Y. S. Lincoln, 273–85. Thousand Oaks, CA: Sage.

Taylor, Dorceta. 2007. "Employment Preferences and Salary Expectations of Students in Science and Engineering." *Bioscience* 57. doi: 10.1641/B570212.

Taylor, Dorceta. 2014. *The State of Diversity in Environmental Organizations.* Green 2.0. July. http://vaipl.org/wp-content/uploads/2014/10/ExecutiveSummary-Diverse-Green.pdf.

Taylor, Dorceta. 2017. "Racial and Ethnic Differences in the Students' Readiness, Identity, Perceptions of Institutional Diversity, and Desire to Join the Environmental Workforce." *Journal of Environmental Studies and Sciences* 8:152–68. doi: 10.1007/s13412-017-0447-4.

Taylor, Dorceta. 2018. "Racial and Ethnic Differences in the Students' Readiness, Identity, Perceptions of Institutional Diversity, and Desire to Join the Environmental Workforce." *Journal of Environmental Studies and Sciences* 8:152.

Taylor, Dorceta, Kit Price, and Ember McCoy. 2018. *Diversity Pathways: Broadening Participation in Environmental Organizations.* doi: 10.13140/RG.2.2.19473.68963/1.

TeApatu, H., N. Liddell, and S. Shahtahmasebi. 2014. "Indigenous Research: Culturally Relevant Methodologies." *Dynamics of Human Health* 1(1): 1–8.

Thomas, D. A. 2001. "The Truth about Mentoring Minorities: Race Matters." *Harvard Business Review* 79(4): 98–107, 168. https://europepmc.org/abstract/med/11299697.

US Department of Education (ED). 2016. *Advancing Diversity and Inclusion in Higher Education—Key Data Highlights Focusing on Race and Ethnicity and Promising Practices."* November. Washington, DC: Office of Planning, Evaluation and Policy Development. www2.ed.gov/rschstat/research/pubs/advancing-diversity-inclusion.pdf.

US Department of Interior (DOI). N.d. "Minority Serving Institutions Program." Office of Civil Rights. Accessed August 21, 2019. https://www.doi.gov/pmb/eeo/doi-minority -serving-institutions-program.

US Forest Service (FS). 2004. *Underrepresentation of African Americans in the Forest Service—Problems and Solutions.* African American Strategy Group, Resource Group. August 6. https://www.fs.fed.us/people/aasg/Docs/edAASG%20White%20Paper1.doc.

US Office of Personnel Management (OPM). N.d. Accessed October 25, 2020. https://www .opm.gov/.

Young, C. A., B. Haffejee, and D. L. Corsun. 2017. "Developing Cultural Intelligence and Empathy through Diversified Mentoring Relationships." *Journal of Management Education* 42(3): 319–46.

PART II

Diverse Perspectives and Practical Acts

Listening and Learning for an Inclusive and Equitable Future

Serra J. Hoagland, Crystal Ciisquq Leonetti, Jenny Bryant, and Joanne Bryant Ditsuh Taii

Perspectives of Native Women in Wildlife

Building a Diverse Future

Native people were the last ethnic minority given the right to vote in the United States. Humans were already launched into space before Indigenous people could vote for what happens in what used to be their country.

The year of the Cuban missile crisis and the release of the Beatles' first single, "Love Me Do," 1962, was also the year that the final state, Utah, allowed Native Americans to vote. Nearly six decades later, Native Americans are still disenfranchised from what most Americans agree is a fundamental human right. In 2018, more than 2,600 surveys conducted on Tribal lands in Nevada, Arizona, New Mexico, and South Dakota showed remote Tribal communities had limited access to polling stations, lacked information about registering and voting options, had poor internet access, and experienced mistreatment, mistrust, and intimidation by polling staff and local officials (Native American Voting Rights Coalition 2018). Native Americans, the first stewards of this land, continue to uphold their spiritual commitments and moral obligations to care for wildlife, which are viewed as their direct relatives, yet Native people were the last ethnic minority allowed to vote in this country. The United States sent people into space before all Indigenous people could vote for what happens in what used to be their country.

This chapter provides an overview of Indigenous peoples within the United States, discusses terminology relevant to Tribal communities, and provides justification and background regarding land acknowledgments that are becoming common practice in opening large events. We also cover several common myths about Tribal communities and Indigenous people to help build mutual understanding and prevent misinformed stereotypes from plaguing our profession and relationships between Native and non-Native communities. Furthermore, we spend significant energy reviewing important historical events relevant to Indigenous communities with regard to land stewardship, and we provide an overview of traditional ecological knowledge with an emphasis on Tribal wildlife management in the 21st century. This chapter focuses immense effort on the history of Native American and Alaska Native communities. Our collective history is a large part of who we are, where we come from, and what defines us, and it provides insights into the lives of Indigenous people today. Our intent in focusing on historical events is to recognize the true resiliency of Native communities and the strength of our traditions and practices that persist through time. Lastly, the authors do not intend to place guilt, but in understanding the current situation of Indigenous women in wildlife, we offer some difficult truths. We also are not making sweeping generalizations about gender or race—in every demographic there are allies and detractors.

The marginalization of Native women in the wildlife profession reflects the eradication of Indigenous practices in wildlife conservation and natural resource management more broadly. Indigenous women in the 21st century carry generations of historical trauma and loss of culture, language, and traditions and have experienced grave neglect by mainstream society; thus, Native women are extremely underrepresented in the wildlife profession. Fortunately, with new eras of Tribal self-determination and an increased recognition of Tribal sovereignty, Native people are beginning to play important leadership roles at local, regional, and national scales. Native women, as many Indigenous societies around the world emphasize matriarchal lines, have tremendous potential to serve the wildlife profession and bring holistic ways of knowing to improve our ability to conserve biodiversity. This chapter focuses on several key Indigenous women leaders in the wildlife community who are breaking barriers.

Throughout this chapter we encourage readers to recognize the value that Indigenous people, Indigenous historical and contemporary practices, as well as Indigenous perspectives and cultural teachings can inform current wildlife management and conservation. In this chapter we focus on Native American and Alaska Native communities. Although there are many parallels and similarities between Alaska Native Tribes and Tribes of the lower 48, unfortunately, we do not specifically cover Kānaka Maoli (Native Hawaiian) or Pacific Islander culture and environmental ethics in this chapter. For good references on these topics, we suggest *Native Hawaiian Law: A Treatise*, by MacKenzie, Serrano, and Sproat, and *Native Land and Foreign Desires: Pehea La E Pono Ai? How Shall We Live in Harmony*, by Kame'eleihiwa.

Indigenous peoples rely on gender identity roles, including for individuals with more complex gender identity, and it is noted that many Indigenous languages do not have pronouns (she/her, he/him). Individuals who identified as something beyond male or female were respected as much as anyone else for their unique talent whether it be hunter, healer, leader, builder, sewist, or some other skill set. We focus on women in this book but want to acknowledge the important roles of other gender identities in Indigenous societies.

Lastly, we recognize collecting information for 574+ unique Tribal communities is a massive undertaking and is beyond the scope of this chapter. There is likely an infinite number of traditional and cultural variations among Tribes, clans, families, and generations regarding cultural taboos and gender traditions. Indeed, this information is not appropriate for sharing in a book that sits on a shelf. Our knowledge and traditions are sacred and therefore are to be shared respectfully in person so that they are not taken out of context. However, we encourage future in-person, respectful conversations and pursuits on this topic when Tribal communities are interested in sharing their ways of knowing.

Indigenous Peoples of the United States 101

Indigenous peoples of North America comprise a diverse set of individual communities with a wide range of cultural practices that represent deeply rooted traditions of land stewardship and wildlife conservation. Indigenous people, since time immemorial, have depended on wildlife for their subsistence practices and lifeways. For most Indigenous communities, wildlife are seen as relatives, ancestors, and sacred beings. Wildlife are always given utmost respect and are treated as one would treat a grandparent.

At the time we write this chapter in the United States there are 574 federally recognized sovereign Tribal nations, 229 of which are in Alaska (see map at https://www.bia.gov/sites/bia.gov/files/assets/bia/ots/webteam/pdf/idc1 -028635.pdf). This total number has the potential to increase over time as Tribes formally obtain federal recognition. For instance, the Little Shell Chippewa began their battle for their federal recognition in 1930 and in January 2020 they became the 574th federally recognized Tribe in the United States. In the lower 48 states, these Tribes manage approximately 56 million acres of land.

In addition to the federally recognized Tribes, there are more than 60 state-recognized Tribes (*Federal Register* 2018; NCSL 2022). The range of cultural diversity primarily originates and is dependent upon the geographical landscape and the resources that were historically or currently available to Native people. Within Indigenous communities they are typically organized by family lines, clans, or villages. As of late 2019, there were more than 5 million Native Americans in the United States, representing less than 2% of the population. The 10 states with highest proportion of Native Americans relative to the total state population are (in alphabetical order) Alaska, Arizona, Idaho, Montana, New Mexico, North Dakota, Oklahoma, South Dakota, Washington, and Wyoming. The states with highest proportion of Native Americans relative to the total state population are Alaska (19.5%), Oklahoma (12.9%), and New Mexico (10.7%) (National Congress of American Indians 2020).

Alaska Native or Indigenous peoples of Alaska have nine main cultural or language groups, including Ungangan/Aleut, Sugpiaq/Alutiiq, Yup'ik/Cup'ik, Inupiaq, Athabascan (Diné), Eyak, Tlingit, Haida, and Tsimshian. In Alaska there are 229 federally recognized Tribes, including village-based and regional Tribal governments. Many rural Alaska Native villages are accessible only by small aircraft, boat, or snow machine. Many Alaska Native people speak their respective Indigenous languages before learning English. Tribal land ownership in Alaska has numerous complexities. Tribes own very little land, interpreted by the US Supreme Court as "sovereigns without territorial reach" (Case and Voluck, 2002). One reservation remained in the state after the 1971 Alaska Native Claims Settlement Act (ANCSA): the Metlakatla Indian Reserve located in the southeastern portion of the state. Alaska Native Corporations formed under ANCSA own approximately 44 million acres of fee simple lands, which does not constitute Indian country for Tribal jurisdictional purposes. Alaska Native Corporations must not be confused with Tribal governments. The Tribes are sovereign, with elected officials and Tribal citizens for whom the Tribe provides services such as health care, education, and environmental protection. The corporations are for-profit entities with diverse portfolios, from telecommunications to oil and gas development to federal military contracting, with an elected board, and Alaska Native shareholders who were designated at the time the act passed.

Terminology and Appropriate Land Acknowledgment

Throughout this chapter we generally use the term "Indigenous" as an overarching classification of Native peoples throughout North America who were the original inhabitants of the land prior to European or other foreign settlement. In certain instances, throughout this chapter we respectfully acknowledge individual Indigenous communities by their specific Tribal name when appropriate and possible. "Alaska Native" generally refers to Indigenous peoples from Alaska and the Tribes and corporations as distinct entities with unique land status and governing rights. "American Indian" represents, like the term "Indigenous," Tribal people who are Native to the Americas and is used interchangeably with "Native American." These broader terms clump all Indigenous people into one category for the United States, although they represent hundreds of Tribes, bands, rancherias, and pueblos. **When working with Indigenous communities it is most appropriate and respectful to be as accurate as possible when referencing individual groups and people, such as a specific Tribe.** An even more respectful approach, for instance, is to use the terminology from the language of the people and what they call themselves, such as "Diné" (meaning "the people" in their native language) for the Navajo Tribe. In some cases, the Native name for a community can be used interchangeably with the non-Native name (Yazzie et al. 2019). The definitions of commonly used terms (Table 6.1) may be helpful when working with Indigenous communities. This list will also help the reader understand the sections that follow.

Land acknowledgment is one respectful way to recognize the original inhabitants of the land and honor the Indigenous communities and people that still exist today. As stated by the Northwestern University Native American and Indigenous Initiatives program, land acknowledgment "*is a formal statement that recognizes and respects Indigenous Peoples as traditional stewards of this land and the enduring relationship that exists between Indigenous Peoples and their traditional territories*" (Northwestern 2019). Land acknowledgments have become standard practice for large organizations and gatherings. This process has even been accepted by large universities such as the following example: "*We recognize the University of Wisconsin–Stevens Point occupies lands of the Ho Chunk and Menominee people. Please take a moment to acknowledge and honor the ancestral Ho Chunk and Menominee land and the sacred land of all Indigenous peoples.*"

Land acknowledgment is also one method to correcting inaccurate history lessons taught in schools, like the one used by Native Movement (n.d.), based in Fairbanks and Anchorage, Alaska: "*We recognize that we work throughout the unceded territories of the Indigenous Peoples of Alaska. Our offices are located on the traditional territories of the lower Tanana Dene Peoples and the Dena'ina Peoples. We acknowledge and honor the ancestral & present land stewardship and place-based knowledge of the peoples of these territories.*"

The goal of a land acknowledgment is to honor the sacred ties Indigenous communities once had and often maintain despite centuries of colonialism, assimilation, and marginalization. Another example, in 2017, The Wildlife Society (TWS) with coordination provided by the Native Peoples' Wildlife Management

TABLE 6.1 Common terms and definitions regarding Indigenous land status, membership, and recognition.

Term	Definition
Blood quantum	The amount of an individual's Native blood determined by the biological parent's heritage. Many Tribes delineate a specific blood quantum at which to deliver and administer services to Tribal members. Many Tribes in the lower 48 recognize Tribal membership as 1/4 while others recognize up to 1/64. Blood quantum is a sensitive subject for many Tribal members as Indigenous people are the only group subject to validating their identity in such a way. Many Indian people consider it partly a tactic of genocide because over many generations through assimilation into the dominant society, Native blood quantum rarely increases.
BIA (Bureau of Indian Affairs)	The primary US government agency responsible for carrying out the services and benefits for Indian people. Housed within the Department of Interior.
Cultural appropriation	Adopting part or an element of one culture or identity by members of another culture or identity. It becomes controversial when members of a dominant culture appropriate from disadvantaged, underrepresented cultures.
Federal recognition	Tribes that are federally recognized by the United States are provided government services and benefits in exchange for land and resources.
Fiduciary trust responsibility	The US federal legal obligations to uphold treaty rights, executive orders, and other laws and policies affecting federally recognized Tribal nations.
Historical trauma	Also described as inter- and multigenerational trauma, experienced by a group of people. American Indians and Alaska Native communities have experienced multiple generations of traumatic events, oppression, and marginalization, which over time can negatively impact the health and well-being of communities.
Indian reservation	Typically land that was formed through treaty or executive order for the benefit of Indian people. Most Tribal lands are not open to the general public, and they often require access permits that are approved by the Tribal government.
Treaty	Supreme law of the land. International law agreed upon by two independent nations.
Tribal membership	An individual is recognized by a particular Tribe through that Tribe's governing body and requirements for Tribal membership (see *blood quantum*). Many Tribes issue Tribal identification cards similar to a driver's license or identification card. The Bureau of Indian Affairs also distributes identification cards for Tribal membership that is commonly referred to as a CIB (certificate of Indian blood).
Tribal sovereignty	A sovereign nation is free, independent, and autonomous with a unique governing body, culture, laws and policies, education system, health care, etc. with the primary function of fulfilling the needs of its own membership.
Tribal governance structure	The governing organization and leadership for the nation. It is typically set out in a Tribal constitution or other guiding document. Various departments, programs, and projects (e.g., health care, education, environmental protection) within the Tribal government can range from compact, contract, or direct services.

Working Group organized a land acknowledgment and Tribal blessing in Albuquerque, New Mexico, at its annual conference. The land acknowledgment included a local Tribal representative, Tim Smith (Chiricahua Apache from Mescalero and Kewa Pueblo) to welcome the attendees to that place, recognize the historic and contemporary Tribal presence in the region, and pay respect to past and future generations of Indigenous people who hold ties to that landscape (Figure 6.1). Out of respect for Smith, TWS provided complimentary

FIGURE 6.1. Tim Smith (Chiricahua Apache from Mescalero and Kewa Pueblo), at the time this photo was captured, served as a biological technician for the Pueblo of Sandia Environmental Department. Here, he is shown providing the Tribal welcome and land acknowledgment to The Wildlife Society conference attendees in Albuquerque, New Mexico.
Photo courtesy of The Wildlife Society.

registration to honor his participation and encourage Tribal involvement in all aspects of the meeting. **Encouraging land acknowledgments by local Tribal representatives** and **offsetting the financial burden to participate at conferences** are two meaningful mechanisms for organizations to engage with Tribal communities.

Myths and Stereotypes about Indigenous Peoples of the United States

First of all, we all do not all live in tepees or igloos! We do not all receive payments from our Tribes, Native-owned corporations, or Indian casinos. Myths and stereotypes (both positive and negative) about Indigenous people remain common today in the media and our culture. These misperceptions will unfold in different circumstances and leave the average citizen a complex image to sort out. The diversity of Indigenous groups within North America makes it even more challenging to formulate an educated understanding of issues that Indigenous people face. Indigenous demographics are as varied as the rest of the country.

Reclaiming Native Truth, a project dedicated to dispelling myths and misconceptions about Native Americans, highlighted society's dual views about Indigenous people. Natives are portrayed as noble but savage, poor yet wealthy, and creatures of nature who trash their reservations. Interestingly, the greatest biases toward Indigenous people are found in communities closest to Tribal reservations, but overall, biases vary by region. Nearly 60% of Americans believe cultural genocide occurred in the United States, and 40% of survey respondents believe that Native Americans are extinct. Indigenous people are rendered invisible by this kind of misinformation (Reclaiming Native Truths 2018b).

A widely broadcast commercial shown on Earth Day in 1971 featuring a "Native American," portrayed by an Italian American, fueled some of the stereotypes we see half a century later. The Ad Council developed the commercial as part of the Keep America Beautiful advertisement to prevent litter and environmental pollution. This impressed upon many Americans notions about the "noble savage," among other duplicities (Dunaway 2017). The commercial is also a classic example of cultural appropriation. Some cultural appropriation cases eventually end up in court, such as the *Navajo Nation v. Urban Outfitters* (2016) when the retailer sold items with "Navajo" design and labels. **Actively preventing cultural appropriation** allows Tribal communities to maintain their own identity and prevents inappropriate use of cultural symbols, art, and traditions.

One author found herself sitting on an airplane listening to a man talk about how he always sees Native people intoxicated hanging around the laundromat. Even after attempting to inform him about historical trauma (Table 6.1 and sections below), he continued to disparage "those people" he sees regularly "just using their casino money for booze." We have also experienced countless misrepresentations of Native identities, even by other Natives, for instance, assuming all Native people living in Arizona are from the Navajo Nation simply because it is the largest and most populous Tribe. There are 21 federally recognized Tribes in Arizona, and some are offended when not referred to correctly. These types of interactions are detrimental to relationships and collaborative partnerships. Interestingly, we have even encountered these types of interaction over email, which is often even more offensive because there is adequate time to look up information and respectfully initiate a dialogue. Fortunately, these slip-ups can be ameliorated if respect is in place, and we can use them as a teaching opportunity. We can make changes in our contemporary times by **referring to Native scholars' Tribal affiliations** appropriately and accurately.

Historical and Contemporary Context of Indigenous Peoples of North America

A brief historical context of Indigenous peoples in North America is necessary to understand the current conditions we face today. Indigenous peoples' history before European settlement varies by Tribal community, but most have specific creation stories that discuss their historical ties to a specific place. Native creation stories often cite inhabitation in the Americas since "time immemorial." Recent publications show evidence of humans near White Sands National Park in New Mexico around the Last Glacial Maximum, 23,000 to 21,000 years ago, indicating that ancient Native peoples shared the landscape with megafauna such as the mammoth (Bennett et al. 2021). New anthropological research shows 56 million Indigenous people were killed by European settlers in North, Central, and South America. Decimating this population led to global changes in climate (Koch et al. 2019) and was a large contributing factor in the Little Ice Age (less carbon dioxide in atmosphere due to reforestation and forest change).

Post–European settlement, Indigenous communities went through various eras that affected their livelihoods and well-being. First was the "Doctrine of Discovery," or **Manifest Destiny**, circa 1493, which became the basis of all European claims in the Americas as well as the foundation for the United States' western expansion. This 15th-century principle of "international law" gave

Christian explorers the right to claim lands they "discovered" for their leaders. The second major era is known as the **Treaty-Making Era**, spanning more than two centuries, from 1600 to 1871. Treaties were intended to act as the supreme law of the land, an agreement between independent nations, but instead they resulted in significant land, territories, and resources of Indigenous people being lost. An example is the Hellgate Treaty of 1855 between the US government and Tribes in western Montana who are today more commonly referred to as the Confederated Salish and Kootenai Tribes. Although their original territories spanned 22 million acres, the treaty conveyed 12 million acres to the US government and left the Tribes with 1.25 million acres known today as the Flathead Indian Reservation. Sadly, however, of this original land, "*over half a million acres passed out of Tribal ownership during land allotment that began in 1904*" (CSKT n.d.). Other examples of treaties can be found in the National Archives' American Indian Records (Native American Heritage 2021). This era represented the international treaties signed with American Indian Tribes, none of which have been fully upheld.

These government-sanctioned laws and policies continued during what are generally known as the **Indian Removal Era** (1830–1850), **Reservation Era** (1850–1880), **Allotment and Assimilation Era** (1887–1930), **Indian Reorganization Era** (1930s-1945), **Termination and Relocation Era** (1945–1961), and most recently the **Indian Self-Determination Era** (1970s–present). Many of these early policies were highly detrimental to Indigenous communities, such as the boarding schools of the Allotment and Assimilation Era where Indigenous youth were taken far from their homes and prohibited from speaking their language and practicing their culture. This sequence of events and the trauma that evolved and magnified over time and was passed down from one generation to the next is known as historical trauma (Whitbeck et al. 2004; Evans-Campbell 2008), which may be characterized as intergenerational posttraumatic stress disorder. "*Many of the dynamics in effect in the Jewish experience are like those in the Native American experience, with the crucial exception that the world has not acknowledged the Holocaust of Native people in this hemisphere. This lack of acknowledgement remains one of the stumbling blocks to the healing process of Native American people. The inherent denial keeps the colonial perpetrators trapped in an aura of secrecy and continuing alienation, since their acts continue to haunt them with guilt and existential emptiness*" (Duran and Duran 1995). This is why truth, healing, and reconciliation commissions in Canada have been helping First Nations people (Indigenous communities of Canada) alongside the colonizers, by allowing both to let go of afflictions they may not realize both are carrying. This applies just as much in the research and management of wildlife. There have been historic missteps and they must be reconciled.

First contact in Alaska occurred in the middle of the 18th century by Russian explorers, but time of settlement with European or American explorers varied by location, ranging from 1750 to 1870 (Langdon 2014). In 1867, Russia sold Alaska to the United States for $7.2 million. Tlingit and Haida from southeast Alaska rightly protested that Russia could not sell what it didn't own. Russians did not occupy Alaska, nor did they conquer the people already there. But Manifest Destiny was too powerful, and the sale went through. At that point, government agents, gold prospectors, fish processors, and whalers began wandering north from the United States. The federal government joined with religious organizations to assimilate Alaska Natives and run government-funded schools and churches (Hensley 2009). Similar to what occurred in the lower 48, mission and boarding schools stripped most children from their parents and ancestral knowledge, culture, and joy of harmony with the land. The introduction of disease, processed foods and alcohol, and these schools inflicted posttraumatic stress syndrome on several generations (Duran and Duran 1995). The Allotment and Assimilation, Indian Reorganization, Termination and Relocation, and Indian Self-Determination eras during the time periods described above apply in Alaska after its purchase during the territorial years. Alaska was admitted to the union in 1959 as the 49th state.

On two remote islands in the Bering Sea, where the Russians had enslaved Unangan people for almost a century to harvest and process fur seal pelts, the US government took over right where the czar left off. It wasn't until a century later with passage of the 1966 Fur Seal Act that the Ungangans of St. Paul and St. George Islands were granted independence.

The fight for land settlement in Alaska was negotiated by oil tycoons and white politicians in the 1960s behind the backs of most Alaska Natives (Case and Voluck 2002). Tribes were not consulted, and individual Alaska Native people had no say, except for several young Alaska Native recent college graduates who led the charge on behalf of Native interests. Alaska Native land claims were "settled" in 1971 by the congressional passage of the Alaska Native Claims Settlement Act. Alaska Natives kept only 44 million of 365 million acres in Alaska through this act. ANCSA established corporations, for-profit businesses chartered by the Alaska state government, as vehicles to acquire title to the lands. ANCSA set up a capitalist, business-centered land ownership system that was entirely new to Alaska Native people. It was a deliberate social engineering project, a departure from agrarian assimilation (as was the case through the Allotment Act in the lower 48), to business assimilation—the

US vocation of the day. The experiment is still at play, sometimes pitting Tribes against corporations. In some areas, there have been concerted efforts to reunite, where Tribes and corporations meet regularly to discuss purpose and priority, and a resulting revival of cultural values. Because of provisions of the ANCSA, Alaska Tribes own very little of what is known legally as Indian country.

The historical context described above has led to many negative situations that Native communities encounter. Today Native communities experience historical trauma, gaps in education, poor high school graduation rates, poverty, high suicide rates, poor infant health, sexual violence including MMIW (missing and murdered Indigenous women), drug and alcohol abuse, and more due to loss of culture and genocide. In many ways, Indigenous people are the canary in the coalmine for human behavior on a global, international scale. Things like racism (both external and internal [i.e., self-inflicted]), self-image, stereotypes, misconceptions, cultural and language barriers, and discrimination are detrimental to Native people and our communities. All these historical and contemporary issues that Native people have experienced has led to the severe underrepresentation we see in the wildlife profession.

Besides critical social factors, Indigenous communities are experiencing environmental stresses and are highly vulnerable to wildfire (Davies et al. 2018), thawing permafrost, thinning sea ice, rising sea levels with more extreme tidal fluctuations (Inuit Circumpolar Council 2012), and other natural disasters. The potential risks combined with limited adaptive capacity of Indigenous communities creates a dire state. Sadly, Native communities are often referred to as 4th world countries, like 3rd world countries landlocked within a 1st world country.

Although the historical context is depressing, there is hope for today in contemporary times. "*The legitimization of Native American thought in the Western world has not yet occurred, and may not occur for some time. This does not mean that the situation is hopeless in the Native American community*" (Duran and Duran 1995). Indigenous people envision this revitalization as the **7th generation**, the time when Indigenous people will lead once again. Despite challenges that Tribal nations faced over the generations, Indian people are reasserting, reforming, and redefining their rightful sovereignty. Some purport that as soon as Native communities are viewed as political entities rather than ethnic minorities, there is room for change and adaptation (Gary Morishima, personal communication, November 2018). Government apologies around shared wildlife management history are creating space for more open government-to-government dialogue, increased empathy, and policies that work for, instead of against, Native communities. Even mainstream icons notice this trend. Former NBA player Joakim Noah stated, *"We're going to need the Indigenous people to lead the way,"* at the grand opening of the Standing Rock Sioux solar farm in North Dakota in 2018.

Indigenous Environmental Ethics

Indigenous people have traditionally lived in harmony with the world around them, through ecosystem changes, weather pattern changes, and drastic climate changes. They have an intricate worldview, with complex cultural mandates governing interaction of the human, natural, and spiritual worlds (Kawagley 1995). Successful hunters and gatherers exhibit a sense of responsibility, appropriate attitude, and even care in thought so as not to injure the minds of humans or animals. Indigenous people have been called the "original ecologists" and first stewards of the land because of their connection to, understanding of, and participation in the ecosystem (Fienup-Riordan 1991). Even today, many Indigenous people who now live in modern houses remain in proximity and intimacy with the land and water because of their high cultural, social, and nutritional reliance on wild plants and animals.

Indigenous communities, broadly, are place-based societies with spiritual connections to the land. The dominant society tends to associate Native American environmental values as whimsical, mythical fallacies. This is often spurred by our educational institutions' inability to reclaim Native truths. For instance, in her work, Stephanie Fryberg at the University of Washington discusses how the psychology of omission (omitting the fact that Native people were on the landscape) fuels the cycle of bias against Native Americans. They are not viewed as present on the landscape because they are not covered in course curricula. In reality, Native traditions are based on and borrow from an understanding of ecology and adaptations to living on the landscape for hundreds of generations. One of the chapter authors, Crystal Leonetti, describes:

> From a very young age, I understood that all natural things, living and non-living, have a spirit, and that in another life and world, I may come back as one of those things. I knew that all things must be treated with respect. I don't remember anyone explaining this to me, but I believe it is part of my ancestral memory. A Dena'ina friend of mine from Tyonek, Alaska, remembers tearing the leaves of grass with her hands as she walked down the trail. Her auntie stopped and ask if she'd like someone to tear her arms off. That is what she was doing, tearing the arms off the grass. "Only take it if you intend to make it into something beautiful and respect its spirit," she explained.

Traditional Ecological Knowledge as a Way of Knowing

Traditional ecological knowledge (TEK) tends to focus on environmental anomalies, the outliers over long time periods. In some ways the norm is mundane and easier to predict with hundred-year data sets held in the stories and traditions of Tribal communities. TEK focuses on experiential learning, where our elders and stories are our textbooks. Knowledge is held by the community, and individual members have a responsibility that is community driven. Our elders and cultural practitioners hold the knowledge, whereas Western science tends to be derived from individual experts with formal education. TEK is abstract, qualitative, inclusive, and holistic, whereas western science can be portrayed as concrete, quantitative, reductionist, and exclusive (Mason et al. 2012).

Scientific and traditional knowledge systems have much in common, especially in the environmental sciences. Both rely on observations and look for patterns in weather, physical conditions of the land and sea, and animal behavior. They both value reliable information that can be replicated. For traditional knowledge bearers, the ability to replicate results of an experiment is the ability to safely gather food for the family and community. A hunter's life literally relies on this information (Huntington and Noongwook 2012).

Tribal Wildlife Departments

Tribes can assert their sovereignty through wildlife conservation because wildlife is a trust resource. In the lower 48, many Tribes have their own wildlife departments that manage habitat improvement projects, distribute big and small game permits, manage external grants, and establish management practices that meet the needs of their communities. In general, Tribal natural resource departments are significantly underfunded compared to their federal counterparts (Gordon et al. 2012), which affects capacity and ability to recruit a talented, highly skilled workforce. Furthermore, Tribes often face disproportionate burden in conserving threatened and endangered species because their lands have low human densities, low development, and provide lots of habitat for rare species. To "*avoid or minimize the potential for conflict and confrontation,*" Secretarial Order 3206 was enacted in June 1997, recognizing Tribal sovereignty in managing their own trust resources, including threatened and endangered species.

Tribes are funded primarily by the Department of the Interior Bureau of Indian Affairs (BIA) and other government and nongovernment grants and

BOX 6.1 **Rhonda Pitka, Koyukon Athabascan and Inupiaq**

Rhonda Pitka is the first chief of the Beaver Village Council (Tribal government), where she has served since 2011 in Beaver, Alaska (Figure 6.2). Chief Pitka's parents are Ron Yatlin and Antoinette of Pitka of Beaver. Her siblings are George and Charlie Yatlin and Elizabeth Blackbird. She is a single mother to Alaina. Her maternal grandparents are Elsie Pitka and the late Elman Pitka of Beaver. Her paternal grandparents are Tony Sam Sr. and the late Emily Sam of Huslia, the late George Frank of Galena, and the late Minnie Yatlin of Huslia.

She credits her grandmother, Elsie Pitka, for her dedication to learning and service to her community. Her grandmother was the teacher aide at the Beaver School for 25 years and instilled a lifelong love of reading and learning in her life. Grandma Elsie had only a third-grade education. She told Rhonda at a young age, "You know you're going to school. It's so much easier for you. You can do this. You need to do this. You need to go to school, and you need to get your education." Chief Pitka says, "*The strength and tenacity and stubbornness of the women in my life motivates me to do more because they are really tough. They have had challenges in their own lives that they have been able to heal. The health and wellness of my family depend on our ability to hunt and fish. We grew up in fish camp in the summers. Our family is one of a handful of families in our part of the river that still lives in fish camp in the summer. My parents instilled a work ethic in our family that has served me well.*"

Chief Pitka's mission is to help empower the Indigenous people of Alaska to steward the lands and wildlife, while exercising self-determination. She explains, "*The disenfranchisement of Alaska Native people in a complex regulatory and management regime is an injustice. The management of subsistence continues to be dominated by federal and state governments. The traditional hunting and fishing practices, including the ceremonies that accompany these practices, provide for the well-being and survival of our people and communities. Our traditional practices and resources are in continued jeopardy. The decision-making processes are often inaccessible to our traditional hunters and fishers.*"

FIGURE 6.2. Rhonda Pitka (Koyukon Athabascan and Inupiaq).
Photo courtesy of Sheila Vent, Tanana Chiefs Conference.

Among many other roles, secretaries of the interior and agriculture appointed Chief Pitka as a public member of the FSB in 2017. *"My role on the Federal Subsistence Board as a public member has been challenging and rewarding,"* she says. Chief Pitka is one of three public members of the FSB and the first Alaska Native woman on the board. Chief Pitka says, *"It is a blessing to listen to hundreds of hours of public testimony from people who want to feed their families and not be criminalized for it."*

agreements. Tribes have elected councils of varying sizes and titles. They are immensely diverse in governance structure and priority workload. For example, in Alaska, Yup'ik Tribal governments on the Yukon Kuskokwim Delta are traditionally an egalitarian society, but to obtain a funding stream from BIA, they had to develop a constitution and hold elections. Their priorities range from health care and education to subsistence hunting and fishing advocacy.

Federally recognized Tribes in the lower 48 with a land base have legal authority over wildlife resources on trust lands. Unfortunately, although Alaska Native people have stewarded wildlife and land for millennia, they do so now under governance from state and federal management. Tribal governments in Alaska range in capacity. Some have the ability to sustain a wildlife professional or department, but most lack the resources. Many Tribes rely on council members to advocate for hunting and fishing rights at state board of game or fish meetings, Federal Subsistence Board meetings, the North Pacific Fisheries Management Council, the Alaska Migratory Bird Co-Management Council, or various marine mammal commissions. There are few places where Tribes have been able to manage lands and wildlife in a customary and traditional way under these new systems and institutions. One successful structure is the Alaska Migratory Bird Co-Management Council, which manages migratory birds for the spring/summer subsistence harvest. The council has three equal votes—one federal, one state, and one Alaska Native (Migratory Bird Subsistence Harvest in Alaska 2003).

On St. Lawrence Island, where the land is closer to Russia than the United States, and Indigenous culture more closely identified with Indigenous Russians, traditional laws remain the most important hunting regulations of the people who live there. The two Tribal governments on St. Lawrence Island have enacted sovereign governing structures for many purposes, including managing wildlife and their harvest. They rely on Pacific walrus (*Odobenus rosmarus*), bowhead whale (*Baleana mysticetus*), polar bear (*Ursus maritimus*), waterfowl, sea ducks, and fish and shellfish as their mainstay. They wrote down

BOX 6.2 Patty Schwalenberg, Gii-wii-aa-si-no-kwe (Lac du Flambeau Ojibwe, Wisconsin)

Patty Schwalenberg (Figure 6.3) is described as having conservation in her DNA by longtime collaborative partner and coauthor of this chapter, Crystal Leonetti, Alaska Native affairs specialist, US Fish and Wildlife Service (USFWS). One of her favorite projects over the years has been helping to transform federal wildlife agency culture to one that is more culturally aware and tribally collaborative. Growing up on her traditional lands, her passion for the outdoors grew until it became a life mission—to work in conservation, fight for the rights of Tribes to sit at the decision-making table, and, most importantly, build the technical capacity of Tribes so that they can fully participate at all levels in the natural resource decisions that affect them. Currently, Schwalenberg is the executive director of the Alaska Migratory Bird Co-Management Council and collaborates with the USFWS to provide a week-long, immersion training for USFWS employees to work more effectively with Alaska's Tribes. She has worked with Leonetti at the Fish and Wildlife Service to transform this course into an online 13-class series educational and cultural experience, providing training to up to 100 students per course. When able, the class is held in person, immersion-style with elders present at all times, making traditional medicines, eating Native foods, beading, and feeling the drum beat in one's chest through the week. Schwalenberg's teaching talent extends into mentoring, having worked with many young Alaska Native men and women over the years. Her work in this area continues inspired by her late husband, Dewey's, positive influence on their lives. Schwalenberg is an integral part of the story, shaping Tribal fisheries and wildlife stewardship in Alaska today.

FIGURE 6.3. Patty Schwalenberg, Gii-wii-aa-si-no-kwe (Lac du Flambeau Ojibwe, Wisconsin).
Photo courtesy of the Chugach Regional Resources Commission.

BOX 6.3 Karen Linnell "Kaawool.ge" (Ahtna and Tlingit)

Growing up, Karen Linnell (Figure 6.4) watched and learned from her parents and elders; she saw how they worked for the people. On several occasions, the president of her Alaska regional corporation asked her how she would handle a situation at work, getting her to think outside of her role and to think like a leader. Linnell has a long history of working for the Ahtna people. She firmly believes that you cannot change things by complaining about them, but that you can effect positive change by getting involved. Linnell now serves as executive director of the Ahtna Intertribal Resource Commission (AITRC), whose vision is *"to manage our own resources on our own homelands."* The AITRC's board believes that the creator put us on this land to be stewards of it. *"I love my family and my people. That's why I am working to secure our hunting and fishing rights for generations to come. It's much more than putting food on the table, it is part of who we are. We shouldn't have to fight for that last moose, caribou, or salmon. We need to work to ensure these resources have healthy populations and are here for our grandchildren's grandchildren. We need to quit managing for a single species or piece of land but start looking at the whole landscape, just as our ancestors did."*

FIGURE 6.4. Karen Linnell "Kaawool.ge" (Ahtna and Tlingit).
Photo courtesy of Ahtna Inter Tribal Resource Commission staff.

their traditional walrus hunting laws in the 1920s, which have since been made into a Tribal ordinance. The walrus ordinance must be obeyed by hunters, and should someone violate it, they must report to the Tribal council and be dealt with by the council and other hunters. Should the offender also violate federal law for wasteful take, the Tribe reports them to federal authorities, who will investigate the violation under the Marine Mammal Protection Act, which may result in a year in jail and up to a $100,000 fine.

Wildlife as Relatives in Native Societies

Traditional Native American stories were not only to entertain but also to instruct, guide, and inspire. Many Western or non-Indigenous people might refer to these as "mythical" tales, but to Indigenous people, they recount events that took place in the Distant Time, before the world as we know it was created. These stories explain the natural history of the Indigenous people's environment, including land formations, geographic climate, hydrology, vegetation and animal distribution, behavior, physiology, and life history. We are taught that animals could "speak" in ancient times, and characters that appear to be human usually became animals by the end of the story. Many characters can transform themselves from human to animal and back again, while others who die as a human are transformed into animals, maintaining their human characteristics. It is said that many patterns for the future behavior of both humans and animals were established during this time before and during creation. After the world came into its current form, animals lost the ability to transform and along with it their human soul, but they retained a protector soul to enforce taboos and rituals that should be observed when they are killed. There is much power in the intricate functioning of the natural world, and maintaining the balance and flow of that power was crucial to the health and survival of Indigenous people. The intimate relationship between the natural world and humans is not one of separation and dominance but of reciprocal interdependence and intense emotional connection at a most basic level. Indigenous people live in a natural world that is aware, sensate, and personified. The land and animals are always cognizant, they can feel, and they can be offended, so they must always be treated with proper respect. All things in nature have a special kind of life, something powerful that cannot be directly altered to serve the needs of humanity. The practical challenges of survival by hunting, fishing, and resource harvesting require a deep objective understanding of the environment and the methods for effectively and efficiently utilizing its resources. As such, Indigenous people were innately sophisticated natural historians, especially well versed in animal behavior and ecology. But nature's spiritual and empirical aspects exist equally, creating an elaborately inseparable worldview in which environmental phenomena or events are often caused or influenced by supernatural forces. Indigenous behavior toward nature is governed by a complex belief system of which ideology is a fundamental element, as important as the more tangible practicalities of rationalistic or scientific harvesting and utilizing natural resources.

In such a worldview, there are reasons for a particular taboo or harvesting practice based on the stories of the Distant Time:

> Three men set out together for a hunting journey—two of them were rich and influential chiefs, and the other was poor but very smart. After they had wandered far and wide without success, the smart man returned home. But the other two kept on, becoming thin and ragged. Their skins tanned and their hair grew long; and finally they began to live like animals. Then they were transformed, one into a wolf and the other into a wolverine. This is why the wolverine's ceremony begins with an announcement that the chief has arrived, and why these animals are presented with a "banquet" remembering their hardships long ago. The smart man, who left the others and returned home, became the first shaman. And since then, a shaman always officiated at the ritual. (Jetté and Jones 2000)

Stories can even explain characteristic of plants, such as three widows who pinched or cut their thighs while mourning their lost husband and became spruce (*Picea glauca*, rough bark), alder (*Alnus crispa*, red dye bark), and balsam poplar (*Populus balsamifera*, long grooved bark).

The inanimate also were able to transform, and some creation stories explain the origins of land formations such as solitary hills or rock outcrops. A Koyukon Distant Time story tells of a race of giants who were starving. One giant, along with his wife, left their people in the mountain range to the south and started walking toward a range of mountains on the other side of the river valley in search of food. Having traveled to the Yukon River, which was only halfway, they succumbed to starvation and fell down exhausted, the husband in front of his wife, dropping his pack in front of him. There they remained as Pilot Mountain and Bishop Rock.

These stories can also offer anecdotal insights into historic range expansion or contraction of particular species for which extensive baseline scientific information is lacking. In 2004, biologists began a study to identify the extent of overlap between tundra swans (*Cygnus columbianus*) and trumpeter swans (*Cygnus buccinator*) nesting in the Koyukuk National Wildlife Refuge. Because swans cannot be identified reliably from the air, all swans counted in the Koyukuk area during statewide trumpeter swan surveys were considered trumpeters with a qualifier recognizing that some portion contained tundra swans. Earlier attempts to elucidate the extent of overlap produced a 90% tundra swans and 10% trumpeter swan nesting distribution in the late 1980s and early 1990s. Trumpeter swans were making a strong comeback from near extinction at the turn of the century, but historic baseline data of swan distribution on the Koyukuk before 1968 was lacking. Refuge biologists went to visit Koyukon elders Catherine and Steven Attla in the remote rural village of Huslia, Alaska, in the middle of the Koyukuk Refuge. Steven Attla was nearly 80, and he knew every waterfowl call that occurred in his area from hunting and living among these birds his whole life. He was a master imitator of any call they asked him about . . . except that of the trumpeter swan. He knew the tundra swans instantly and could make the call perfectly but didn't know the horn-like trumpeter call. Curious about this, biologists looked up the word "swan" in the Athabascan dictionary (compilation began in 1898 by a Jesuit anthropologist and completed by a Koyukon Athabascan author in 2000) and could only find the whistling swan (tundra swans) referenced. But also nestled in the text was a small note taken from the priest's notes relaying the Athabascan story of the swan from distant times, "*At the time that beasts were men, the swan fought against K'etsetl (teal) and was beaten, whereupon the swan lost his fine voice and now has but a pitiful cry*" (Jetté and Jones 2000). This indicated that the trumpeter swan may have occupied the Koyukuk region prior to its decline. The healthier population of tundra swans occupying the tundra region of the Selawik valley to the north most likely moved south into the newly vacated and available periphery habitat during the 1835–1897 period of intense market hunting by the Hudson's Bay Company for trumpeter swan skins. Protection of trumpeter swans and their habitats along with transplant breeding programs enabled the species to make a successful comeback between the 1950s and the 1990s. The population has been steadily increasing throughout its historic range in North America, and the distribution of nesting trumpeter and tundra swans on the Koyukuk Refuge in 2004 was 50/50.

Women in Native Societies and Wildlife Conservation

Here we present an example from Joanne Bryant Ditsuh Taii, Gwich'in, who is the Tribal communications and outreach specialist for the US Fish and Wildlife Service in Alaska. She was born and raised with very little outside influence in remote Arctic Village, the northernmost village of the Athabascan

(Diné) people. She speaks and translates the Gwich'in language. Here is her description of Indigenous women's relationship to wildlife and conservation:

I grew up traditionally, with parents and grandparents teaching me about the subsistence way of life, survival skills, language, culture, Gwich'in heritage. The basic methods and necessities to survive.

Back in the 1960s, there was limited transportation for passengers and supplies. During my birth, there was no airplane, so I was born naturally in a single-bedroom log cabin with no running water or electricity and only a wood stove for heat. Back then, there was no medicine, but I had eight experienced midwives who took care of me and Mom. Having been born in the village, Mom and others have to endure the pain of child bearing but also take responsibility taking care of a large family. Raising children is a shared responsibility by the parents and grandparents.

Hunting and trapping is a Gwich'in man's job. After the hunters kill the animal, they butcher the meat on tree branches or willow for cleanness. They make sure the blood is cleaned as much as possible so as not to scare other animals away. The stomach and guts are a delicacy and are given to the elders.

It is Gwich'in women's job to take care of the meat. Native women display great strength to wash, cut, hang meat, and keep the smoke going in the smoke house so the bugs won't disturb the dry meat and the meat aroma is of smoke. It is their role to clean, preserve, and cook the meat. They are not allowed to hunt with their spouse because it is considered a taboo. After the hunt, the meat and skin are taken to the cache, and the women cut the meat and work on the skin. The skin is tough and excess fat is removed and stretched out with all the necessary skin tanning tools. The men hunt and trap, and the women prepare the meat and skin.

Gwich'in women can't be around hunters and animals during their menstrual period because it is considered "bad luck" and taboo. When a woman has her period, she is placed in a tent for one week with a big decorative hat to cover her face. She is not to communicate with anybody but sits in the corner quietly and sews. This is to clear her complexion and purify her. The Gwich'in were very strict in following the protocol and rules. This is for survival purposes.

Women take care of animals after harvest except for wolf (*Canis lupus pambasileus*), fox (*Vulpes vulpes* and *Alopex lagopus*), lynx (*Lynx canadensus*), martin (*Martes americana*), muskrat (*Ondatra zibethicus*), ground squirrel (*Spermophilus parryii*), or rabbit (*Lepus arcticus*). The hunter takes care of these furbearers. Once the hide is removed and dried on a board, the women sew hats and gloves from the fur. Very seldom, during lean times, a woman will harvest small game such as ground squirrel, muskrat, or rabbit. Fishing is considered appropriate for both men and women, especially ice fishing or throwing lures in the river.

Underrepresentation of Native Women in Wildlife

Of the 229 Tribal governments in Alaska, at time of this writing we know of 3 that have a dedicated wildlife professional, and only one is a woman. Across the board, American Indians and Alaska Natives represent only 0.7% of wildlife biologists (Zippia n.d.), and if we extrapolate those numbers, it is likely that Native women represent less than 0.3% of wildlife biologists, 1 out of 300. Out of all the Tribes in Alaska, there are 5–10 that have wildlife ordinances that designate hunting areas, practices, and limits. Most, if not all Tribes, work on wildlife research and management issues, but these issues compete with the many duties of multitalented (and often overworked) Tribal staff. For instance, the Tribal forester may also serve as the wildlife biologist and as the grant writer and program manager for the natural resource department. Funding may come from competitive grants from foundations or the government. Most often the grants have guidelines and questions that the funders want answered, which do not necessarily correspond to the Tribe's priorities or questions. In summary, not only are Native women underrepresented in the wildlife profession, but they often serve multiple roles in their departments and often are not provided ample staffing and financial resources to conduct their work.

Top Native American female leaders the authors have viewed as role models in the wildlife and natural resource profession include Elveda Martinez, Julie Thorstenson, Paige Schmidt, Gloria Tom, Robin Kimmerer, Sally Carufel Williams, Karen Linnell, Seafha Ramos, Rhonda Pitka, Molly Chythlook, Stephanie Quinn-Davidson, Patty Schwalenberg, and Brooke Wright.

Focusing efforts on **Indian education** can help bridge the gap and increase representation of Native women in the wildlife profession. Unfortunately, Alaska Natives face extreme challenges, and the state has one of the most expensive K–12 education systems in the nation and consistently ranks at the bottom for performance at all levels. Alaska faces education challenges including the vastness of the state, a very limited road infrastructure, lack of STEM teachers, and a high teacher turnover in rural communities. Alaska Natives have the lowest annual household income in the country, and they have the lowest four-year high school graduation rate in Alaska and second lowest in the United States. STEM industries are a critical share of Alaska's economic vitality, and the majority of STEM occupations (75%) require at least an undergraduate degree. Though Alaska Natives make up 19% of Alaska's population and 10% of the workforce, they are only 6% of the state's workers in computer, engineering, and science occupations. Specific figures are not available because national data sets combine Alaska Natives with American Indians, but the college dropout rate among American Indian / Alaska Native (AI/AN) students in public universities is the highest of any other student group. AI/AN students have the lowest percentage of public and private high school graduates of any race or ethnicity who took Algebra II, Analysis/Pre-calculus, or Statistics/ Probability in high school. In 2016, the Alaska Native Science and Engineering Program (ANSEP) graduated 47 Alaska Natives with BS degrees in STEM, and 19 were Alaska Native women. Of these 19 women, 11 earned degrees in sciences including engineering, mathematics, and physical sciences. The current mitigation attempts, which often involve highly successful but locally focused efforts, have not resulted in increased graduation rates. The National Science Foundation has found that fragmented or isolated efforts provide an insufficient response to the acknowledged scope and scale of the lack of women in STEM, especially those of minority backgrounds (email communication with ANSEP at University of Alaska). **More Native role models in positions of leadership** plus **culturally congruent curriculum** support Natives in higher education (Gervais et al. 2017).

So what can we do to foster this demographic? Making it possible to go home often is one strategy to promote Native American student success rates (Waterman 2012). This connection to home allowed Native students to stay culturally centered. Many Native women are a critical component to the family dynamic and often require ample time at home with small children, elderly parents, and other caregiving, which can have an impact on their professional, career and leadership development in the workplace. Furthermore, leaving your Tribal community and then coming back home can be challenging when your family members think you have "sold out" or left for too long and forgotten your traditions.

Lastly, what are the barriers for Native women to join the workforce? Often remote work locations and low pay (Tribal jobs pay less than federal equivalent) require frequent relocation. Plus, how does one do this with commitments to community, family, and cultural ceremonies? All this comes with privilege. There are few biological technician jobs available in places where early career professionals want to call home, resulting in brain drain from Indian country.

Fortunately, there are programs that work to support Native students, such as The Wildlife Society Native Student Professional Development program, the Native American Fish and Wildlife Society (NAFWS) youth practicum, ANSEP, and individual Tribal outdoor programs such as the Navajo Nation Outdoors Women's and Friends program.

BOX 6.4 Laurel James (Yakama Nation Wildlife Program Manager and Yakama Tribal member)

Laurel James steps off the jet bridge and poses on the tarmac before a Boeing 747 that would be the first Alaska Airlines flight powered by jet fuel composed of Douglas-fir forest slash and tree debris from the Flathead and Muckleshoot Tribal forests (Figure 6.5). The Tribal biomass project was the first of its kind, with goals of restoring forest health and mitigating fire risk by implementing fuel reduction projects. Exactly one decade prior to the photo, Laurel found herself directing a helicopter crew on the Yakama Nation to use chainsaws to take the tops off trees to develop artificial nest cavities for the northern spotted owl (*Strix occidentalis caurina*). It was a day in the life of Laurel James, soon to be Dr. Laurel James. In her tenure at University of Washington as a master's and doctoral student, Laurel has served as mentor for more than three dozen Native students in natural resources. She was also one of three Native students to serve on the congressionally mandated Indian Forest Management Assessment Team in 2013. Laurel was nominated to be one of the inaugural cohort members of the University of Washington Husky 100 in 2016 for her contributions to the institution. She is a single mother and active in community ceremonies, gatherings, and traditions. In her position with the Yakama Nation, Laurel works until midnight releasing pronghorns (*Antilocapra americana*) and is back at the office before dawn to prep trailers for next load. Her position as a role model for Native youth will continue to impact wildlife populations in the Yakama Nation and beyond for generations to come.

FIGURE 6.5. Landing in Washington, DC, from Seattle, Washington, Laurel James (Yakama, *second from right*) poses with Congresswoman Suzan DelBene (*far left*), Tribal Councilman Louie Ungaro (Muckleshoot, *second from left*), and Secretary of Agriculture Tom Vilsack (*far right*). Photo courtesy of Indroneil Ganguly.

BOX 6.5 The Birth of the Native American Fish and Wildlife Society

Shortly before 1980, in a dorm room at the New Mexico State University Las Cruces campus sat four Native men from various Tribes in the Southwest, Norman Jojola, Butch Blazer, John Antonio, and Joe Jojola. On note paper they wrote preliminary ideas for building the first society dedicated to wildlife conservation in Indian country. Later that year, 20 Tribal representatives met in Gallup, New Mexico, to formalize the idea. By 1983 the NAFWS was incorporated as a 501(c)(3) organization with a mission of assisting Native American and Alaska Native Tribes with the conservation, protection, and enhancement of their fish and wildlife resources (pers. communication with Julie Thorstenson, NAFWS executive director). Today, the NAFWS works diligently to pass resolutions in support of policy that benefits Tribal wildlife and natural resource programs. Currently it is working toward policy that specifically redirects $97.5 million in funding for Tribal wildlife conservation through the Recovering America's Wildlife Act. Additionally, the NAFWS supports Native American students through scholarships, as well as building and maintaining a network of Native American wildlife professionals. The organization has a strong commitment to supporting Native women in natural resources. As an example, in 2019 the NAFWS National Conference, hosted by the Gila River Indian Community was themed Native Women: Through the Hands of our Grandmothers, Mothers, and Daughters—We Are the Stewards of Mother Earth.

How to Be an Ally

Being an ally is important for enhancing Tribal sovereignty and capacity to manage wildlife populations for future generations. One way to support Indigenous peoples is to educate yourself about the historical and current contexts that Indigenous communities face. A good way to be an ally is to **learn about the Tribal communities** close to your home or workplace. Learning more about local Indigenous cultural practices, their history, values, and beliefs is one way to better understand their needs and what may drive their governing bodies, land management practices, and educational and health care systems, and so on. Never assume that you know everything about a community, and approach the learning process with humility and humble inquiry. **Do more**

listening and much less talking. Listen without carrying on an inner dialogue, composing your next question or response. Often, a response is not necessary. As always, respect for others is a fundamental component of being an ally for any marginalized community. Little things go a long way. Introduce yourself as a whole person (where you grew up, who your ancestors are, the names of parents/spouses/children, your title comes last), visit often, follow through, don't make false promises, attend community events (both happy and sad).

Indigenous women in wildlife are here on our own merits, our own hard work, complete with setbacks and successes, just like anyone else in the wildlife field. To be an ally means to not assume that we got here by any other method than our own merits.

Please note that you will not be an ally by casually mentioning a past lineage or heritage from some Tribe that you know nothing about. Speaking from personal experience, this is a pet peeve for most Native people and hits some serious heart strings when it happens—and it happens often. Interestingly, 36% of survey respondents from a random population thought they had Native ancestry (Reclaiming Native Truth 2018a). The first question a Native will ask is, "What Tribe?" If you don't know, you have damaged both your credibility and your relationship. If you believe you have some Native American or Indigenous heritage, take the initiative to look it up, find a way to identify the actual community you claim ties to, and then be proactive about your enrollment or contributions to the cultural preservation of that community. Fortunately, a great resource with more information on how to be an ally for Indigenous communities is available from Reclaiming Native Truths: *Changing the Narrative about Native Americans: A Guide for Allies* (2018b).

In addition to personal behavior, research, and actions, it is also good to get your organization thinking on a policy level. Organizations and government agencies alike can be allies by holding land recognitions and acknowledgments, volunteering with Native programs, fostering welcoming environments for Native people to thrive, and supporting Native youth initiatives such as the American Indian Science and Engineering Society or the National Congress of American Indians youth programs, or ANSEP. Organizations can be allies by educating others about the limited public access on most Indian reservations, treaty rights, and federal-Tribal trust responsibility. They can acknowledge and respect the vast Indigenous knowledge system about the wildlife around them. Government agencies especially can involve Tribes meaningfully by placing high value on local knowledge and inviting them to jointly design research studies and management plans (Brooks and Bartley 2016).

Organizational allyship also includes sharing research ideas and outcomes, as well as sharing management responsibility. Big changes through large policy initiatives help build stronger relationships with Tribes, like the US Fish and Wildlife Service Native American Policy update in 2016 and a supplemental Alaska Native Relations Policy.

Conclusion

Indigenous women in STEM and natural resources are rising. There are more role models today in positions of leadership likely due to the American Indian Movement in the 1970s. Indigenous communities are finally reaping the benefits of the Indian Self-Determination Act one to two generations later. Native languages are making a resurgence, with positive impacts in many Indigenous communities. Language preservation helps preserve culture and traditions and our relationship to place.

Indigenous peoples have high level of identity with and responsibility to care for wildlife and human communities. Indigenous peoples also have a unique ability to balance conflicting perspectives and navigate complex relationships. Today we see a strong passion and level of dedication in Native youth for environmental concerns. Indigenous groups are focused on inclusivity and parity. If trends in student demographics are any indication for where the wildlife profession is headed, our future is in good hands.

Discussion Questions

1. On your phone or computer, go to the Tribal Connections Viewer and Map provided by the USDA Forest Service available online here: https://usfs.maps.arcgis.com/apps/webappviewer/index.html?id=fe311f 69cb1d43558227d73bc34f3a32. Read through and accept the use policy, then locate your home and click on the map to identify what Tribes are listed. Once you find the names of some Tribes, do some internet searches on those Tribal communities. Do they have a local Tribal government office, cultural center, wildlife department, or museum nearby? How would you propose reaching out to this community if you wanted to learn more about their views on local state wildlife conservation or habitat improvement plans? How would you appropriately plan, conduct, and reference a land acknowledgment for these Indigenous communities for a large public gathering? Here are some examples to

get you started: https://native-land.ca/, http://landacknowledgements.org/, and https://usdac.us/nativeland.

2. Identify wildlife species that are of great significance to the Tribal communities in your region. How would you incorporate their stories and traditions (TEK) into wildlife management decisions? For example, you could establish game species harvest seasons or limits that are informed by the cultural committee or the Tribal Historic Preservation Office. Read McCorquodale, *Cultural Contexts of Recreational Hunting and Native Subsistence and Ceremonial Hunting: Their Significance for Wildlife Management* (1997) for details on the differences between recreational hunting and Native practices.

3. Read about the 2019 Supreme Court *Herrera v. Wyoming* case of the Crow Tribal member who harvested an elk within ancestral Crow territory but off reservation lands. What year was the initial treaty signed? Were wildlife incorporated into the treaty? Do states have higher authority than treaty laws? What sort of impacts does this have for other Tribes and Indian people?

Additional Resources

Fish and Wildlife Biologist Demographics and Statistics in the US: www.zippia.com/fish-and-wildlife-biologist-jobs/demographics/
Native American Fish & Wildlife Society: www.nafws.org/

Activity: Cross-Cultural Dialogue Experience

Learning Objectives: Participants will identify and appreciate that their communication style reflects their culture, understand that cultural groups differ in how they communicate and listen, and learn that effective interactions with others can be improved by appreciating these cultural differences.

Time: 30–45 min.

Description of Activity

Materials needed: None

Process:

Part 1 (10–15 min.): This exercise works well early in the academic term when students are just meeting each other. Participants form pairs or groups of three and are provided with time for a short discussion (5–10 min.). Each person in the pair or group will have an opportunity to share with each other their life story. Begin the discussion with "You will tell your life story to your group."

Part 2 (10 min.): After the discussion ends, groups are presented with the following list when considering their own and their group members' communication styles.

1. Eye contact: How much eye contact did the speakers make with the listeners and vice versa?

2. Physical distance: How close was the speaker to other people in the group?

3. Hand gestures: How much gesturing did the speaker use when telling their story?

4. Direction of story: How was the knowledge transferred? Was it linear or circular? Did the speaker get straight to the point explicitly and clearly, with few deviations? Or did the speaker talk around the main point telling anecdotes and stories? How did people classify their life story? Did it start from their home and upbringing or was it based on their accomplishments?

5. Interruptions or questions: Did the listeners ask questions or interrupt the speaker while they told their origin story? Was it uncomfortable speaking the entire time about your story?

Part 3: Debriefing and Discussion Questions (10–15 min.)

1. How did you feel when interpreting your communication style?
2. What did you learn?
3. What was most surprising about the experience?
4. What are potential outcomes from a conversation when cultural communication style is ignored?
5. How would you use the information you learned or your experience from this activity in your life as a wildlife professional, student, or citizen?

Teaching Notes

Different verbal and nonverbal communication styles are used and interpreted differently across cultures. The five communication styles participants discussed are interpreted differently across the globe and even among different cultural groups within the same country.

1. Eye contact: Some cultures consider it respectful to minimize eye contact while others show respect by maintaining eye contact. The level of eye contact can also be interpreted by some as a signal of

trustworthiness about the story. Still other cultures consider eye contact embarrassing and avoid it after making initial eye contact with another speaker. For instance, many Native American youth are taught to not look elders in the eyes because it is a sign of disrespect.

2. Physical distance: Creating an arm's length of space between speakers is more comfortable for some while for other cultures, physical closeness is preferred.

3. Hand gestures: Many cultures use arm and hand gestures when speaking to accentuate their spoken words while others find it to be impolite or a sign of dishonesty.

4. Direction of story: Some cultures consider not getting to the main point as wasting another person's time while other cultures consider it insulting to the other person if you have to tell them something explicitly.

5. Interruptions or questions: Interrupting to ask questions can either be interpreted culturally as showing respect and interest in the speaker by engaging them or be interpreted as disrespectful. For instance, interrupting, especially someone who is older than you, in Tribal communities can be seen as highly disrespectful.

The more communicators understand potential differences in communication styles, the more likely they are to improve communication with others outside of the own culture. One easy solution for people to consider is that matching your communication style to that of the person you are speaking with can create a more positive communication experience for everyone.

References and Resources

Korac-Kakabadse, Nada, Alexander Kousmin, Andrews Korac-Kakabadse, and Lawson Savery. 2001. "Low- and High-Context Communication Patterns: Towards Mapping Cross-Cultural Encounters." *Cross Cultural Management: An International Journal* 8(2): 3–24.

Lewis, Richard D. *When Cultures Collide: Leading across Cultures*. London: Nicholas Brealey, 2018.

Literature Cited

Bennett, M., D. Bustos, J. Pigati, K. Springer, T. Urban, V. Holliday, S. Reynolds, M. Burdka, . . . and D. Odess. 2021. "Evidence of Humans in North America during the Last Glacial Maximum." *Science* 373(6562): 1528–31. doi: 10.1126/science.abg7586.

Brooks, J. J., and K. A. Bartley (2016). "What Is a Meaningful Role? Accounting for Culture in Fish and Wildlife Management in Rural Alaska." *Human Ecology*, 44(5): 517–31.

Case, David S., and David Avraham Voluck. 2002. *Alaska Natives and American Law*. 2nd ed. Anchorage: University of Alaska Press.

CSKT History and Culture. N.d. "Honoring the Past to Ensure the Future." Accessed March 29, 2020. www.csktribes.org/history-and-culture.

Davies, I. P., R. D. Haugo, J. C. Robertson, and P. S. Levin. 2018. "The Unequal Vulnerability of Communities of Color to Wildfire." *PLoS ONE* 13(11): e0205825. doi: 10.1371/journal.pone.0205825.

Dunaway, F. 2017. "The 'Crying Indian' Ad That Fooled the Environmental Movement." *Chicago Tribune*. November 21. https://www.chicagotribune.com/opinion/commentary/ct-perspec-indian-crying-environment-ads-pollution-1123-20171113-story.html.

Duran, Eduardo, and Bonnie Duran. 1995. *Native American Postcolonial Psychology*. Albany: State University of New York Press.

Evans-Campbell, T. 2008. "Historical Trauma in American Indian/Native Alaska Communities: A Multilevel Framework for Exploring Impacts on Individuals, Families, and Communities." *Journal of Interpersonal Violence* 23(3): 316–38.

Federal Register. 2019. 84 FR 1200. Indian Entities Recognized by and Eligible to Receive Services from the United States Bureau of Indian Affairs. February 2. www.federalregister.gov/documents/2019/02/01/2019-00897/indian-entities-recognized-by-and-eligible-to-receive-services-from-the-united-states-bureau-of.

Fienup-Riordan, Ann. 1991. *The Real People and the Children of Thunder: The Yup'ik Eskimo Encounter with Moravian Missionaries John and Edith Kilbuck*. Norman: University of Oklahoma Press.

Gervais, Breanna K., Chase R. Voirin, Chris Beatty, Grace Bulltail, Stephanie Cowherd, Shawn Defrance, Breana Dorame, . . . and Serra J. Hoagland. 2017. "Native American Student Perspectives of Challenges in Natural Resource Higher Education." *Journal of Forestry* 115(5): 491–97.

Gordon, J., J. Sessions, J. Bailey, D. Cleaves, V. Corrao, A. Leighton, L. Mason, M. Rasmussen, H. Salwasser, and M. Sterner. 2013. *Assessment of Indian Forests and Forest Management in the United States*. www.itcnet.org/issues_projects/issues_2/forest_management/assessment.html.

Hensley, William L. Iggiagruk. 2009. *Fifty Miles from Tomorrow*. New York: Sarah Crichton Books.

Huntington, Henry P., and George Noongwook. 2012. "Traditional Knowledge and the Arctic Environment: How the Experience of Indigenous Cultures can Complement Scientific Research." PEW Charitable Trusts. https://www.pewtrusts.org/~/media/legacy/oceans_north_legacy/attachments/arctic-traditional-knowledge-brief--artfinal_v4_web.pdf.

Inuit Circumpolar Council. 2012. *Food Security in the Arctic*. Inuit Circumpolar Council—Canada. www.inuitcircumpolar.com/wp-content/uploads/2019/01/icc_food_security_across_the_arctic_may_2012.pdf.

Jetté, Jules, and Eliza Jones. 2000. *Koyukon Athabascan Dictionary*. Fairbanks: Alaska Native Language Center, University of Alaska.

Kawagley, Oscar A. 1995. *A Yupiaq Worldview: A Pathway to Ecology and Spirit*. Prospect Heights, IL: Waveland Press.

Koch, A., C. Brierley, M. M. Maslin, and S. L. Lewis. 2019. "Earth System Impacts of the European Arrival and Great Dying in the Americas after 1492." *Quaternary Science Reviews* 207: 13–36. doi: 10.1016/j.quascirev.2018.12.004.

Langdon, Steve J. 2014. *The Native People of Alaska: Traditional Living in a Northern Land*. 5th ed. Anchorage: Greatland Graphics.

Mason, Larry, Germaine White, Gary Morishima, Ernesto Alvarado, Louise Andrew, Fred Clark, Mike Durglo Sr., . . . and Spus Wilder. 2012. "Listening and Learning from Traditional Knowledge and Western Science: A Dialogue on Contemporary Challenges of Forest Health and Wildfire." *Journal of Forestry* 110(4): 187–93.

McCorquodale, S. 1997. "Cultural Contexts of Recreational Hunting and Native Subsistence and Ceremonial Hunting: Their Significance for Wildlife Management." *Wildlife Society Bulletin* 25(2): 568–73.

Migratory Bird Subsistence Harvest in Alaska. 2003. 16 U.S.C. 703–712, tit. 50, § 92, 67 FR 53517, Aug. 16, 2002, as amended at 68 FR 43027, July 21.

National Congress of American Indians (NCAI). 2020. "Indian Country Demographics." Updated June 1. www.ncai.org/about-tribes/demographics#R2.

National Conference of State Legislators (NCSL). 2022. Federal and State Recognized Tribes. www.ncsl.org/research/state-tribal-institute/list-of-federal-and-state-recognized -tribes.aspx#State.

Native American Heritage. 2021. "Viewing American Indian Treaties." National Archives. Last reviewed March 8. www.archives.gov/research/native-americans/treaties/viewing -treaties.

Native American Voting Rights Coalition. 2018. *Voting Barriers Encountered by Native Americans in Arizona, New Mexico, Nevada and South Dakota*. January. www.narf.org /wordpress/wp-content/uploads/2018/01/2017NAVRCsurvey-summary.pdf.

Native Movement. N.d. Accessed June 16, 2022. www.nativemovement.org/.

Navajo Nation v. Urban Outfitters, No. 12-195 (D.N.M. September 19, 2016). https://narf .org/nill/bulletins/federal/documents/navajo_nation_v_urban_outfitters_2016.html.

Northwestern University. 2019. "Land Acknowledgment." Native American and Indigenous Initiatives. www.northwestern.edu/native-american-and-indigenous-peoples/about /Land%20Acknowledgement.html.

Reclaiming Native Truths. 2018a. *Changing the Narrative about Native Americans: A Guide for Native Peoples and Organizations*. www.firstnations.org/wp-content/uploads/2018/12 /MessageGuide-Native-screen.pdf.

Reclaiming Native Truths. 2018b. *Changing the Narrative about Native Americans: A Guide for Allies*. www.firstnations.org/publications/changing-the-narrative-about-native -americans-a-guide-for-allies/.

Waterman, S. J. 2012. "Home-Going as a Strategy for Success among Haudenosaunee College and University Students." *Journal of Student Affairs Research and Practice* 49(2): 193–209, doi: 10.1515/jsarp-2012-6378.

Whitbeck, L. B., Adams, G. W., Hoyt, D. R., and Chen, X. 2004. "Conceptualizing and Measuring Historical Trauma among American Indian People." *American Journal of Community Psychology* 33(3/4): 119–30.

Yazzie, J., Fulé, P., Y-S Kim, and A. Sánchez Meador. 2019. "Diné Kinship as a Framework for Conserving Native Tree Species in Climate Change." *Ecological Applications* 29(6): e01944. doi: 10.1002/eap.1944

Zippia. N.d. Fish and Wildlife Biologists Demographics and Statistics in the US. Accessed November 30, 2021. https://www.zippia.com/fish-and-wildlife-biologist-jobs/demo graphics/.

CHAPTER

7

Rena Borkhataria, Mamie Parker, Camya Robinson,
Maria Johnson, Sheila Minor Huff, Princess Hester,
Audrey Peterman, Charisa Morris, Modeline Celestin,
and Samantha Miller

Women of Color in Wildlife

Stories of Support and Setbacks

At some really basic level you are seeing how we are significantly shifting in terms of racial and ethnic diversity, and what does that mean for the conservation community and organizations like the Park Service that want to stay relevant? They are actually going to have to address the issue, and have that be addressed at a lot of different levels in terms of stewardship, engagement, leadership. Who is getting jobs, who is taking part?
                                                        —Carolyn Finney

When we began this chapter, our intention was to focus on lifting up and exposing the experiences and life lessons of Black women pioneers in the conservation profession. What we found was that some experiences and lessons pertain directly to Black women, but many experiences are common to women of color in general, and to women as a whole, as they navigate the predominately white- and male-dominated conservation space.

Diversity initiatives within conservation and preservation organizations have been successful in promoting and supporting white women (Taylor 2014) but not at recognizing the intersectional challenges associated with gender and race. Intersectionality refers to the net effects or outcomes of two or more social categories being greater than the sum of their parts. The overlap of categories creates an experience that is distinct from that of members of either category alone (Crenshaw 1994). The experiences of women of color are unique to those identities and cannot be considered to be the experiences of all women coupled with the experiences of all people who are racially diverse. The intersectional effects of gender and race in conservation are only beginning to receive attention. Studies of the topic may be hampered, however, by the scarcity of women of color in our field. In many cases it is difficult, if not impossible, to maintain the anonymity of study participants (Miriti 2020). For this reason, in some instances we have taken an additive approach despite its limitations.

Intersectional supports are needed to ensure women of color have equal opportunities entering, persisting, and advancing in wildlife and conservation professions. The authors of this chapter, both those who are relatively new to the field and those who have been working in conservation for 30- to 40-plus years, can attest that some of the biggest barriers they have faced could have been ameliorated by leaders who actively embraced diversity, equity, and inclusion rather than just paying it lip service. We hope recent efforts by agencies, organizations, academic programs, and professional societies to diversify the field of conservation will ease some of the challenges specific to women of color.

Throughout this chapter we've embedded our own experiences as women of color in the field of conservation in boxes. Personal narratives related to experiences of women of color in conservation are rare yet important for expanding awareness of institutional injustices and racism (Miriti 2020) and of our accomplishments in our field (Karp 2019). Here we provide our open and honest thoughts on the challenges and rewards of being pioneers in our field while dealing with racism, sexism, and discrimination. Throughout this chapter you will find our stories highlighted. These are stories from Black, Asian American, and multiracial women. They are the stories of recent graduates, mid-career professionals, and pioneers who established prominent and inspirational careers in a field that was not always welcoming. In many cases we have shared experiences that are both complicated and painful to relive and to recount. In sharing such vulnerability, we hope these stories will resonate with our readers and will transcend a purely academic approach to the subject matter. However, to provide context for these accounts we include background information on how women of color enter careers in wildlife, examine barriers more generally, provide suggestions for creating more welcoming spaces, and challenge all readers to take on the work of dismantling racism in our field.

Entering the Wildlife Profession:

Exposure

The decision to pursue a career in ecology and natural resources is often influenced by people's childhood backgrounds and learning experiences. In particular, childhood exposure to nature is a strong predictor of which students will choose to pursue natural resource careers (Moreno et al. 2020). Family attitudes toward nature, influential adults, and nature- or science-related media also play a strong role in decisions by women of color to pursue a career in natural resources (Haynes and Jacobson 2015; Prevot et al. 2018; Bal et al. 2020; Dokes et al. 2020; Miao and Cagle 2020; Szczytko et al. 2020).

For too long, these types of experiences were out of reach for most women and girls of color. However, this is changing. It is encouraging to see nonprofits like Outdoor Afro, Latino Outdoors, Soul River, Soul Trak, and others working to connect racially diverse children and families to nature and the outdoors. Some include opportunities and programs specifically for girls and women.

Mamie Parker shares her journey from the banks of Arkansas's Bayou Bartholomew and Lake Enterprise to the one of the highest-ranking positions in the US Fish and Wildlife Service as a pioneering Black woman in fisheries (Box 7.1). She attributes her decision to pursue a career in fisheries and with the USFWS to the influence of her mother, an avid outdoorswoman who loved to fish. She ensured that young Mamie grew up spending time outside and in particular, fishing. These activities, along with the encouragement of a high school biology teacher, were a major influence in her decision to pursue a career in fisheries and eventually led to her becoming the first woman of color to serve as regional director and head of fisheries (Figure 7.1).

Education

An undergraduate degree, and in many cases a graduate degree, in a wildlife-related major is generally required of anyone who wants to pursue a career in wildlife. Master's degrees are "preferred" for even entry-level biologist jobs with the US Fish and Wildlife Service (management; OPM N.d.). Once in these majors, women of color may find the field to be alienating, even if the subject matter is interesting, because of their underrepresentation in the field (Puritty et al. 2017). Traditionally, wildlife classes have been composed almost exclusively of white students, primarily men (Dokes et al. 2020), so women of color often stand out in relation to their white peers. As "the only" nonwhite person in a given setting and possibly the only woman, women of color are "hypervisible" and are even more likely to feel excluded or unwelcome in classes, wildlife-related majors, and the field as a whole (Blosser 2020; Dickens et al. 2019).

For women of color pursuing wildlife degrees, their university as a whole may feel alienating as well. Nine of the top 10 wildlife programs at a large national college information clearing house (College Factual 2021) are at "land grant" universities—large public universities that were established by the Morrill Act of 1862. Compared to the national average, these universities lack

BOX 7.1 **Mamie Parker**

As I look back on my career, I think of where my journey started. Born in the Mississippi Delta to a single mother, my early life was marred by the realities of the segregated South. A little more than a month before I was born, President Dwight D. Eisenhower sent army troops to Arkansas to enforce a federal court order and escort and protect African American students integrating Little Rock Central High School. It would take approximately a decade for our small town in the southern part of the state to end segregation, and I was the first little African American girl to enter Wilmot Elementary School's 3rd-grade class. While this was a historic milestone and a proud moment for our family and the community, it was a very difficult experience for me.

Eventually, a teacher of mine became one of my strongest allies. He often reminded me that one day these difficult experiences would help me survive in any environment. All of my teachers encouraged me to study hard, keep my core values, and remember who I was and what I could become by being a pioneer. As the environmental movement of the 1970s grew, that high school teacher and the words to Marvin Gaye's hit song, "Mercy Mercy Me," which described the many ways the earth's environment was being degraded, inspired me to love biology and to contribute to the ongoing cultural shift toward a more environmentally conscious society.

FIGURE 7.1. Dr. Mamie Parker earned a BS in biology from the University of Arkansas, and an MS in fisheries and wildlife management and PhD in limnology from the University of Wisconsin–Green Bay. She served as the head of fisheries and as a regional director for the US Fish and Wildlife Service and was the recipient of a Presidential Rank Award. She is now a sought-after success coach and speaker. Here is Mamie Parker in 1985, plowing her way through cattails, bogs, isolation, racism, sexism, and other challenges, with the dream of inspiring other Black girls.
Photo courtesy of the subject.

While the support of my teacher helped inspire me to dedicate myself to a career in science, my strongest influence throughout my childhood was my mother. My mother loved the outdoors and spent what free time she had gardening and fishing. Her love for the outdoors became imprinted on me during these times. She taught me life lessons and survival tactics on our trips to the banks of Bayou Bartholomew and Lake Enterprise. She was certain that this pioneer experience would help me to help others. She was so right. My childhood experiences felt unbearable at times, but they served me very well in the years to come as I moved through the ranks of the US Fish and Wildlife Service, from a budding biologist to the first African American regional director and head of fisheries in this country.

When I first came to Washington during the Clinton administration to work in the USFWS Director's Office, I noticed that the eyes of the janitorial staff would follow me as I moved through the hallways of the Department of the Interior. I didn't know why. Then one afternoon I ran into one of the janitors in the restroom and she said to me, "I've been here for many years and I have never seen another African American woman in that office that you are occupying who wasn't there to clean it. We are so proud of you for representing us." This experience highlights the fact that other members of my race were counting on me to make them proud.

For many years, I continued to struggle with the isolation, lack of role models, exclusion and various racial and gender biases, particularly unconscious biases. However, the various assignments in the field offices, exceptional training, and exposure in the USFWS gave me the experiences, excitement, and joy of a lifetime. I was fortunate to move through national fish hatcheries and field offices in Wisconsin and Minnesota, fields and farms in Missouri, streambanks in Georgia and the Southeast, briefings at the Executive Office of the White House, testimonies on Capitol Hill, and board meetings throughout the country. Wherever I went I was the only one. It was rewarding but challenging, at times, to find the strength and stamina to push through these barriers to success.

When I struggled to find support, or felt that I was excluded within the workplace, I found support from the words of my mother during our outdoor adventures in addition to the support of my family and encouragement from my allies and mentors. My strength came from previous reminders from my mother, my siblings, and my teachers and from solace in awareness that my struggles could only uplift women who came before me and provided support and inspiration to those that would follow. When I was selected as the USFWS Northeast regional director, my photograph was placed along the wall next to every other person who had previously held that position. This was the first year the portrait would be in color to show the very first African American regional director in the history of the agency.

The journey toward being a "first" was a lonely one and a very difficult one, but it was worthwhile when I hear how it has inspired other women, particularly little African American girls. I have spent an entire career sharing my story and trying to blend these ideals, life lessons, and survival tactics into my conservation work and speeches.

Now, with years of experience under my belt, I can say that where we are now marks another great cultural shift within conservation. I, along with many other pioneers, created spaces in the conservation field for Black women and other women of color because our hard work, dedication, and stamina spoke louder than our race and gender differences. For a long time, we were just struggling to find a seat at the table. Now, we are in influential positions that allow us to build upon the skills, lessons, and success of our collective careers to support greater access and a smoother road for young women of color coming after us. The success of these women will drive how environmental organizations address issues of diversity, racism, and sexism.

diversity (de Brey et al. 2019; College Factual 2021; Figure 7.2). Institutional diversity is important to racially diverse undergraduates (Haynes et al. 2015). So although racially diverse students pursuing environmental careers generally take the same types of undergraduate classes as their white peers (Taylor 2018), women of color attending small and racially diverse universities may lack the programs and resources to prepare them fully for wildlife careers.

Camya Robinson shares with us her story as an undergraduate in one of the top wildlife programs in the nation (Box 7.2).

Internships

In addition to an appropriate education in natural resources, environmental science, or other wildlife-oriented topics, many jobs in conservation require prior experience (Bodin 2020). At some point, women of color who aspire to careers in wildlife should complete a relevant internship to gain practical experience in the field. Yet internships and fieldwork opportunities are very often unpaid and frequently require students and recent graduates to move long distances without compensation. People who cannot afford to take unpaid opportunities, including many people of color, are effectively cut off from viable paths to careers in the field (Fournier and Bond 2015; Vercammen et al. 2020; Bodin 2020). Owing to the gender and racial disparities in compensation that permeate our society, women of color are least likely to have the resources to take advantage of unpaid opportunities.

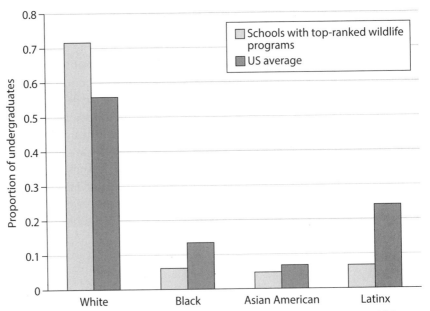

FIGURE 7.2. Racial diversity at land grant universities housing 9 of the top 10 wildlife programs compared to the national average for universities. Land grant universities were chosen based on 2017-2018 ratings for wildlife programs by College Factual (2021) and included the University of Michigan, the University of Florida, the University of Tennessee, Knoxville, South Dakota State University, the University of Wisconsin–Madison, Mississippi State University, and Virginia Tech. National averages were taken from de Brey et al. (2019).

Furthermore, poorly planned internships can be particularly isolating, uncomfortable, hostile, or unsafe for women of color (as well as other non-majority or at-risk groups) (Demery and Pipkin 2021). While intersectional data are not available in this area, sexual harassment and assault targeted primarily at women trainees are common in fieldwork settings, especially when rules and consequences are unclear (Clancy et al. 2014; Nelson et al. 2017). Internships involving physical demonstrations of strength and gendered division of labor can also be alienating (Nelson et al. 2017). Negative experiences like these can affect the mental health, productivity, and professional development of interns (Demery and Pipkin 2021). While the effects of such experiences on women of color have not been reported in an intersectional context, the additive effects of age, gender, and race, including stereotypes of sexual availability among women of color (Rosenthal et al. 2020), suggest that women of color in internship settings face additional challenges.

Here, Maria Johnson shares her experience in a federal internship program (Box 7.3).

BOX 7.2 Camya Robinson

As a first-generation college student, it was difficult navigating through college because I had to do most things without the help of my family. They were able to give emotional support and constant words of encouragement, but I couldn't rely on them to help me through the university experience because they had not gone through it themselves. I also come from a low-income family, so there was little financial support that my family could give. Luckily, I was able to receive a full scholarship which lifted the financial burden of college off my shoulders. Still, I was not in a position where I could take advantage of unpaid internships, no matter how good the experience was because I had to pay rent and other bills for the summer that my scholarship did not cover.

I applied for a minority scholarship that was offered to people from marginalized groups in the wildlife field. The whole application process was somewhat tactless. It seemed as if I had to "prove" myself to the selection committee and show how I was going to fight for diversity and inclusion in the field. When the time came for me to accept the award, it was obvious that the minority scholarship was an afterthought to the main, original scholarship, which was awarded to a white person. Although I am proud of my achievements and proud to win an award recognizing them, I also had to learn that some organizations are only interested in using me as proof that they're diverse without providing actual structural changes to better include and accommodate people of color.

There are not many Black people in the wildlife field, let alone Black queer women. It is hard to connect with people who share my identities and to see myself represented in this field. Moreover, it is difficult to find people to look up to when the only environmental figures that are talked about in my class are all rich, white, men who also held beliefs of white supremacy in some capacity (Theodore Roosevelt, Gifford Pinchot, Charles Darwin). It is disheartening to be expected to look up to these people and praise their achievements, while also completely overlooking their racism. There hasn't been one class that I have had where marginalized people were highlighted and celebrated for their contributions to the field of wildlife. It can sometimes feel isolating being one of maybe two Black women in most of my classes, at conferences, and at the jobs that I've had. Sometimes discrimination can happen in the form of not being included and not feeling seen.

What helped me succeed in completing my degree was mentorship, paid research positions, and therapy. The Doris Duke Conservation Scholars

FIGURE 7.3. Camya Robinson received a BS in wildlife ecology and conservation from the University of Florida in 2020. Here she sets a bear hair snare trap while working a field job in Idaho in 2018. Photo courtesy of the author.

Program provided me with the former two things and through that program I was able to meet others who shared the same passions as me and meet most of the people of color that I know in this field. The other students in the program and the people that mentored us are a big source from which my inspiration comes from. They give me hope that I can truly succeed in this field and in life.

Careers

Taylor (2018) found that the desire of women of color to pursue a postgraduate career in environment organizations was no less than those of other cultural groups and they were similarly prepared for such careers academically. Yet she found that fewer are actually entering those careers. The top two factors in deciding whether to advance to a career in an environmental profession included opportunities to take on leadership roles and opportunities for advancement. These factors were not as important to white students. Other important factors included diversity programming within the organization, costs of living at the job location, and the racial diversity of the organization's leadership.

Haynes and Jacobson (2015) and Balcarcyk et al. (2015) conducted interviews with recent hires to the US Fish and Wildlife Service, the National Park Service, and the US Geological Survey to document barriers they had encountered as they pursued a natural resource career. While many barriers were shared by racially diverse and white employees, some were unique to people of color (Figure 7.5). Financial factors were of particular concern to people of color related to unpaid internships, low pay, and student loans. Structural and institutional barriers included universities that weren't aware of career options in wildlife, organizational structure and barriers to advancement, relocation expectations, and lack of skills or previous coursework to be eligible for certain jobs. And women of color reported discrimination based on sex and race to also have been a discouraging factor (Balcarczyk et al. 2015; Morales and Jacobson 2020; Haynes and Jacobson 2015).

Confronting Racism and Sexism in Conservation

There are many ways that women of color in conservation experience acts of racism and sexism in the workplace, including overt acts of harassment and abuse as well as more subtle acts of discrimination. Subtle forms of discrimination

BOX 7.3 **Maria Johnson**

I grew up in a fairly large city, one with a lot of racial diversity, and I had never spent time in the Midwest. So it was a big adjustment to go to a remote wildlife refuge near a small town and not see anyone who looked like me or the people in my family. And the thing that was really shocking to me is that I'd never realized that parts of the country were so, so far behind. The male chauvinism was really excessive. I couldn't open my own door, I couldn't be the last person to walk out of a room, things like that. If a man and a woman came into a room to talk to someone, the staff would always talk to the man, and the woman would just sit there ignored. It seems really tiny, but when it's day after day, it starts to feel really degrading.

And in the community I experienced the same thing. Plus, because I'm a woman of color, everyone would ask me where I was from, because in a small town it's clear that you're not from around there. And I don't think they saw me as someone in a professional role.

Even though they'd requested an intern for the summer, the office really wasn't prepared for a young person, or a person of color, or a woman, or a young woman of color. I felt like this was supposed to be a really great thing, a great opportunity, but instead they made it clear to me that I was "just an intern." They didn't treat me like someone who might be moving into a position with the agency, it seemed like they didn't understand the goals of the internship program or didn't care.

There were other things too. I spent the summer living in a bunkhouse with a bunch of guys. I was a little freaked out when I got there and found out that would be the living arrangement. It turned out fine, but it seems like the sort of thing they should tell someone beforehand so they can make an informed decision on whether it's something they'd be comfortable with.

I did really like the work I did there. I traveled a lot into the nearest big city, where I worked with communities of color trying to make conservation more relevant, especially for young people. And I realized that talking about conservation in terms of native plants or monarch butterflies didn't mean much to people, but when you presented the same topics through an environmental justice lens—how it relates to water quality, for example, people understood why it mattered for them, and then they started to care.

I've thought a lot about whether I would do it again, and I would, because the work I want to do involves broadening access and bringing diversity, equity, and inclusion to this field. If I hadn't gone there, I wouldn't know that this is something that happens, that this is what people are dealing with out there.

FIGURE 7.4. Maria Johnson received a BS in natural resources from the University of Arizona in 2018 and is currently the program coordinator for resilience internships in the Arizona Institute for Resilient Environments and Societies (AIRES) at the University of Arizona. In this picture, she is showing young kids how to fish during an Urban American Outdoors event in Kansas City in 2018. Photo courtesy of Jonathan Bell.

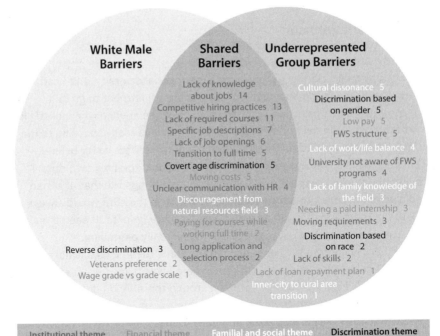

White Male Barriers   **Shared Barriers**   **Underrepresented Group Barriers**

Lack of knowledge about jobs 14
Competitive hiring practices 13
Lack of required courses 11
Specific job descriptions 7
Lack of job openings 6
Transition to full time 5
Covert age discrimination 5
Moving costs 5
Unclear communication with HR 4
Discouragement from natural resources field 3
Paying for courses while working full time 2
Long application and selection process 2

Reverse discrimination 3
Veterans preference 2
Wage grade vs grade scale 1

Cultural dissonance 5
Discrimination based on gender 5
Low pay 5
FWS structure 5
Lack of work/life balance 4
University not aware of FWS programs 4
Lack of family knowledge of the field 3
Needing a paid internship 3
Moving requirements 3
Discrimination based on race 2
Lack of skills 2
Lack of loan repayment plan 1
Inner-city to rural area transition 1

Institutional theme   Financial theme   Familial and social theme   Discrimination theme

FIGURE 7.5. Barriers to a natural resource career perceived by FWS recent hires. Numbers represent the number of hires who discussed each of the themes. The four main thematic categories of barriers were institutional, financial, familial and social, and discrimination. Reproduced with permission from Balcarczyk et al. (2015).

include a lack of representation, toxic stereotypes, barriers to advancement, and microaggressions.

Harassment and Abuse

Several studies have documented the prevalence of sexual harassment and assault of women doing fieldwork (Clancy et al. 2014). In addition to the psychological and physical costs of such experiences, they can affect productivity and professional development (Demery and Pipkin 2021). Potential outcomes include aversion to or avoidance of their field sites because of traumatic memories or posttraumatic stress, exclusion (self-imposed or otherwise) from professional settings and activities, discomfort and alienation associated with continued professional interactions with their abuser, and decisions to alter career paths or leave the profession (Nelson et al. 2017).

While violence specific to women of color was not reported, women of color are more likely than white women to be targets of sexual harassment and assault (Cassino and Besen-Cassino 2019; Crenshaw 1994) and to be perceived as sexually available (Rosenthal and Lobel 2016; Rosenthal et al. 2020; Rosenthal and Lobel 2020). For example, women employed by the US Forest Service were more likely than men to report problems with sexual harassment in the workplace along with nonsexual verbal harassment and bullying based on a variety of factors, including sex and race (USDA 2017).

Representation

In 2018, a picture went viral of a lone Black woman standing with a group of prominent white male ecologists at the 1971 International Conference of the Biology of Whales in Virginia. While every man had his name in the caption, the woman wasn't identified. This led to an internet furor, reported in media sources as prominent as CNN (Williams 2018) and the *New York Times* (Fortin 2018). This woman was Sheila Minor Huff, a Black scientist who had worked with numerous agencies and administrations on wildlife and environmental projects. This sort of erasure denies women and girls of color of important role models. Sheila shares some of her survival strategies below.

Representation in images related to the profession are important for establishing a sense of belonging or "fit." Bal and Sharik (2019a) found that while women and people of color are often featured on the homepages of 11 major natural resource professional societies, there was a paucity of images of women and people of color shown among society leadership or engaging in professional activities in the field. Similar dynamics were shown for the homepages of university natural resources departments (Bal and Sharik 2019b). This lack of meaningful representation of natural resource professionals can influence decisions by women of color of whether to pursue careers in this field.

Sheila Huff shares the lessons of a long career as a pioneering Black woman in natural resources (Box 7.4).

Toxic Stereotypes

In general, the stereotype of women in the workplace is that they are community-minded, passive, and essentially weak, while men are expected to be dominant, assertive, and essentially strong (Rosette et al. 2018). This creates a quandary for women aspiring to advance in the field of wildlife. Working in wildlife at times requires physical strength and the ability to respond quickly and change plans as circumstances require, for example, if someone is injured, a vehicle breaks down, or the path to a field site is obstructed. It has been our experience, however, that when women assert themselves, they are often overruled, ignored, or have their

BOX 7.4 Sheila Minor Huff

I began my career as a biological research technician. This was because I refused to be hired as a secretary. Instead, I told my future supervisor that he would have to offer something else, so he created an entry-level position in the field of animal sciences. I continued to work in research and the management of natural resources and went on to earn a master's degree in biological science. In 1979, I became one of the first female TWS Certified Wildlife Biologists, and in my 35-year career, I worked at various governmental agencies including the Smithsonian Environmental Research Center. I also wrote environmental impact statements for the Federal Energy Regulatory Commission. In 2006, I retired from the US Department of the Interior as a GS-14, Step 10 environmental protection specialist.

My parents were born in the 1920s. They made sure that I grew up aware of the challenges the world would have to offer. And they went on to be beacons of encouragement and light to me. They provided a home of respite and support but never sheltered me. My parents were honest with me about the world and allowed me to experience real world issues. They taught me at an early age that experiences, positive and negative, build you up.

Throughout my career, I developed many strategies for survival and empowerment. I share these below in the hope that they will support and empower women in our field, whether they are just entering the field or near the end of their career.

FIGURE 7.6. Sheila Minor Huff received her BS in biology from American University in 1970 and her MS in biology from George Mason University in 1979. She earned her certification as a wildlife biologist in 1979. She retired in 2006 with 35 years of federal experience in natural resources management. In this photo, she is live-trapping small mammals within the Rhode River watershed in Edgewater, Maryland, circa 1973, while employed at the Smithsonian Institution. Photo courtesy of the subject.

- Develop and practice enjoyable outlets throughout the day.
- Practice positive self-talk.
- Practice deep breathing exercises (possibly try a guided app).
- Take an exercise class to break up the workday or go for a brisk walk alone.
- Call friendly, inspiring, and uplifting people throughout the day.
- Attend faith-based groups during your lunch period.
- Identify and utilize your "safe place" (e.g., a library, museum, under a tree, or just out in nature).
- Listen to your inner protector and do what makes you feel peaceful.
- Identify people who help you feel refreshed and around whom you can feel natural. Choose carefully whom you trust. Conversely, take note of those who appear angry or argumentative.
- Ask and repeat verbatim all questions/comments that sound hurtful or offensive to you.

- Dress for success and your own personal empowerment.
- Avoid "being right."
- Be aware of nonverbal attitudes and communication techniques when interacting with staff.
- Steel yourself to be the only one (at least at first). Remember it is okay to feel sad, anxious, or depressed, but don't let it consume you. Be sure to identify therapy or counseling services that suit your needs.
- Become involved with groups outside of your job. You can tap into your creative side, give back to the community, and take time for your own professional development.
- When it's time to leave, leave with grace.

---

ideas co-opted by the white men in their group. After a certain point, they may stop trying because it seems futile, reinforcing the stereotype of passivity. This creates a no-win situation in which women are less likely to be viewed as competent and dominant enough to lead in the field because they were ignored or faced reprisals for the very behavior necessary to advance.

When seen through an intersectional lens, these sexist dynamics become more complicated. Historical narratives that uphold white systems of oppression have led to persistent preconceptions about women of color. Black and Latina women are sexualized and face stereotypes of being unintelligent, poorly educated, less sensitive, and less dependent than white women and of being more volatile, aggressive, and loud (Ghavami and Peplau 2013; Rosenthal and Lobel 2016; Rosenthal et al. 2020). In contrast, Asian American women are expected to be intelligent, capable, and submissive (Ghavami and Peplau 2013; Choi et al. 2021). Negative experiences and stigmatization related to these and other stereotypes have far-reaching personal and professional effects and can cause profound damage to physical and mental health.

Many women of color face the additional burden of being expected to act as representatives of their race or serve as a "model minority" to dispel negative stereotypes (Apugo 2019). This may prevent them from bringing their authentic selves to the workplace. They may also try to manage their identities in ways that will increase their agency or mitigate the negative consequences of belonging to a marginalized group (Apugo 2019; Buchanan and Settles 2019; Liu et al. 2019). Maintaining these inauthentic identities, often referred to as "acculturation" or "code-switching" can be traumatic—both psychologically harmful and emotionally draining (Liu et al. 2019). It may also result in decreased productivity or attrition that again can negatively reinforce stereotypes of these marginalized groups (Daniels 2018).

BOX 7.5 Princess Hester

I am a product of an impoverished community. If there was a checklist of all the possible obstacles one could face in life, my card would be full, yet because of God's grace here I stand. I became a nature enthusiast at an early age because being outside allowed me to escape. It is my love for the environment and all things in nature that prepared me to make a more significant impact. But I didn't know that I would have a career in science until later in life.

To get here, I engrossed myself in everything science so that I could excel in the requirements of my job. I am proud of my many accomplishments, but the one I am most proud of is a robust youth environmental

FIGURE 7.7. Princess Hester earned a degree in business management from Pepperdine University in 1998 and an MS in public administration from California State University in 2007. She now serves as director of administration at the Riverside County Habitat Conservation Agency (RCHCA) in Riverside, California. Photo courtesy of Christian Lomeli, Promeli Videoworks

education program focused on sharing the importance of preserving our natural resources. The activities are STEAM-based (Science, Technology, Engineering, the Arts and Mathematics), geared toward providing opportunities of inclusion for students in underserved communities.

Every year I attend conferences, symposiums, and meetings of wildlife professionals. In most cases I am the only African American in the room. We work in an environment where we constantly must prove our worth. Even when we are the expert in the room, our opinions are looked upon as invalid or non-intellectual. I must admit that sometimes I really get frustrated when this happens. No, I should tell the truth, I get angry. It's been good to have Sheila's advice to fall back on.

I have written hundreds of reports and papers on different topics and themes, yet this contribution to this chapter was one of the most difficult. My generation was told, "You have a great job, you should be glad you are in the position you are in; keep your head down and stay quiet; let your work speak for you and you will be recognized." I have learned that is the opposite of what you should do.

My advice to those who come after me is to speak out, speak loudly and denounce injustice, join together, create circles of trust and share your experience. When you are in the position and you have the opportunity elevate another woman of trust, work to lift each other and stand together against the racial prejudices that have plagued us. And for those who are on the receiving end of this connection, be responsible, recognize the importance of continuing such a practice and elevate others. Each one should teach one, each one should bring one; everyone brought should evolve.

Here Princess Hester discusses some of the triumphs and challenges she's experienced in her career (Box 7.5).

Barriers to Advancement

There are other ways that women in general, and women of color in particular, who violate white male norms are subject to backlash. Negative performance reviews, incivility, and social penalties all serve as barriers to advancement (Kabat-Farr et al. 2020; Rosette et al. 2018). Because unjust punishments and incivility in the workplace can lead to negative attitudes and decreased productivity, such patently unjust treatment can create a self-reinforcing dynamic, again with the potential to result in attrition and turnover of women of color (Mooijman and Graham 2018) that perpetuates false stereotypes.

Microaggressions

Women of color regularly experience microaggressions in the workplace in general. Microaggressions are subtle, continual, and commonplace snubs, insults, or acts of discrimination that stigmatize women of color or culturally marginalized groups and remind them of their subordinate status in society and their "otherness" (Sue et al. 2007). Microaggressions can also lead to and exacerbate imposter syndrome (McGee et al. 2021).

The cumulative stress associated with microaggressions results in a wide range of negative outcomes for women of color. These include a lower quality of life, lower emotional well-being, increased depression, and worse physical and mental health (Douds and Hout 2020; Williams 2020; Williams et al. 2020). Furthermore, uncertainty around what constitutes a microaggression (Midgette and Mulvey 2021) can increase self-doubt in interpreting interactions with others.

The Consequences of Discrimination

Justifications for diversity and inclusion efforts in natural resources often appear to be instrumental or performative, designed to meet goals related to representation and, as is the case in many organizations (Chang et al. 2019; Glass and Cook 2020), to give the appearance of diversity without creating meaningful change or antiracist policies (Bal and Sharik 2019a, 2019b; Batavia et al. 2020). Yet there are many reasons to do the work required to diversify our field. One of

BOX 7.6 Audrey Peterman

In the age when climate change poses an imminent, existential threat to the future of humanity, it is abominable that any American should feel excluded from being responsible by dint of ephemeral characteristics such as race or gender. In almost a quarter century of environmental advocacy, I have many times been met with the fallacy that "you (Americans of color) have no interest in protecting the environment because you have so many 'survival' issues to deal with—paying your rent and light bill. Besides, so many of you are afraid of nature / the woods." The heaping of scorn with which I've met such comments has taken detractors aback momentarily but seems to have done little to change their minds and attitudes.

FIGURE 7.8. Audrey Peterman is a national award–winning conservationist and a leader in the movement to engage more Americans of color in the enjoyment, care, and protection of our national parks and public lands system. She is the founder of the Diverse Environmental Leaders Speakers Bureau and holds a diploma in mass communications from the University of the West Indies, Jamaica, and a bachelor of arts in communications from St. Thomas Aquinas College, New York. Here she is in Yosemite Valley in 2005, celebrating the 10th anniversary of the hard-copy newsletter she and her husband published, *Pickup & GO!* Photo courtesy of Frank Peterman.

WHY are Americans so invested in maintaining the fallacy that nonwhite people lack the intelligence, soul, and capacity to care for the natural world that is both our birthright and our life support system? To paraphrase the great American writer Upton Sinclair, it is hard to get someone to understand something when his livelihood depends upon him not understanding it. Sadly, in my experience, the majority of the environmental/conservation community in America is structured around maintaining the idea that white Americans have superior sensibilities and must "try, so hard," to engage a sliver of the unwashed masses of nonwhites in this noble cause. This has the effect of concentrating an entire sector and significant sums of money in the hands of a small minority, with a pittance doled out to "others."

In our 2009 book, *Legacy on the Land: A Black Couple Discovers Our National Inheritance and Tells Why Every American Should Care*, my husband Frank and I ask the question, "What will happen when Americans find out that the worst effects of climate change that could have been mitigated with an 'all hands on deck approach' were instead stymied by foolish, self-serving leaders?" That question may be moot now that leaders at the highest levels of our government dismiss science and deliberately advance policies that accelerate climate change. Still, as one who believes in the power of "We, the people," I have to work mightily to avoid focusing on the road not taken, one in which millions more Americans would know about our land treasures in our national parks, forests, and wildlife refuges and be inspired to strive for balance between conservation and consumption.

The greatest irony of all, to me, is how the historic contributions of Americans of color are overlooked or summarily dismissed as little more than a novelty. As a speaker, one of my favorite presentations to give is titled, "Our National Parks: The Glue That Holds America Together." For the past decade I've seen audiences spellbound by the story of Sir Lancelot Jones, scion of an African American family that owned two and a half islands in Biscayne Bay, Florida, since 1897 and whose insistence that "it is good to have somewhere that people can go and leave the hustle and bustle behind and get in the quietness of nature" led to the creation of Biscayne National Monument and eventually the national park that today is the largest marine unit in the National Park System.

As a proud naturalized American of Jamaican origins, I am so grateful that I grew up in a majority-Black country where our motto is *"Out of many, one people."* My formative years were spent outdoors with my friends. The ease, conviviality, and sense of belonging in the world that I thus absorbed have been invaluable in helping me combat the race prejudice that's so prevalent in my adopted country.

I think it's past time for a discussion of the word "leader" to describe people in positions of authority who reinforce our challenges instead of finding ways to alleviate them. To my mind, a leader looks ahead, sees the trends, and works to devise ways to head them off. In the case of the conservation sector, blind, self-serving attitudes and a lack of accountability have placed us on the edge of a precipice, literally teetering on the brink. Our youngest people are having to take the lead abdicated by the fat cats pretending to the title. It's time to stop the unnatural separation of people and unite to meet the challenge of our time. All hands on deck!

---

the most compelling is to create an inclusive profession that reflects the complex identities of a diverse workforce (Cook and Glass 2014). The intersectionality of race, gender, LGBTQ+ identity, mental and physical differences, and abilities that make each person unique brings with it a diversity of perspectives, approaches, and solutions that will fundamentally transform our field (Parker 2019; Batavia et al. 2020). Furthermore, racial justice demands it.

The environmental and conservation challenges we face require the intellect, energy, and professional contributions of everyone with a desire to contribute. We must engage all of our global citizenry as supporters of conservation, which will require input and messaging from people who represent their cultures and interests.

Because conservation efforts have been guided by the interests of capitalist hegemony and white supremacy, climate change, environmental degradation, unsustainable use of natural resources, and loss of biodiversity have a disproportionate impact on communities of color.

Here Audrey Peterman describes ways in which discrimination and exclusion damage the conservation mission we share and the planet we live on (Box 7.6).

How to Be an Ally

Many conservation professionals and professional societies have offered ideas and strategies for including and supporting women of color in the profession. Topics addressed include recruitment and hiring practices, mentorship, and confronting one's own racial attitudes and unconscious bias.

Recruitment

Position announcements and job openings should be worded carefully. Women of color lack confidence in many areas of STEM relative to white men and may not view themselves as qualified when required to perform a subjective self-assessment (Ellis et al. 2016). In general, women are stereotyped as less competent in math and science than men and Black and Latine individuals stereotyped as less competent than whites and Asian Americans (Eaton et al. 2020). These racist and sexist stereotypes can amplify the self-doubt or fear of failure commonly known as "imposter syndrome" (McGee et al. 2021). Therefore, as Gewin et al. (2018) suggest, job announcements that sound "as if only superheroes need apply" should be avoided.

Hiring

Selection committees should be diverse and should be able to show they took concrete actions to attract a diverse applicant pool (Gewin et al. 2018). They should also be careful not to set their sights on "ideal employees." The problem with this is approach is how people in positions of influence imagine the "ideal employee." People tend to "recognize" or categorize people based on their most extreme attributes—those most stereotypically associated with the category. Therefore, the imagined "ideal" candidate will possess those characteristics most stereotypically associated with the position being hired for, and people who do not possess those characteristics will be overlooked (Brown-Iannuzzi et al. 2012). It seems likely that for as long as white male conservation heroes set the default image of a wildlife professional, the "ideal employee" will be a stereotypically masculine white man. This results in a self-reinforcing cycle that perpetuates white men as the stereotypical wildlife biologist and therefore the ideal candidate. When this is the case, women of color are much more likely to be overlooked. Furthermore, stereotypes related to deficits in math and science skills in women of color may also work against applicants in the evaluation stage if superlatives related to these qualities are used.

To counter this bias in how applicants are evaluated and hiring decisions made, Brown-Iannuzzi et al. (2012) recommend having those involved with the hiring process imagine a variety of people they might hire for the position as they write job descriptions and evaluate candidates. They found that when people were instructed to imagine a variety of potential candidates, they were more likely to imagine nonwhite people in those positions. Requiring people who will evaluate job candidates to describe a broad range of characteristics that a candidate could possess can reduce bias in the hiring process.

Safety

In addition to basic safety protocols for fieldwork, supervisors should have plans for addressing discrimination, sexual harassment, and sexual assault both in the field and in other work settings. Supervisors, teams, and interns should educate themselves about risks inherent to the area in which they're working and the people they may encounter (Clancy et al. 2014; Demery and Pipkin 2021). Other best practices for supporting women of color (and others) include having clearly stated rules and consequences for discriminatory behaviors and sexual harassment and assault, egalitarian working conditions, women in leadership positions, and intentional and safe workplaces (Nelson et al. 2017; Demery and Pipkin 2021). Both Demery and Pipkin (2021) and Nelson et al. (2017) provide a comprehensive list of resources and best practices for supervising at-risk interns and employees in the field. Best practices for non-field-based antiracist and antisexist support are also available in the literature (Chaudhary and Berhe 2020).

Representation

Another way to create a more welcoming field for women of color is to produce, promote, and support media featuring them actively engaging in wildlife careers rather than tokenizing them or relegating them to the sidelines (Bal and Sharik 2019a, 2019b). Contemporary, racially diverse ecologists who maintain an active social media presence and work to increase representation in our field are proliferating. These women serve as powerful role models for girls and women of color aspiring to a career in wildlife.

BOX 7.7 Charisa Morris

My comfort in the workplace as a Black woman working in wildlife is directly correlated with the behavior of my team and supervisory chain. A strong support team and supervisor can help me be resilient when facing barriers and challenges. The absence of this support can be devastating.

From the start, well-intentioned decisions made to increase diversity in the workplace actually put me at a disadvantage. I was a strong graduate of a high-ranked natural resources college, but I was brought on in a cohort of "diversity hires." My recruitment into the agency was clearly associated with my race, which raised intense skepticism about my qualifications. Though I was well qualified and had a fervent desire to contribute, I was perceived as a burdensome obligation rather than an asset.

I resolved to overcome preconceptions about my aptitude by distributing my transcripts and resume to coworkers and campaigning to be included on teams. While my coworkers became more supportive after this active engagement, my supervisory chain remained antagonistic. I was engaged in a constant battle to be treated with dignity and respect. I was told I could not do fieldwork because I was a young woman and might get injured or be attacked.

FIGURE 7.9. Charisa Morris earned a BS in forestry and wildlife in 1999 and an MS in fisheries and wildlife in 2006 from Virginia Tech. She has worked as a wildlife biologist for the US Fish and Wildlife Service since 1998 and has served as an endangered species biologist (in the Northeast and nationally). Since then, her assignments have all been national—she has served as chief of migratory bird conservation, chief of staff, deputy of the Science Applications program, and is currently the US Fish and Wildlife Service national science advisor. Here she is shown trapping a federally endangered Delmarva fox squirrel (*Sciurus niger cinereus*) in 2006, prior to this species being recovered and delisted in 2015. Photo courtesy of the subject.

I wanted to play an active role as a wildlife professional but was instead left on the sidelines by a supervisory chain that valued the optics of diversity but didn't engage or support candidates after the hiring was complete.

The effect this disenfranchisement had on me was profound. I became despondent. The confidence, capability, and commitment to conservation my university professors and mentors had cultivated within me withered. I even wrote a letter to the Department of Interior describing the draining effect the experience had on my soul in an attempt to end the chain of mindless recruitment. I found myself unconsciously adopting the discriminatory behaviors I was experiencing, valuing input less if it originated from a woman or minority. When I realized I had this reaction, I recall being struck by how powerfully those negative messages had infiltrated what had previously been my rather strong-willed mind.

Luckily, the management in my program eventually changed, and our new supervisor was a woman of color who supported and encouraged innovation and development of her team members. With her leadership, trust, and support, I joyfully excelled in the field, gained regional recognition for my work, and secured a position at our agency's headquarters. Yet I continued to note and experience challenges and barriers in this profession associated with my race and gender.

Even on the healthiest teams, I feel pressure to assimilate. I would pretend to know the music my colleagues listened to or would simply stay silent as they talked about their shared hobbies. I would quietly accept that I didn't truly belong and was a cultural interloper being allowed to live out my career dreams. I noticed many of my female role models often had gender-neutral clothing and styles; to be seen as professionals, they had to take away any part of their identity that could distract the men at the table from their message. During the times I worked under management that was undeniably racist, sexist, or both, I too would try to remove or hide any characteristic of myself that might trigger their behavior. I would eventually suppress so much that there wouldn't be much left to give. The effect of those situations on both my physical health and well-being were immediate. After joining one particular team with a notably overt discriminatory dynamic, my health declined drastically within the first six weeks. I developed severe allergies, a chronic cough, started losing my hair, and eventually developed a heart condition within the first six months. When I moved on to my next position several years later, my health challenges resolved within three months.

The single most important factor affecting my comfort as a Black woman in wildlife is the support of my supervisor and team. Although a great team does not mean I will always be comfortable being different, when they accept and celebrate me as I am, I can, too. However, a team that embraces discrimination and harassment can completely crush my sense of self, my ability to be a contributing employee, and my health. A truly strong, understanding, and welcoming support team and supervisor makes all the difference in the world.

Mentorship

Perhaps the most valuable way to be an ally for women of color in wildlife is through mentorship. Mentoring is key to identity development in underrepresented populations. While women of color benefit from mentors who are similar to them, shared values and strong support can be as or more important as simply sharing race or gender (Atkins et al. 2020). Women of color value personal closeness to their mentors, but they also want them to provide the kind of accountability that will help them navigate the challenges they face and that support them in their aspirations (Allen et al. 2005; Atkins et al. 2020). Sensitivity to race-specific stereotypes faced by women of color (e.g., Asian American "model minority," the "angry Black woman") is important in helping them navigate challenges unique to their race (Chin and Kameoka 2019). Mentoring best practices are too numerous in the literature and online to list here.

Women of color with mentors receive substantial benefits both for their careers, in terms of salary, promotions, and job satisfaction, and for their well-being, in terms of role models, acceptance, confirmation, counseling, and friendship (Allen et al. 2004; Hund et al. 2018). Benefits also accrue to mentors (Allen et al. 2006).

Charisa Morris relates how she benefited from professional mentorship throughout her career (Box 7.7).

Conclusion

Creating a welcoming and supportive environment for women of color in wildlife will be challenging and will take time. We hope the stories we've provided will open people's eyes to some of the intersectional racism and sexism women of color encounter in wildlife. By validating the existence of race- and sex-based bias and discrimination in our field, we hope to reassure the women of color who've experienced them that they are not alone.

We must all continue to educate ourselves about the history of white supremacy and racism with respect to the environmental movement as a whole, the conservation profession in general, and in the institutions and organizations we interact with. Too often it falls upon women of color to educate and advise people on ways to welcome and support them, despite lacking the power and stature to create change. We hope that all our readers will reflect upon the ways each of you, regardless of race, can act as allies to help dismantle racism, discrimination, injustice, and the supremacy of white norms and values in our field.

Doing this work is difficult and requires open and honest dialog. Glenn Singleton's Courageous Conversations model (2022) stresses the need to stay engaged, speak your truth, experience discomfort, and expect and accept non-closure. There are no quick fixes to the systemic racism and sexism that permeate our society and profession, and we will all make mistakes at times. It is crucial that we persevere with our commitment to listen, learn, and take action. By holding love, compassion, conviction, and courage in our hearts, together we will make a difference.

BOX 7.8 Modeline Celestin

The town of Belle Glade, Florida, is my hometown. Our acclaimed motto is "Her soil is her fortune." Yet the fortune does not trickle down to the working-class residents, many of whom work in the agriculture sector. It is also where my parents immigrated 30+ years ago from Haiti, with nothing but fortitude in their pockets. At my core, I identify as a Black Haitian American woman, an identity that is integral to the cultivation of my self-worth. As a young child tethered to the land, I inherited the intrinsic values my parents harvested from their manual labor as migrant farmworkers. We fostered an ecological ethic, only to have it desecrated by land fragmentation.

My upbringing fueled my intense passion for biodiversity, particularly community and ecosystem responses to human-induced environmental stressors. Ecology is a multifaceted discipline that has nurtured my adaptive thinking and guided my transformative learning experiences. From mountains to islands, I have worked multiple field seasons in remote rural conditions with extreme weather that were both arduous, isolating, and labor intensive. Those experiences of conducting field-based research were physically and mentally challenging but have also shaped me into becoming the resilient ecologist I am today. I harness the principles and skills I have experienced from working in wildlife ecology and conservation into my everyday life. I would not be me without ecology.

A major motivator for pursuing a career within the wildlife field for me is to leverage human capital to dismantle systemic barriers and eradicate the institutionalized oppression that creates intergenerational trauma within disadvantaged communities. Ultimately, I know that I will aid in the creation of a more just and equitable conservation field and a more just and equitable future. I look forward to building upon the legacies of pioneers like the ones in this chapter.

FIGURE 7.10.
Modeline Celestin earned a BA in sustainability studies with a minor in wildlife ecology and conservation at the University of Florida in 2019. She is currently working toward a doctoral degree in ecology and evolutionary biology at Rice University. Here she is conducting amphibian surveys in Idaho in 2017.
Photo courtesy of Daniel Choi.

As we draw our chapter to a close, Modeline Celestin shares her vision for the future as part of the next generation of women of color in wildlife (Box 7.8).

Discussion Questions

1. Who do you picture when you think of a wildlife professional? Why? What are the most important attributes of a wildlife professional and how can those be elevated in hiring and selection decisions?
2. Does the culture of your institution or organization feel equitable to you? What would be needed to make it better?
3. How can *you* help build bridges and advocate for people of color in our field, both personally and institutionally?

Additional Resources

Field Safety:

Clancy, K. B. H., R. G. Nelson, J. N. Rutherford, and K. Hinde. 2014. "Survey of Academic Field Experiences (SAFE): Trainees Report Harassment and Assault." *PloS One* 9(7): 9. https://journals.plos.org /plosone/article?id=10.1371/journal.pone.0102172.

Demery, A. J. C., and M. A. Pipkin. 2021. "Safe Fieldwork Strategies for At-Risk Individuals, Their Supervisors and Institutions." *Nature Ecology & Evolution* 5(1): 5–9. www.nature.com/articles/s41559-020-01328-5.

Employers:

https://diversityinconservationjobs.org/resources-for-employers/

Job-seekers:

https://diversityinconservationjobs.org/resources-job-seekers/

Inclusive Mentoring:

https://www.mentoring.org/resource/national-mentoring-resource-center/

Activity: Identity in the Wildlife Profession

Learning Objectives: Participants will engage in dialogue to learn about the personal experience and identity of other groups in the wildlife profession.

Time: 1.5 hr.

Description of Activity

Materials needed:

- Signs or table tents for each group/table
- Large paper or multiple sheets of paper
- Markers or pens

This activity would work best in a room with small tables that would facilitate small group discussions.

Process:

Activity preparation: On each table, create a sign or table tent with one of the following identities (these are suggestions—more or different identities can be included):

- Native American
- African American
- Latine
- European American
- LGBTQ+
- Women of color
- White men
- Other

Part 1 (20 min.): Open the activity by describing to the participants the basics of dialogue: For successful dialogue, everyone should be conscientious both of their words and of listening to others. Dialogue is different from a debate in that the goal is to discover common ground among participants, explore a diversity of perspectives, listen for strengths rather than flaws in other people's statements, and to work as a group. During this activity, it is fine to focus on listening to other's positions and to gather new ideas (i.e., being quiet is okay).

Next, encourage everyone to select one of the groups for discussion. Most people belong to more than one group, but participants should select a group that they would like to focus on today. If participants would prefer to join a group that is not included, they can join the "Other" group and rename it.

Once in their groups, instruct participants to begin a dialogue with their group members about what it means to be in that group as a wildlife professional. Ask participants to spend the next 15 minutes addressing the following questions and writing notes, key words, or even drawing images on the paper on their tables:

1. What does membership in this group mean to you?
2. How does being in this group affect you as wildlife professional?
3. How does your group affect the wildlife field?

It might be easiest to print these questions on a sheet of paper or have them on a PowerPoint slide for participants to reference.

Part 2 (40 min.): Next ask that the larger group rotates together among the tables and have one or two representatives from each table spend 4–5 minutes retelling the general stories from the group and explaining notes and words written by the group. While the representatives are speaking, the remaining participants should be asked to sit or stand quietly and focus on listening; there are no questions or comments solicited during this time. Facilitators should not summarize at the end of each visit to a table—leave all speaking to the group representatives. Once everyone has visited each table and group, acknowledge everyone for sharing and for listening. Then have the larger group congregate again for a debrief.

Part 3: Debriefing and Discussion Questions (15+ min.)

1. How did you feel when thinking about your group membership?
2. How did you feel when hearing about others' group membership?
3. What did you learn?
4. What was most surprising about the experience?
5. What are potential outcomes for the larger group?
6. How would you use the information you learned or your experience from this activity in your life as a wildlife professional, student, or citizen?

Teaching Notes

This is a risky activity for some people. Facilitators should be aware and recognize that this activity can make some people nervous or uncomfortable; therefore, modeling a general attitude of being open and calm will support participants.

References and Resources

This activity was modified from Bill Withers and Keami D. Lewis, "Diversity Dialogue," in *The Conflict and Communication Activity Book: 30 High-Impact Training Exercises for Adult Learners* (New York: AMACOM American Management Association, 2003).

Literature Cited

Allen, T. D., R. Day, and E. Lentz. 2005. "The Role of Interpersonal Comfort in Mentoring Relationships." *Journal of Career Development* 31(3): 155–69. doi: 10.1177 /0894845305031003001.

Allen, T. D., L. T. Eby, M. L. Poteet, E. Lentz, and L. Lima. 2004. "Career Benefits Associated with Mentoring for Proteges: A Meta-analysis." *Journal of Applied Psychology* 89(1): 127–36. doi: 10.1037/0021-9010.89.1.127.

Allen, T. D., E. Lentz, and R. Day. 2006. "Career Success Outcomes Associated with Mentoring Others: A Comparison of Mentors and Nonmentors." *Journal of Career Development* 32(3): 272–85. doi: 10.1177/0894845305282942.

Apugo, D. 2019. "A Hidden Culture of Coping: Insights on African American Women's Existence in Predominately White Institutions." *Multicultural Perspectives* 21(1): 53–62. doi: 10.1080/15210960.2019.1573067.

Atkins, K., B. M. Dougan, M. S. Dromgold-Sermen, H. Potter, V. Sathy, and A. T. Panter. 2020. "'Looking at Myself in the Future': How Mentoring Shapes Scientific Identity for STEM Students from Underrepresented Groups." *International Journal of Stem Education* 7(1): 15. doi: 10.1186/s40594-020-00242-3.

Bal, T. L., M. D. Rouleau, T. L. Sharik, and A. M. Wellstead. 2020. "Enrollment Decision-Making by Students in Forestry and Related Natural Resource Degree Programmes Globally." *International Forestry Review* 22(3): 287–305. doi: 10.1505/146554820830405627.

Bal, T. L., and A. L. Sharik. 2019a. "Image Content Analysis of US Natural Resources-Related Professional Society Websites with Respect to Gender and Racial/Ethnic Diversity." *Journal of Forestry* 117(4): 360–64. doi: 10.1093/jofore/fvz023.

Bal, T. L., and T. L. Sharik. 2019b. "Web Content Analysis of University Forestry and Related Natural Resources Landing Webpages in the United States in Relation to Student and Faculty Diversity." *Journal of Forestry* 117(4): 379–97. doi: 10.1093/jofore/fvz024.

Balcarczyk, K. L., D. Smaldone, S. W. Selin, C. D. Pierskalla, and K. Maumbe. 2015. "Barriers and Supports to Entering a Natural Resource Career: Perspectives of Culturally Diverse Recent Hires." *Journal of Forestry* 113(2): 231–39. doi: 10.5849/jof.13-105.

Batavia, C., B. E. Penaluna, T. R. Lemberger, and M. P. Nelson. 2020. "Considering the Case for Diversity in Natural Resources." *Bioscience* 70(8): 708–18. doi: 10.1093/biosci/biaa068.

Blosser, E. 2020. "An Examination of Black Women's Experiences in Undergraduate Engineering on a Primarily White Campus: Considering Institutional Strategies for Change." *Journal of Engineering Education* 109(1): 52–71. doi: 10.1002/jee.20304.

Bodin, M. 2020. "Call for Change: Why Unpaid Work Experience Must Stop." *Nature* 586 (7829): 463–64.

Brown-Iannuzzi, J. L., B. K. Payne, and S. Trawalter. 2012. "Narrow Imaginations: How Imagining Ideal Employees Can Increase Racial Bias." *Group Processes & Intergroup Relations* 16(6): 661–70. doi: 10.1177/1368430212467477.

Buchanan, N. T., and I. H. Settles. 2019. "Managing (In)Visibility and Hypervisibility in the Workplace." *Journal of Vocational Behavior* 113:1–5. doi: 10.1016/j.jvb.2018.11.001.

Cassino, D., and Y. Besen-Cassino. 2019. "Race, Threat and Workplace Sexual Harassment: The Dynamics of Harassment in the United States, 1997–2016." *Gender Work and Organization* 26(9): 1221–40. doi: 10.1111/gwao.12394.

Chang, E. H., K. L. Milkman, D. Chugh, and M. Akinola. 2019. "Diversity Thresholds: How Social Norms, Visibility, and Scrutiny Relate to Group Composition." *Academy of Management Journal* 62(1): 144–71. doi: 10.5465/amj.2017.0440.

Chaudhary, V. B., and A. A. Berhe. 2020. "Ten Simple Rules for Building an Antiracist Lab." *PLoS Computational Biology* 16(10). doi: 10.1371/journal.pcbi.1008210.

Chin, D., and V. A. Kameoka. 2019. "Mentoring Asian American Scholars: Stereotypes and Cultural Values." *American Journal of Orthopsychiatry* 89(3): 337–42. doi: 10.1037/ort0000411.

Choi, S., S. Z. Weng, H. Park, J. Lewis, S. A. Harwood, R. Mendenhall, and M. B. Huntt. 2021. "Sense of Belonging, Racial Microaggressions, and Depressive Symptoms among Students of Asian Descent in the United States." *Smith College Studies in Social Work*: 27. doi: 10.1080/00377317.2021.1882922.

Clancy, K. B. H., R. G. Nelson, J. N. Rutherford, and K. Hinde. 2014. "Survey of Academic Field Experiences (SAFE): Trainees Report Harassment and Assault." *PLoS One* 9(7): 9. doi: 10.1371/journal.pone.0102172.

College Factual. 2021. "2021 Best Colleges for Wildlife Management." https://www
.collegefactual.com/majors/natural-resources-conservation/wildlife-management
/rankings/top-ranked/.

Cook, A., and C. M. Glass. 2014. "Analyzing Promotions of Racial/Ethnic Minority CEOs."
Journal of Managerial Psychology 29(4): 440–54. doi: 10.1108/jmp-02-2012-0066.

Crenshaw, K. W. 1994. "Mapping the Margins: Intersectionality, Identity Politics, and Violence
against Women of Color." In *The Private Nature of Public Violence*, edited by Martha
Albertson Fineman and Rixanne Mykitiuk, 93–118. New York: Routledge.

Daniels, J. R. 2018. " 'There's No Way This Isn't Racist': White Women Teachers and the
Raciolinguistic Ideologies of Teaching Code-Switching." *Journal of Linguistic Anthropology* 28(2): 156–74. doi: 10.1111/jola.12186.

de Brey, C., L. Musu, J. McFarland, S. Wilkinson-Flicker, M. Diliberti, A. Zhang, C.
Branstetter, and X. Wang. 2019. *Status and Trends of Education of Racial and Ethnic
Groups 2018* (NCES 2019-038). US Department of Education. Washington, DC:
National Center for Education Statistics.

Demery, A. J. C., and M. A. Pipkin. 2021. "Safe Fieldwork Strategies for At-Risk Individuals,
Their Supervisors and Institutions." *Nature Ecology & Evolution* 5(1): 5–9. doi: 10.1038/
s41559-020-01328-5.

Dickens, D. D., V. Y. Womack, and T. Dimes. 2019. "Managing Hypervisibility: An Exploration of Theory and Research on Identity Shifting Strategies in the Workplace among
Black Women." *Journal of Vocational Behavior* 113:153–63. doi: 10.1016/j.jvb.2018.10.008.

Dokes, T. J., G. J. Roloff, K. F. Millenbah, B. H. K. Wolter, and R. A. Montgomery. 2020.
"Natural Resource Undergraduate Students in the New Millennium." *Wildlife Society
Bulletin*. October 8. doi: 10.1002/wsb.1128.

Douds, K. W., and M. Hout. 2020. "Microaggressions in the United States." *Sociological
Science* 7:528–43. doi: 10.15195/v7.a22.

Eaton, A. A., J. F. Saunders, R. K. Jacobson, and K. West. 2020. "How Gender and Race
Stereotypes Impact the Advancement of Scholars in STEM: Professors' Biased
Evaluations of Physics and Biology Post-doctoral Candidates." *Sex Roles* 82(3/4):
127–41. doi: 10.1007/s11199-019-01052-w.

Ellis, J., B. K. Fosdick, and C. Rasmussen. 2016. "Women 1.5 Times More Likely to Leave
STEM Pipeline after Calculus Compared to Men: Lack of Mathematical Confidence a
Potential Culprit." *PLoS One* 11(7). doi: 10.1371/journal.pone.0157447.

Fortin, J. 2018. "She Was the Only Woman in a Photo of 38 Scientists, and Now She's Been
Identified." *New York Times*, March 19. https://www.nytimes.com/2018/03/19/us
/twitter-mystery-photo.html.

Fournier, A. M. V., and A. L. Bond. 2015. "Volunteer Field Technicians Are Bad for Wildlife
Ecology." *Wildlife Society Bulletin* 39(4): 819–21. doi: 10.1002/wsb.603.

Gewin, V., B. Gaensler, D. Taylor, I. Uchegbu, E. V. Scarpetta, L. Michel, and G. Valentine.
2018. "It Takes More Than a Vow." *Nature* 558(7708): 149–51.

Ghavami, N., and L. A. Peplau. 2013. "An Intersectional Analysis of Gender and Ethnic
Stereotypes: Testing Three Hypotheses." *Psychology of Women Quarterly* 37(1): 113–27.
doi: 10.1177/0361684312464203.

Glass, C., and A. Cook. 2020. "Performative Contortions: How White Women and People of
Colour Navigate Elite Leadership Roles." *Gender Work and Organization* 27(6):
1232–52. doi: 10.1111/gwao.12463.

Haynes, N. A., and S. Jacobson. 2015. "Barriers and Perceptions of Natural Resource
Careers by Minority Students." *Journal of Environmental Education* 46(3): 166–82. doi:
10.1080/00958964.2015.1011595.

Haynes, N., S. K. Jacobson, and D. M. Wald. 2015. "A Life-Cycle Analysis of Minority
Underrepresentation in Natural Resource Fields." *Wildlife Society Bulletin* 39(2):
228–38. doi: 10.1002/wsb.525.

Hund, A. K., A. C. Churchill, A. M. Faist, C. A. Havrilla, S. M. Stowell, H. F. McCreery, J. Ng, C. A. Pinzone, and E. S. C. Scordato. 2018. "Transforming Mentorship in STEM by Training Scientists to Be Better Leaders." *Ecology and Evolution* 8(20): 9962–74. doi: 10.1002/ece3.4527.

Kabat-Farr, D., I. H. Settles, and L. M. Cortina. 2020. "Selective Incivility: An Insidious Form of Discrimination in Organizations." *Equality Diversity and Inclusion* 39(3): 253–60. doi: 10.1108/edi-09-2019-0239.

Karp, B. 2019. "Theme 4: What Does Diversity Require of Us?" *Fisheries* 44(8): 380–81. doi: 10.1002/fsh.10309.

Liu, W. M., R. Z. Liu, Y. L. Garrison, J. Y. C. Kim, L. Chan, Y. C. S. Ho, and C. W. Yeung. 2019. "Racial Trauma, Microaggressions, and Becoming Racially Innocuous: The Role of Acculturation and White Supremacist Ideology." *American Psychologist* 74(1): 143–55. doi: 10.1037/amp0000368.

McGee, E. O., P. K. Botchway, D. E. Naphan-Kingery, A. J. Brockman, S. Houston, and D. T. White. 2021. "Racism Camouflaged as Impostorism and the Impact on Black STEM Doctoral Students." *Race Ethnicity and Education* 21. doi: 10.1080/13613324.2021.1924137.

Miao, R. E., and N. L. Cagle. 2020. "The Role of Gender, Race, and Ethnicity in Environmental Identity Development in Undergraduate Student Narratives." *Environmental Education Research* 26(2): 171–88. doi: 10.1080/13504622.2020.1717449.

Midgette, A. J., and K. L. Mulvey. 2021. "Unpacking Young Adults' Experiences of Race- and Gender-Based Microaggressions." *Journal of Social and Personal Relationships*. doi: 10.1177/0265407521988947.

Miriti, M. N. 2020. "The Elephant in the Room: Race and STEM Diversity." *Bioscience* 70(3): 237–42. doi: 10.1093/biosci/biz167.

Mooijman, M., and J. Graham. 2018. "Unjust Punishment in Organizations." In *Research in Organizational Behavior: An Annual Series of Analytical Essays and Critical Reviews*, vol. 38, edited by A. P. Brief and B. M. Staw, 95–106. Amsterdam: Elsevier.

Morales, N., and S. Jacobson. 2020. "Student Perceptions of Environmental and Conservation (EC) Careers: Exploring Perspectives of Diverse University Students." *Environmental Management* 66(3): 450–59. doi: 10.1007/s00267-020-01304-6.

Moreno, B., C. Crandall, and M. C. Monroe. 2020. "Factors Influencing Minority and Urban Students' Interest in Natural Resources." *Journal of Forestry* 118(4): 373–84. doi: 10.1093/jofore/fvaa008.

Nelson, R. G., J. N. Rutherford, K. Hinde, and K. B. H. Clancy. 2017. "Signaling Safety: Characterizing Fieldwork Experiences and Their Implications for Career Trajectories." *American Anthropologist* 119(4): 710–22. doi: 10.1111/aman.12929.

Parker, M. 2019. "Theme 1: Why Diversity?" *Fisheries* 44(8): 357–58. doi: 10.1002/fsh.10322.

Prevot, A. C., S. Clayton, and R. Mathevet. 2018. "The Relationship of Childhood Upbringing and University Degree Program to Environmental Identity: Experience in Nature Matters." *Environmental Education Research* 24(2): 263–79. doi: 10.1080/13504622.2016.1249456.

Puritty, C., L. R. Strickland, E. Alia, B. Blonder, E. Klein, M. T. Kohl, E. McGee, M. Quintana, R. E. Ridley, B. Tellman, and L. R. Gerber. 2017. "Without Inclusion, Diversity Initiatives May Not Be Enough." *Science* 357(6356): 1101–2. doi: 10.1126/science.aai9054.

Rosenthal, L., and M. Lobel. 2016. "Stereotypes of Black American Women Related to Sexuality and Motherhood." *Psychology of Women Quarterly* 40(3): 414–27. doi: 10.1177/0361684315627459.

Rosenthal, L., and M. Lobel. 2020. "Gendered Racism and the Sexual and Reproductive Health of Black and Latina Women." *Ethnicity & Health* 25(3): 367–92. doi: 10.1080/13557858.2018.1439896.

Rosenthal, L., N. M. Overstreet, A. Khukhlovich, B. E. Brown, C. J. Godfrey, and T. Albritton. 2020. "Content of, Sources of, and Responses to Sexual Stereotypes of Black and Latinx Women and Men in the United States: A Qualitative Intersectional Exploration." *Journal of Social Issues* 76(4): 921–48. doi: 10.1111/josi.12411.

Rosette, A. S., R. P. de Leon, C. Z. Koval, and D. A. Harrison. 2018. "Intersectionality: Connecting Experiences of Gender with Race at Work." *Research in Organizational Behavior* 38:1–22.

Singleton, G. E. *Courageous Conversations about Race: A Field Guide for Achieving Equity in Schools and Beyond.* 3rd ed. Thousand Oaks, CA: Corwin Press, 2022.

Sue, D. W., C. M. Capodilupo, G. C. Torino, J. M. Bucceri, A. M. B. Holder, K. L. Nadal, and M. Esquilin. 2007. "Racial Microaggressions in Everyday Life: Implications for Clinical Practice." *American Psychologist* 62(4): 271–86. doi: 10.1037/0003-066x.62.4.271.

Szczytko, R., K. T. Stevenson, M. N. Peterson, and H. Bondell. 2020. "How Combinations of Recreational Activities Predict Connection to Nature among Youth." *Journal of Environmental Education* 51(6): 462–76. doi: 10.1080/00958964.2020.1787313.

Taylor, D. E. 2014. *The State of Diversity in Environmental Organizations.* Ann Arbor: University of Michigan, School of Natural Resources and Environment.

Taylor, D. E. 2018. "Racial and Ethnic Differences in the Students' Readiness, Identity, Perceptions of Institutional Diversity, and Desire to Join the Environmental Workforce." *Journal of Environmental Studies and Sciences* 8(2): 152–68. doi: 10.1007/s13412-017-0447-4.

US Department of Agriculture (USDA). 2017. *Survey of the Forest Service Region 5 Regarding Sexual Harassment.* Office of Inspector General. https://www.oversight.gov/report/USDA OIG/Survey-Forest-Service-Region-5-Regarding-Sexual-Harassment.

US Office of Personnel Management (OPM). N.d. "Classification & Qualifications: Wildlife Biology Series, 0486." Accessed August 30, 2021. www.opm.gov/policy-data-oversight /classification-qualifications/general-schedule-qualification-standards/0400/wildlife -biology-series-0486/.

Vercammen, A., C. Park, R. Goddard, J. Lyons-White, and A. Knight. 2020. "A Reflection on the Fair Use of Unpaid Work in Conservation." *Conservation & Society* 18(4): 399–404. doi: 10.4103/cs.cs_19_163.

Williams, D. 2018. "The Identity of the Lone Woman Scientist in This 1971 Photo Was a Mystery: Then Twitter Cracked the Case." CNN, March 28. www.cnn.com/2018/03/20 /health/woman-scientist-1971-twitter-mystery-trnd/index.html.

Williams, M. T. 2020. "Psychology Cannot Afford to Ignore the Many Harms Caused by Microaggressions." *Perspectives on Psychological Science* 15(1): 38–43. doi: 10.1177 /1745691619893362.

Williams, M. T., J. W. Kanter, A. Pena, T. H. W. Ching, and L. Oshin. 2020. "Reducing Microaggressions and Promoting Interracial Connection: The Racial Harmony Workshop." *Journal of Contextual Behavioral Science* 16:153–61. doi: 10.1016/j.jcbs.2020.04.008.

CHAPTER

8

Katherine M. O'Donnell, Brenda McComb, Travis L. Booms, Claire Crow, and Alexander Novarro

Breaking the Binary
LGBTQ+ Wildlifer Perspectives

If you dare to let go, if you dare to stop thinking about what box you fit into and just start being who you are and letting yourself want what you want, then I think you'll wake up one day and find yourself sitting in the right box, which might not be a box at all.
                                                                          —*Riese Bernard*

Breaking the Binary

At an early age, we learn how to use basic characteristics to categorize animals. Birds have feathers, reptiles have scales, mammals have fur. When we learn about a new animal, we feel confident that we know what box it belongs in. A Swedish botanist with a childhood fondness for flowers, Carl Linnaeus, likewise recognized categories into which flora and fauna fell; in 1753, he created the neat and orderly system we scientists still use to classify species. As we grow, we learn about exceptions to these rules. Pangolins have scales, but they are not reptiles. Dolphins do not have fur, but they are still mammals. We begin to develop a more nuanced view of biology. The lines between categories become blurry, and we realize that the "rules" are much more complicated than we thought as children. As we mature academically and professionally, we see that the neat little Linnaean boxes we learned in Biology 101 are simplified versions of reality. Our modern nomenclature (both cultural and scientific) hides the infinite complexity present in the natural world adeptly described by J. B. S. Haldane (a pioneer in modern population genetics) as "*not only queerer than we suppose, but queerer than we can suppose*" (Haldane 1927).

Many human characteristics, including gender and sexuality, are no different. Though traditionally treated as simple binaries (either/or), aspects of gender and sexuality exist as spectrums with much more complexity than our cultural norms suggest. The tendency to treat these concepts as binary is a societal construct that is oversimplified and exclusionary. If there are only two boxes but neither one feels true, then the message is clear: you do not belong. Each aspect of sexuality and gender occurs on a continuum, and where a person falls on one spectrum (e.g., sexual orientation) does not dictate where they land on the others (e.g., gender identity). For many people, where they are today may not match where they are in the future—these sexuality and gender characteristics can be fluid, or it may take time for someone to find where they fit. In the United States and much of Western culture, for most people whose sexual orientation or gender identity does not neatly fit in a binary box (Figure 8.1), coming to terms with our identities is a process that takes years of often painful self-reflection and doubt.

The list of terms for describing gender and sexual identities continues to expand. With that growth, it has become common to see the familiar initialism LGBT with additional letters, including Q (queer, questioning), I (intersex), A (asexual, agender), and P (pansexual, polyamorous). In this chapter, we use LGBTQ+ (lesbian, gay, bisexual, transgender, queer, and related identities) as an umbrella term for those who identify as gender and sexual minorities (i.e., non-cisgender, non-heterosexual). Though the term LGBTQ+ is widely used, we acknowledge that it is not perfect, as many non-cisgender and non-heterosexual people do not identify as L, G, B, T, or Q (Box 8.1). Furthermore, two people may identify with the same term but define it differently, or two people may have similar sexual orientations or gender identities but use different terms to describe themselves. Terminology also changes over time, often

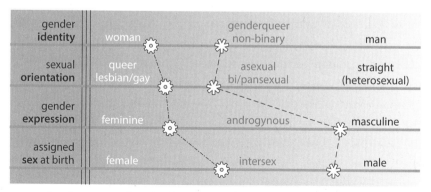

FIGURE 8.1. Spectrums of gender identity, sexual orientation, gender expression, and biological sex. Two sets of dashed lines illustrate possible intersections of gender and sexual identities, but infinite combinations exist.

BOX 8.1 **LGBTQ+ Terminology**

- -

We concede that these are simplified definitions and that this list is far from comprehensive. We also acknowledge that some may define these terms differently—even two people who identify with the same word. It is best to allow someone to share their identity and how they define it, if and when they feel comfortable doing so.

GENERAL

Queer: an umbrella term often used to refer to anyone who is not hetero-sexual or cisgender (a gender identity that conforms to the sex assigned at birth); increasingly used in place of "LGBTQ" (e.g., "the queer commu-nity" instead of "the LGBTQ community")

SOGI: an acronym for "sexual orientation and gender identity"; terms like "SOGI minorities" are sometimes used instead of "LGBTQ" or similar initialisms

SEXUALITY

Sexual orientation: a deep-rooted emotional, romantic, or sexual attraction to other people

Asexual: lack of sexual attraction/desire for other people (though some may feel romantic attractions)

Bisexual: emotional, romantic, or sexual attraction to more than one sex or gender

Gay: emotional, romantic, or sexual attraction to people of the same gender

Lesbian: woman who is emotionally, romantically, or sexually attracted to other women

Pansexual: someone who is emotionally, romantically, or sexually attracted to people regardless of gender

GENDER

Gender identity: a person's self-identification as man, woman, a blend of both (e.g., **genderqueer**, **nonbinary**), or neither (**agender**)

Gender expression: presentation of one's gender identity through behavior, clothing, hairstyle, or voice; categorized on masculine-to-feminine spectrum

Gender nonconforming: broad term for those whose behavior does not fit conventional gender expectations or whose gender expression does not fit neatly into a traditional archetype (related terms: genderqueer, nonbinary, **gender fluid**)

Androgynous: identifying or expressing as neither distinctly masculine nor feminine

Genderqueer, trans*, transgender: umbrella terms used to describe a person whose gender identity or expression does not align with their sex assigned at birth (antonym: **cisgender**)

Gender transition: process of aligning gender expression (appearance) with gender identity (internal); may include social transitions (e.g., changing name, pronouns, clothing), physical transitions (e.g., surgical, hormonal interventions), or both

SEX

Sex assigned at birth: also referred to as "biological sex"; involves hormones, chromosomes, and anatomy designated at birth; often only accounts for external anatomy; categorized on male-to-female spectrum

Intersex: a term used to describe a wide range of natural bodily variations in hormones, chromosomes, and anatomy; if apparent at birth, often "treated" by doctors to align all sex markers with either male or female sex, but these medically unnecessary procedures are becoming rarer as awareness increases (Note: "intersex" has replaced the term "hermaphrodite," which is considered insulting)

---

resulting in generational differences in labels. For example, some members of older generations consider the word "queer" offensive, as it was once a commonly used insult. However, many people, especially in younger generations, have reclaimed the word and use it as an umbrella term encompassing many gender and sexual identities.

Reflect on the connotations associated with some labels that wildlifers commonly use to describe other species—"endangered," "pest," "game," "predator"— each word is laden with implicit cultural values. People label other people, too, and those labels can have important personal and political implications. Labels can be used to marginalize or ostracize others, but they can also enable people to find others with similar identities and establish supportive communities. These groups can be incredibly important; the feeling of being seen, understood, and accepted by others is empowering. Labels also foster visibility, which contributes to increased societal acceptance and a broadening definition of the norm.

Understanding LGBTQ+ Identities within the Framework of Difference, Power, and Discrimination

Members of LGBTQ+ communities have long hidden their identities to avoid harassment and discrimination (e.g., James et al. 2016) or have chosen to live and work in locations where supportive communities exist. Still, LGBT* US federal agency employees in science, technology, engineering and math (STEM) fields have reported more negative workplace experiences than their non-LGBT colleagues (Cech and Pham 2017). These disadvantages extend across age and position level for LGBT-identifying women and men (Cech and Pham 2017), and they extend beyond the workplace (Mallory and Sears 2015). In a recent survey of transgender individuals in the United States, 16% of respondents were denied equal treatment within the previous year; 88% were based on gender identity or expression, and 36% on sexual orientation (James et al. 2016). The pervasiveness of inequalities is evident in our media, as evidenced by a sample of headlines we found using the search term "LGBT" in Google News in February and March 2019:

- "LGBT Community in Chechnya Faces 'New Wave of Persecution'" (UN News)
- "South Dakota Leads the Way in Anti-LGBT Bills for 2019 Session" (*Washington Blade*)
- "64 Percent of California LGBT Students Are Bullied" (*Sacramento Bee*)
- "A Member of the West Virginia House of Delegates Is Facing Bipartisan Criticism for a String of Anti-LGBT Statements" (CNN)
- "LGBT Group Sues Arizona over Law Barring HIV, AIDS Instruction in Schools" (*Washington Post*)
- "Google Resists Pressure to Pull LGBT 'Conversion Therapy' App" (*Axios*)
- "United Methodist Church Keeps Ban on Same-Sex Weddings, LGBTQ Clergy" (NPR)

We lack data regarding the prevalence of these discriminatory behaviors in the wildlife profession, but our collective lived experiences[†] as LGBTQ+ wildlife professionals make it clear to us that we continue to face a unique set of challenges. How does a person's gender identity or sexual orientation shape the feeling of inclusion or exclusion in various parts of their lives? Consider the variety of identities that people use to describe themselves (Figure 8.2). Within this conceptual framework, the central horizontal line represents privilege in our society. With privilege comes power, granted to an individual whether they want it or not. Privilege may be reflected by having opportunities that others

---

* Though we use LGBTQ+ in our writing, when referencing other sources, we use the term the authors used in the cited publication.

† In a social justice context, the term "lived experience" refers to firsthand narratives and impressions of people who are minorities or members of oppressed groups.

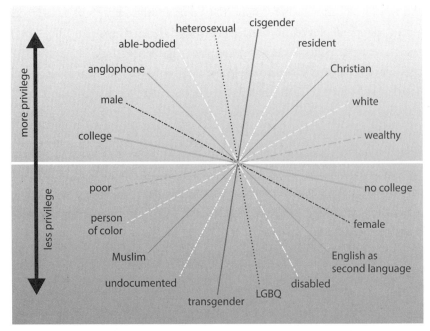

FIGURE 8.2. Illustration of power differentials based on a small sample of identities within the United States. The dashed lines of various types connect related identities above and below the privilege line.
Adapted from Morgan 1996.

do not, having a voice in a decision, or simply by being a majority identity in a population; those differences confer power to a person above the privilege line. People above the privilege line may still face many challenges, but those challenges are not the result of their underrepresented identity, and they often face fewer obstacles to success and advancement than those with less privilege. Based on the lived experiences of the authors of this chapter, those we have observed, and the data provided by many citations listed at the end of this chapter, we recognize that lesbian, gay, bisexual, transgender, gender nonconforming, and other queer people are granted less privilege than cisgender, straight people in current US society and the wildlife profession; consequently, they are placed below the privilege line. The relative distance of identities below the line may reflect the degree of privilege lost by an individual with that identity and the level of discrimination they may experience. For instance, although lesbians and gay men continue to face obstacles in society and the workplace, the hurdles faced by transmen and transwomen place them farther below the privilege line (James et al. 2016).

Context matters in this framework. The magnitude of disparities in societal privilege may be smaller in San Francisco than in rural Idaho, or in social work careers versus those in natural resources agencies. We lack data to quantify where wildlifers fall on the diagram in Figure 8.2, but our experience indicates that the majority of wildlife professionals in the United States today fall above the privilege line in many, if not most respects—most (especially those in advanced career stages) are white, male, cisgender, straight, Christian, able bodied, English speaking, and were born in the United States. The US delegation to the Pacific Flyway Council, a policy-making body composed of the directors or appointees from each state and provincial wildlife agency in the western United States, Canada, and Mexico, provides one especially stark example: not a single woman or person of color was present as of 2019.

Individuals below the privilege line may face obstacles or anxiety about how their work is viewed by those in the majority. Consider the level of societal privilege within the wildlife profession for someone who is a person of color, transwoman, lesbian, Muslim, and disabled. What challenges might she face in a wildlife agency? What voice might she have in decisions about her workplace? At a practical level, would she face backlash for using a women's restroom at her office? If the wildlife profession (or any other STEM profession) is to promote *equity* in the workplace, then a concerted effort is needed to elevate all identities currently below the privilege line to be on par with those above it. That effort does not mean simply treating people *equally*; an individual with a marginalized identity will have faced a lifetime of obstacles that their peers have not. Equity efforts are most successful when those with greater societal privilege recognize inequity and work intentionally to raise privilege by removing barriers and providing opportunities for those with less privilege. Otherwise, the power differential is perpetuated, and the discriminatory environment persists. Progress may be happening in some venues for those who identify as lesbian or gay, but when those identities intersect with other marginalized identities, progress in equity may be assumed but is often overestimated.

An individual's professional success is not solely based on their workplace accomplishments; their ability to be successful depends on their whole lived experience. As a society, we may believe that we have become more accepting of LGBTQ+ identities since Matthew Shepard's murder in 1998, but some data indicate otherwise. In 2017, 7,175 hate crimes were reported in the United States, 1,130 (16%) of which were based on sexual orientation and 119 (2%) on gender identity (DOJ 2018). Crimes targeting LGBTQ+ people increased by about 6% in 2018 (DOJ 2019). Evidence of such explicit violence can be found in the news media daily. Reports of violence against LGBTQ+ individuals continue, and the level of violence is likely greater toward people who are also of minoritized races, religions, or ability status. The Human Rights Campaign began tracking anti-transgender violence in 2013; it reported that at least 26 transgender and gender nonconforming individuals were killed in the United States in 2019 alone—a staggering 91% of whom were Black women (HRC 2019). Now place this information in the context of a wildlife professional, working in remote, often rural locations in communities where marginalized people are potentially at risk; bias may be explicit, confrontational, and deeply personal. LGBTQ+ professionals must include in their decision-making process about research and field work a variable that their cisgender, straight counterparts often do not—is this workplace safe for me?

In addition to explicit bias that can result in unsafe environments for LGBTQ+ people, interpersonal interactions expose implicit bias, often in subtle ways. We may make assumptions about someone's gender or sexual orientation, or their race or religion. These unconscious assumptions influence how we interact with others via our actions, conversations, and decisions. Although the validity of these tests has been questioned (Lopez 2017), implicit association tests may provide some insight into our own implicit biases on many topics, including gender and science, sexuality, and gender and careers (Project Implicit 2011). Whether or not you take these tests, introspection is important if we truly want to advance equity for marginalized individuals. Equity efforts may be particularly important at this time in the United States, as a divisive political environment has further eroded protections for those below the privilege line. Efforts to expel trans people from the military, revoke protection for young undocumented immigrants (DACA), or build border walls increase the differences in privilege between those above and below the privilege line. That may shift in the future with changes in elected representatives, but the divisions in our country seem clearer now than they have been in decades.

Why This Chapter Is Needed

It is impossible to know how many LGBTQ+ people are employed in the wildlife profession in the United States, as those data do not exist. Based on the proportion of the US population that identifies as LGBTQ+ (~4.5% overall, varying widely among states; Newport 2018) and the estimated number of zoologists and wildlife biologists (19,400; BLS n.d.), we conservatively estimate that there are 800–900 LGBTQ+ wildlife biologists in the United States (4.5% of 19,400 = 873), and approximately 500 LGBTQ+ members of The Wildlife Society (4.5% of 11,000 members = 495). However, we do not know whether these numbers are accurate because many LGBTQ+ wildlifers may not want to be labeled as such for reasons described on the next page (Box 8.2). Because

BOX 8.2 **Out at Work**

The act of coming out is often portrayed as a singular decision—an LGBTQ+ person comes to understand and accept their identity, they announce it to the world, and it's done. In reality, LGBTQ+ people must decide to come out (or not) in countless situations that happen every single day. The mental calculations (Is this person safe to tell? Can I mention my partner around these folks?) can be exhausting, especially in professional contexts where the "wrong" choice could harm your career.

Some of the aspects wildlifers most love about their profession—remote study locations, tight-knit field crews, attending conferences—are also aspects that can complicate LGBTQ+ wildlifers' decisions about if and when to disclose their sexual orientation or gender identity. It's different for every person and every situation. Even the most "out and proud" person might go back in the closet when they fear for their safety or their job. The risk of limiting career advancement is real, as are threats to physical safety and security in some situations (e.g., working in remote areas in small groups of people who may not understand LGBTQ+ identities, working in countries where one's sexual orientation or gender identity is illegal). How should an LGBTQ+ wildlife biologist act in a country where their identity is punishable by death? Should they follow their intellectual curiosity (and associated opportunities for professional development) to places they know are discriminatory?

The spectrum concept is again useful here. Rarely is someone either 100% closeted or 100% out (in all situations, to all people, etc.). In social science terminology, identity disclosure is a spectrum from concealment to revealment (Wax et al. 2018). As Figure 8.3 illustrates, an individual may choose to reveal their identity more freely in their personal life or their professional life. To describe some of these intersections of personal/professional "degrees of outness," we borrow terms from a podcast episode, "Out at Work" (Low and Tu 2017). Under each heading below, we share our personal experiences when we found ourselves in one of these categories—each section was written by one of the authors of this chapter. We urge readers to think about where each of these narratives would belong on Figure 8.3.

Out and Proud: Doesn't Feel the Need to Censor Any Part of Their LGBTQ+ Identity

After coming out while a department head at a large university, I applied for a position as dean of a different prominent natural resources college. I was not offered the job, and there were undoubtedly many good reasons for that decision by the provost, but I was curious for feedback on how I could improve on my interviews for future positions. I contacted a colleague who was close to the search process. He was kind and thoughtful in his response, but one comment caught me by surprise, and perhaps it should not have: concerns were raised about how alumni might react to a trans-woman as dean. Among all the reasons for not getting the job offer, this one made me realize that there are limits to acceptance of trans identities, even in academia. I have encountered similar reactions from other academic and development offices since then. I am out. I am who I am. And that is a problem, but it is not my problem. I will not change who I am to get a job. People with the authority to hire others must become aware of the potential for the gender identity or sexual orientation of an applicant to influence their hiring decisions, consciously or subconsciously, and be bold in advancing those individuals who may face challenges because of their identity.

People Can Tell: Not Formally Out at Work, but Their LGBTQ+ Identity is Implicitly Known

In a recent chat with my lifelong friend—the one I first came out to—I mentioned that it was now 11 years since I had told her. She responded, "It seems almost longer to me because when you told me, it made the most perfect sense!" It didn't surprise me to hear her say that. I know many people had a hunch well before I officially came out. Is it the clothes I wear, the way I cut my hair, the way I walk and talk? Or something less tangible? We all give off vibes, and something about mine says "gay" to those around me. Regardless of the reason, I realize that even if I haven't formally come out to a colleague, they can probably tell. Quite honestly, I find it easier that way. When "people can tell," it lays the groundwork for those moments when "implicitly out" becomes "explicitly out." It means people aren't surprised, and I don't have to try to interpret their reactions.

For the past 11 years, my disclosure decisions have meant that I'm generally "out and proud" in my personal life, and I don't actively hide my sexual orientation in work situations. Being in the "people can tell" realm leaves room for people to choose to ignore it if they'd rather. But that's their decision—not mine.

It's Complicated: Might Like to Be Out at Work, but Some Aspect of the Job Prevents It

A close LGBTQ+ colleague of mine felt forced to enter the closet every time they conducted fieldwork in rural villages. Their work consisted of surveying villagers by entering their houses, sitting down at their kitchen tables, and inquiring about their use of wildlife resources. They interacted with complete strangers in pretty intimate settings day in and day out. My colleague just didn't feel safe working as an out LGTBQ+ professional in those settings, even while working in a small group. And their insecurity was well placed, as they often encountered outright racism, xenophobia, and clear homophobia while interviewing people in the comfort of their own homes. To accomplish their work and remain safe, they simply lied about their personal lives when asked (and they were commonly asked if they had a family, spouse, etc.). Such are the compromises we must sometimes make as LGBTQ+ wildlife professionals. Sometimes you don't have the luxury of being out, of being you, and you have to endure the derogatory comments and shame heaped upon you by the people you are there to serve.

Not Out: Closeted on the Job

As a gay wildlife biologist, some of the most challenging situations regarding how "out" I am in the workplace occur in remote field situations. How do I address this topic with new graduate students or employees whom I have not previously met and do not know much about but am within 12 hours of finding myself sharing a one-room cabin or bunkhouse? Ideally, one has access to separate sleeping quarters, but the realities of fieldwork mean that a tiny cabin or cramped bunkhouse may be the only option to escape bears, bugs, rain, and snow. Sharing a room with same-sex subordinates is pretty commonplace in wildlife biology. So what are the guidelines for coming out in such situations? Is it better to state it up front? If so, when? During the interview process? As soon as the technician gets off the plane? During casual conversation over dinner? When you are unpacking your stuff in the shared bedroom? None of these felt like appropriate times to out myself to a new subordinate, and so I often didn't.

There are no clear guidelines for these situations, and I find myself following my gut, making some assumptions, and trying to judge the employee's viewpoints. Correct or not, my focus has always been on minimizing their discomfort, sometimes at my own expense. Thankfully, I've found that most younger wildlife professionals and students are pretty indifferent to the topic; meeting LGBTQ+ people isn't new to them, and they often don't see it as an issue (or they have at least acted convincingly so in my presence). After coming out in an international publication, my hope is that word gets around and people will know I am gay long before we meet. Even so, relying on that assumption is no guarantee, and not outing myself directly does nothing to advance the rights and visibility of LGBTQ+ biologists in our field. Hence, I am left judging each situation individually and responding accordingly, no manual included.

While some of these issues may appear to be personal, the importance of strong working relationships with coworkers, founded on good personal relationships, is extremely important to professional success. This is especially so in remote field camps where just a few coworkers live in an intimate work setting; we rely upon each other for safety, and job success depends heavily on successful teamwork. Consequently, the settings in which LGBTQ+ biologists may be most challenged by revealing their sexual orientation and gender identity are also the places where those challenges may have the biggest impact on job performance, satisfaction, and retention.

It's Irrelevant: LGBTQ+ Identity Has Nothing to Do with the Job

As is the case with many others I know who identify as part of the LGBTQ+ community, I uncovered my gender identity through a series of realizations over time. As I learned more about the gender spectrum, I refined the way I defined myself—first to myself, then to others. I thought of this internal discovery process as being like a personal philosophy or religion, in that it was not connected to my professional life. My commitment to the mission of my employer had not changed. I could continue all aspects of my work with success and satisfaction, regardless of my gender identity. And so, I did not share this self-discovery with my coworkers, colleagues or community; I shared it only with family and close friends. What's wrong with that? No harm, no foul, right?

Except that as my career matured, I found myself supervising others and serving in the roles of coach and mentor. I realized that other LGBTQ+ wildlifers were looking for role models, seeking acceptance and inclusion in a career field that is exciting, rewarding, important, and too often excludes or fails to embrace the marginalized. If my gender identity truly does not matter to my career, I reasoned, then what's the big deal about disclosing it? I was lucky enough to be in a supportive workplace. I was astonished to experience a release of tension, tension that I did not know I had been holding.

---

sexual minority students drop out of STEM fields at a higher rate than their heterosexual peers, the proportion of LGBTQ+ wildlife professionals may be lower than in the general US population (Hughes 2018). In fact, we know numerous highly qualified potential contributors who did not want to coauthor this chapter because outing oneself in this profession still has very real and sometimes serious negative consequences in both professional and personal contexts. In a self-perpetuating cycle, the resulting lack of visibility in our profession compounds the problem by leaving LGBTQ+ wildlifers (especially students and young professionals) with few role models and mentors. It also impedes the building of community, peer support, and acceptance that is needed to end this cycle of suppression and discrimination (Figure 8.3).

These biases are not restricted to the wildlife profession; they span many STEM fields. Neuroscientist Dr. Ben Barres (1954–2017) was a transman who experienced both gender discrimination (before he transitioned) and gender privilege (after transitioning) and, therefore, had a unique perspective about how prevalent these issues are in STEM professions (Allen and Daneman 2018; Barres 2006). He wrote about a revealing incident he experienced shortly after his transition, when a faculty member remarked about his previous work (pub-

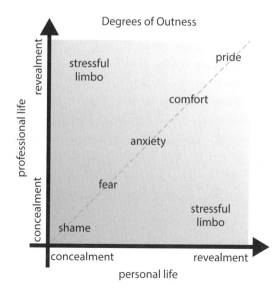

Degrees of Outness

FIGURE 8.3. Because sexual orientation and gender identity can be "invisible" to others, the decision to come out is persistent. LGBTQ+ individuals choose how much of their identity to reveal in both personal and professional contexts—represented by the two axes here. The dashed line indicates congruence between the two contexts. Words within the figure reflect feelings associated with degrees of outness and congruence. We emphasize that these feelings can move around on a given day or in a certain situation. Even if you've been out and proud for years, anxiety can still creep in.

lished as Barbara Barres), *"Ben Barres gave a great seminar today, but then his work is much better than his sister's"* (Barres 2006).

Legal Protections for the LGBTQ+ Community

Nondiscrimination protection for LGBTQ+ people falls into four main spheres of influence: federal law, state law, municipal ordinances, and employer commitments. Intersecting with these spheres of influence are specific policies or practices related to workplace protections, access to housing, public accommodations, access to credit, and public versus private employment. Regardless of where we work and where we live, these laws and policies affect all of us, wildlifers or not. Most antidiscrimination laws related to the LGBTQ+ community are enacted by individual states, but there are a few (currently contested) federal laws that provide a modest level of protection.

Federal Laws

No formal federal legislation specifically provides protection from discrimination for LGBTQ+ people, but executive orders and court decisions have enabled some protections by interpreting existing laws to include LGBTQ+ populations. Such recent changes have provided protections for some aspects of employment, education, marriage, adoption, and military service, but many of them are being challenged in court. For instance, Title VII of the Civil Rights Act was expanded to include LGBTQ+ federal workers and contractors in 2012 (based on gender identity) and 2015 (based on sexual orientation), but President Trump's administration filed a court brief in 2017 reversing those expansions. The reversal was challenged in court, and the administration's interpretation was overturned by the US Court of Appeals for the Second Circuit in 2018. On June 15, 2020, the US Supreme Court ruled 6-3 that the Civil Rights Act protects LGBTQ+ people from discrimination in employment. The ruling did not address protections from discrimination in other areas, such as housing and public accommodation.

In another example, President Obama's administration interpreted the Department of Education's Title IX legislation (which prohibits discrimination based on sex in schools and other institutions that receive federal financial assistance) to include protection based on gender identity. Subsequently, the Trump administration proposed to redefine the legislation to only include protection based on sex assigned at birth. After a legal challenge, the US Court of Appeals for the Second Circuit ruled that gender identity protections do apply under the Civil Rights Act.

These two examples illustrate that while protections at the federal level do exist, enforcement depends on how the executive branch and the courts interpret existing legislation. While the Civil Rights Act does provide protections for LGBTQ+ individuals in employment, it is silent on other areas of potential discrimination. Hate crimes laws provide some additional protection under the Matthew Shepard and James Byrd Jr. Hate Crimes Prevention Act of 2009. Other nondiscrimination protections vary among states, and these issues continue to be debated in court.

State Laws and Local Ordinances

As of 2019, most US states did not provide clear legal protection against discrimination for LGBTQ+ people. In terms of housing and public access, states vary widely in protecting LGBTQ+ people from discrimination. The passage of what are described as "religious freedom" laws also impacts the LGBTQ+ community, particularly where they apply to private schools and businesses. These laws allow individuals or private entities to deny service to someone based on the provider's religious beliefs—effectively allowing religious beliefs to justify discrimination. Note that the pattern of states with employment protections (prior to June 2020) largely mirrors the map of "religious freedom" legislation (Figure 8.4). The degree to which religious beliefs are used to discriminate against the LGBTQ+ population varies regionally but has had broad

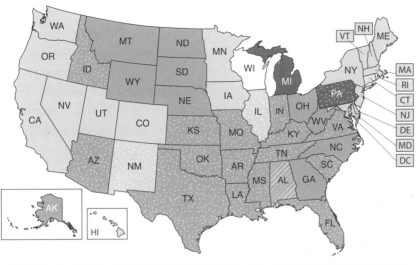

Discrimination laws (base color)	Religious exemption laws
explicitly prohibits discrimination on sexual orientation and gender identity ☐	no pattern ☐ no religious exemption law
no explicit prohibition of discrimination ☐	▦ statutory religious exemption law
explicitly prohibits discrimination based on sexual orientation only ☐	▨ constitutional religious exemption law
interprets existing prohibition on sex discrimination to include SO and/or GI ■	

FIGURE 8.4. Employment nondiscrimination and religious exemption laws in the United States as of April 2019; in June 2020 the US Supreme Court ruled in favor of employment protection for LGBTQ+ employees under Section VII of the Civil Rights Act. Religious exemption laws persist in many states and influence LGBTQ+ people's rights outside of employment.
From MAP (Movement Advancement Project), 2019b, 2019c.

implications nationwide. These sorts of laws make *explicit* discrimination legal in instances of housing and public access, and they may still have implications for hiring LGBTQ+ people despite the June 2020 Supreme Court ruling. Justice Neil Gorsuch, who wrote that majority opinion, recognized that "*the 1993 Religious Freedom Restoration Act as a 'super statute' may offer a potential lifeline to employers who object, on religious grounds*" (Totenberg 2020). What is unseen in these maps are the *implicit* biases that employers, supervisors, advisors, and administrators have that accompany these religious beliefs and how they normalize discrimination.

Many cities and counties have nondiscrimination policies protecting LGBTQ+ people, even if the state does not, covering 2%–100% of LGBTQ+ people per state (Figure 8.4; MAP 2019a). Local ordinances have also been passed to allow or prohibit access to public facilities (e.g., restrooms). Consequently, someone may be protected from employment discrimination, but their lived experience in the workplace, school, or community may be complicated by other barriers depending on where they work, live, and even shop. These barriers affect people's lives, both inside and outside of the workplace.

How Does All of This Relate to LGBTQ+ Wildlife Professionals?

We illustrate how these laws and policies may influence the success of LGBTQ+ people in the wildlife profession by providing examples of policies from states and universities. We assumed that most wildlifers have an educational background in wildlife biology, natural resources, biological sciences, or environmental sciences. Consider the following nondiscrimination policies from two universities that offer undergraduate degrees in biology or related fields.

> Oregon State University, in compliance with state and federal laws and regulations, does not discriminate on the basis of age, color, disability, gender identity or expression, genetic information, marital status, national origin, race, religion, sex, sexual orientation, or veteran status in any of its policies, procedures, or practices. (OSU 2019a)

> Liberty University does not engage in unlawful discrimination or harassment because of race, color, ancestry, religion, age, sex, national origin, pregnancy or childbirth, disability or military veteran status in its educational programs and activities. Liberty University maintains its Christian mission and reserves its right to discriminate on the basis of religion to the extent that applicable law respects its right to act in furtherance of its religious objectives. (Liberty University 2019)

Examine these two statements from the perspective of a potential LGBTQ+ student. How do they differ? Which one specifically calls out gender identity or expression? Sexual orientation? What exactly does "*reserve the right to discriminate on the basis of religion*" mean? What is the lived experience of a student attending either of these two universities with regard to implicit or explicit bias that they may face? How might that experience influence their success as a wildlife professional, and their interactions with other wildlifers?

Now conduct the same analysis for four potential employers:

> The Louisiana Department of Wildlife and Fisheries is an Equal Opportunity Employer. LDWF does not discriminate on the basis of race, color, religion, sex, age, national origin, disability, veteran status or any other non-merit factor in any of its programs, activities, services or employment practices including but not limited to, recruitment, pay, hiring, firing, promotion, job assignment, training, leave, layoffs or benefits. (LDWF 2013)

> It is the policy of the [California] Department of Fish and Game (Department) to provide equal employment opportunity (EEO) to its employees and job applicants

without regard to age, sex, race, ancestry, political affiliation, disability, religion, color, national origin, marital status, sexual orientation, or medical condition. (CDFW 2008)

The Commonwealth [of Massachusetts] provides equal opportunity in state employment to all persons. No person shall be denied equal access because of race, creed, color, religion, national origin, sex, sexual orientation, gender identity, age, or physical/mental disability. (Commonwealth of Massachusetts N.d.)

It is the policy of the State [of Idaho] to ensure equal employment opportunity for all individuals. Discrimination or harassment based on race, color, religion, sex, national origin, age, disability, marital status, citizenship, genetic information, pregnancy, military status, or any other characteristic protected by law is prohibited. (Idaho DHR 2019)

Analyze these four statements through the lens of a lesbian. Where would you feel most protected? Do the same analysis through the lens of a transwoman. What exactly is a "*non-merit factor*"? Which states offer explicit protections for lesbians? For transwomen? And keep in mind that all four of these statements come from state Equal Employment Opportunity websites.

Private companies also vary in the degree to which they are explicit in naming LGBTQ+ identities in their EEO statements. Many companies simply state that they are EEO compliant but do not provide an easily found EEO policy. Here is just one example:

The company does not discriminate in employment opportunities or practices because of a person's race, color, religion, national origin, sex, sexual orientation, gender identity, age, disability, veteran status or genetic information. The company also prohibits harassment of any individual on the basis of any of these characteristics. (Georgia-Pacific 2019)

How does The Wildlife Society (TWS) fare in this regard? In its Code of Ethics, standards of professional conduct, members "*shall studiously avoid discrimination in any form, or the abuse of professional authority for personal satisfaction*" (TWS 2017). Further, TWS's Bylaws state, "(*Section 4): Composition—the Society shall be composed of professionals, students, and others, regardless of age, race, religion, gender, ethnicity, disability, sexual orientation, or nationality who are interested in wildlife resources, and who subscribe to the Society's Objectives and Code of Ethics.*" And further, it calls for "*whistleblower protection for employees*"— reporting of illegal or unethical behavior including "*age, gender or racial discrimination, physical assault, sexual harassment*" (TWS 2012). Could TWS be more explicit in their protections for LGBTQ+ professionals? We believe that they can and should.

How does your school or workplace support LGBTQ+ students, employees and clients? Do some homework. Look for evidence that shows support (e.g., job announcements, EEO information). Does it exist? How might you improve the language presented for your institution? Your community? What resources are offered by your institution? In your community?

Where Do We Go from Here?

For many LGBTQ+ people in the United States, state-level workplace protection becomes one of many priorities when seeking a job. The relative proportion of self-identified LGBTQ+ people per state at least partially reflects the level of nondiscrimination protections in place. For instance, the estimated population of LGBTQ+ adults in Oregon is 5.6% of the total, similar to California and Massachusetts; all three states had workplace protection legislation prior to the Supreme Court ruling in 2020 (MAP 2019c). Compare those proportions to states that did not, such as Mississippi and South Dakota (2.8%–3.5% of the total population). Local nondiscrimination ordinances are most common in more urban areas, but few wildlife jobs are located there. Regardless of whether this pattern indicates a causal relationship, opportunities for wildlifers are limited when the locations with a high proportion of wildlife jobs are in areas lacking protections that could impact career security and personal safety. Based on our own experiences and our observations of other LGBTQ+ wildlifers, securing employment with public universities, some state agencies, federal agencies, and some nongovernment organizations is likely to provide the strongest workplace protections. Employees in states that lack other protections or that have religious freedom legislation are more likely to experience implicit or explicit bias on a daily basis. Until there is federal legislation that provides full nondiscrimination protection in all facets of life, some will feel free to discriminate against members of our LGBTQ+ community. In the meantime, a concerted effort by administrators, managers, colleagues, faculty, and human resources professionals is needed to both minimize instances of bias in the workplace and advance LGBTQ+ people into leadership positions. Establishing bias response protocols (OSU 2018) and search advocates on hiring committees (OSU 2019b) are two structural approaches to begin addressing issues of discrimination and implicit bias in the workplace. Intentional and effective implementation of such practices by well-trained supervisors, staff, and administrators is imperative to make a difference, which to us equates to providing improved learning and career opportunities for LGBTQ+ wildlife professionals.

We find it interesting that the animal subjects that most draw our attention in the wildlife field are often those outside the normal bell curve. The elk, grouse, or wolf that does something completely different from its counterparts is the one that attracts our curiosity. These rare phenotypes are sometimes even afforded special legal protections and often, affection from us and society. For example, rare albino deer are protected from harvest in some states. We look forward to the day when such positive attention and protections are afforded to the rare and unique human phenotypes within our own wildlife profession.

Reasons for Hope

Although there is much more work to be done, perceptions of the LGBTQ+ community in the United States have shifted dramatically in a favorable direction over the past several decades. This shift first became evident in the 1990s. A study by Loftus (2001) showed an 18% increase in the number of Americans who reported that "homosexuality is not wrong at all" between 1987 and 1998. As attitudes about sexual orientation became more liberal, Americans became less willing to restrict civil liberties of the LGBTQ+ community. In the afore-mentioned study, the number of respondents who "would not restrict any civil liberties of homosexuals" increased by 29% between 1973 and 1998 (Loftus 2001). More recent studies show a continued liberalization of US views toward the LGBTQ+ community. Following the 2015 Supreme Court ruling in favor of same-sex marriage, Obergefell v. Hodges, perceived social norms among non-LGBTQ+ Americans shifted toward increased support of gay marriage and gay people (Tankard and Paluck 2017). According to a 2016 Pew Research Center survey, 63% of Americans said that "homosexuality should be accepted by society," compared with 51% in 2006 (Brown 2017). It is no surprise then, that the number of Americans who openly identify as LGBTQ+ continues to rise, with more than 11 million people identifying as LGBTQ+ as of 2017 (Newport 2018).

So far, we have mentioned shifting perceptions and policies toward nonhet-erosexual individuals, but has the country experienced a similar change when it comes to the transgender community? In general, attitudes toward transgen-der people are strongly correlated with attitudes toward gay, lesbian, and bi-sexual individuals (Norton and Herek 2013). However, only 0.6% of US adults identify as transgender, making it substantially less likely for non-LGBTQ+ people to knowingly interact with someone who identifies as transgender

(Flores et al. 2016). This lack of familiarity is relevant because those who report having some type of contact or interaction with a transgender individual are more likely to have a positive attitude toward the transgender community as a whole (Norton and Herek 2013). On the positive side, there has been a clear increase in the visibility of transgender individuals in the media over the past decade, which has been associated with increasingly supportive attitudes toward transgender people and policies (Gillig et al. 2018). Despite the early exclusion of transgender issues from the LGBTQ+ rights movement, they have since been included and have catalyzed an explosion of activism and policy changes (Taylor et al. 2018). These policy changes are far-reaching and include the regulation of identity documentation, discrimination and civil rights laws, health care policy, education policy, and criminal justice.

While LGBTQ+ visibility and advocacy hit society at large like a wave, similar changes have been slower to reach STEM fields. LGBTQ+ life science professionals are much less likely to be "out" to their colleagues (Box 8.3), compared to people in their personal life (Yoder and Mattheis 2016; Barres et al. 2017). This disparity is important because those who reported greater openness at work were more likely to feel safe and welcome there, which undoubtedly affects productivity and the potential for collaborations. This lack of visibility also reduces the number of LGBTQ+ mentors and role models available for future generations. Thus, it is unsurprising that LGBTQ+ students drop out of STEM fields at a disproportionately higher rate than non-LGBTQ+ students (Hughes 2018). In response to this lack of visibility in STEM fields, there has been a movement led by online communities to empower and acknowledge LGBTQ+ scientists. Notably, 500 Queer Scientists (500queerscientists.com) has emerged as a visibility campaign for LGBTQ+ people and their allies working in STEM and STEM-supporting professions. The campaign collects and posts individual, self-submitted biographies and stories to build community connections and to provide role models for the next generation. National organizations with similar goals include LGBTQ+ STEM (lgbtstem.wordpress.com) and oSTEM (ostem.org). Within The Wildlife Society, a new initiative called Out in the Field began in 2019, which aimed to increase visibility of LGBTQ+ wildlifers, build a community of LGBTQ+ wildlife professionals, and identify ways to support all LGBTQ+ wildlifers (Olfenbuttel et al. 2020). This effort was an organic response by a few wildlife professionals (including some authors of this chapter [Travis L. Booms, Katherine M. O'Donnell, Claire Crow]) to an article ("I *Am* One of You") written by Travis Booms in the March/April 2019 edition of the *Wildlife Professional* (Booms 2019). In the article—the first ever in the publication to address LGBTQ+ issues—Booms talked about his experi-

BOX 8.3 Profile of Kelly J., Professional Wildlife Biologist, TWS member

Kelly J. has been a member of The Wildlife Society for 21 years and earned a PhD in Wildlife Ecology from a prominent university in the southern United States. Kelly has authored 16 publications in top-tier journals, has given many presentations at professional society meetings, and now serves as a regional biologist for a state wildlife agency in the western United States. Kelly has received agency awards for their recent work that led to the recovery of an endangered species. Kelly aspires to move into an agency leadership position and is planning to apply for the director position when it is open in a few years. You know Kelly. You have worked with Kelly, attended professional meetings with Kelly, and cited Kelly's papers as the basis for decisions that you make about your own work. But there are some things that you do not know, and will not know, about Kelly.

Kelly is transgender. They transitioned as an undergraduate while earning a BS in biology at a top university in California. Kelly changed all documentation, including their birth certificate. Kelly's doctoral advisor does not know this, nor does their boss, nor do their colleagues. Kelly enjoys running for exercise but has avoided joining their colleagues who go running before work, not wanting to use either of the only two showers in the agency headquarters. Kelly is in a same-sex relationship with Kim, their partner of five years. Kim and Kelly do not live in the same town where either of them works—they live two towns away so that they do not encounter their coworkers while out together. Kim does not accompany Kelly to professional meetings or to social events with coworkers. You may not know these things about Kelly (or the >750 other people like Kelly who work in the wildlife profession in the United States) until they reach a point in their career where they trust that their LGBTQ+ status will not impact their ability to advance in the profession. Then, and only then, may Kelly choose to come out. Or they may not. Many people like Kelly will never come out to their colleagues, which is why we do not know exactly how many LGBTQ+ people there are in our profession. That is why we chose Kelly for this vignette. You may never know who they are. They will not let you.

ence as a gay wildlife biologist and emphasized the need to increase the visibility of LGBTQ+ wildlifers and role models. Recognizing Booms's courage, and seeing something of ourselves in his experience, we were inspired to seek a lasting and positive change in culture starting at the upcoming TWS annual conference. Over the next few months, we planned a social event and created informative tools that supported a safe environment for voluntary visibility of

LGBTQ+ members and allies. We reached out to TWS working groups and leadership, receiving encouragement and financial support. Looking at the 40 places set for the luncheon, we held our breaths, worried about how awkward it might be if only a few people showed up. It was thrilling to watch a crowd arrive that far exceeded the planned space, prompting then-CEO Ed Thompson to order more food and chairs in a clear signal of leadership support. Given the overwhelming support Booms received for the article and the subsequent and continuing success of Out in the Field efforts, it is clear that progress toward LGBTQ+ equity within our profession is moving forward.

Like the broader STEM fields, the wildlife and natural resource professions have room to grow when it comes to increasing the visibility and support of LGBTQ+ individuals. One approach is to create support groups or subcommunities within organizations and professional societies. For example, The Nature Conservancy sponsors an LGBTQ+ employee resources group called Nature's Pride. This group is one of several resources advertised to potential applicants on their website, signaling a welcoming workplace that is likely to attract a diverse applicant pool. Other simple approaches to being more inclusive of LGBTQ+ colleagues at wildlife and natural resource conferences include allowing attendees to have their pronouns (e.g., "they/them", "she/her") indicated on their nametags to reduce misgendering and normalize the practice of including pronouns; sponsoring LGBTQ+ socials to enhance personal and professional networking; choosing host locations that are LGBTQ+ friendly and avoiding those with anti-LGBTQ+ policies (e.g., HB2, a 2016 transgender bathroom bill in North Carolina); and ensuring accessibility to restrooms that align with a person's identified gender, including gender-neutral restrooms. At the annual meeting of the Society for Integrative and Comparative Biology, which largely focuses on zoological research, meeting attendees self-organized an LGBTQ+ social that has grown from a small meetup to a gathering of 40–60 people. Although the road ahead is long, there are many opportunities for creative and engaging ways to support LGBTQ+ wildlife biologists and create an environment that celebrates diverse sexual and gender identities.

The message we leave with you is this—any wildlife professional (and, indeed, any member of our society) who does not identify as LGBTQ+ can be more intentional in their support for those of us who do. We are explicitly asking you to do so for the betterment of our society and our profession (Box 8.4). Out yourself publicly as an ally. Educate yourself about the unique challenges your LGBTQ+ colleagues and friends face. Work with us to overcome those challenges. Speak up about the barriers we face and about your out

BOX 8.4 Changing the Climate

Whether you are an ally or a member of the LGBTQ+ community, these are some actions everyone can take to help change the climate for LGBTQ+ people in the wildlife profession.

- **Know your rights**. Know your employer's policy and what constitutes discriminatory speech, actions, and harassment. Understand whistle-blower protection. Know what resources to use if you or someone at work is harassed.

- **Intervene**. When you witness LGBTQ+ discrimination or harassment, call it out. Whether in a duck blind with coworkers who crack a gay joke or at a conference where a colleague inappropriately remarks about another person's gender identity, identify the statement as unacceptable and show your disapproval, provided you feel safe to do so. Calling out your friends, colleagues, and mentors who venture into the zone of disrespectful language can be terrifyingly difficult. If you fail to do so in the moment, think about how you could have handled it differently and make a plan to do better in the future. Memorize a short, direct statement such as, "That is disrespectful and inappropri-ate." Encountering harassment in the moment can be stressful; memorizing and practicing statements is very useful and effective. Taking bystander intervention training can also help you prepare.

- **Be intentional with words**. Use "spouse" or "partner" in place of "wife" or "husband" when inquiring about a colleague's significant other if you do not know your colleague's sexual orientation. Use the same words to describe a person that they use to describe themselves.
    - Include your pronouns in introductions: "Hi, my name is Casey and I use 'she' and 'her' pronouns." Add your pronouns to your email signature line.
    - If you inadvertently use the wrong pronoun for someone else, quickly correct yourself and go right back to the topic at hand. Mistakes are normal—we are all human. The most respectful action is to acknowledge the mistake without making it the focus of the conversation.
    - Gently correct others when they misgender a person who has told you their pronouns, then go back to the topic at hand. For example, " 'he' is more appropriate," or simply state the correct pronoun.
    - Use gender-neutral language like "everybody," "friends," "person," "they"—language that does not assume anyone's gender. For example, "the person behind the counter" rather than "the woman behind the counter" and "Everybody, may I have your attention" rather than "Ladies and gentlemen, may I have your attention."

- **Network**. If your workplace has an LGBTQ+ employee resource group or network, join it. If your workplace does not have such a group, consider starting one. These groups provide mutual support and inclusive social space, advocate for LGBTQ+ rights and build inclusivity in the workplace culture, and support LGBTQ+ events that connect the workplace with the community (McNaught 2017).
- **Lead with empathy**. We collectively are unaware of the secret struggles, strife, and stories that our colleagues bring with them to work, meetings, and conferences. The facade you meet is false; the truth lies much deeper and is harder to see and address. If a colleague is struggling, recognize the struggle but realize you probably can't appreciate it fully. Simply listening and acknowledging what you hear is incredibly valuable.

Scenario: a coworker quotes a religious text (Bible, Koran, etc.) to support their belief that some sexual orientations or gender identities are unnatural, evil, or worse. Do not quote from a religious text in response. Quote (or paraphrase) from your company's policy on valuing diversity and on acceptable and expected behaviors for every employee in your workplace (McNaught 2017). Use your memorized response to harassment: "That statement is disrespectful and inappropriate."

Actions Specifically for Allies
- **Out yourself as an ally**. Proactively bring up LGBTQ+ issues in group settings and make it clear to those around you that you are an ally. If you know out LGBTQ+ colleagues, friends, or family members, find ways to work them into your conversation in a positive way. Unless you clearly out yourself as an ally and do so regularly, those around you will not know.
- **Ask LGBTQ+ people who are out**. Don't assume you are perceived as an ally simply because you identify as one. Ask, "What can I do to be supportive as an ally?"
- **Don't assume a person you just met is cisgender and heterosexual**. Always speak and behave as if someone in your workplace is lesbian, gay, bisexual, transgender or queer (McNaught 2017). If you believe in statistical probability (we presume that's all wildlife professionals), 1 out of the next 20 people you meet will identify as LGBTQ+. You may never know it, but we are here, listening, and present, nonetheless.
- If you suspect a colleague identifies as LGBTQ+, **do not put them on the spot** by asking them directly. The coming out process is different for everyone, but it is always deeply personal. For many, it is also challenging, painful, and lengthy. It is an intensely personal decision that each of us must make (repeatedly) and should only be made by the person leaving the closet. Rather than asking someone about their LGBTQ+ identity, ensure that they know you accept and support

LGBTQ+ people. Express your support in general terms—perhaps by mentioning a relevant news article or talking positively about an (out) LGBTQ+ relative or friend. Allow your colleague/friend/employee to come out when *they* are ready.

- **Practice using they/them/their to refer to one person**. It may feel normal to say, "Not everyone brings their lunch to the office" rather than "Not everyone brings his lunch to the office." With practice, sentences like "Alicia always brings their lunch to the office" flow easily and sound natural. Use "they" and "their" for anyone who has not told you their pronouns.
- Acknowledge that **some people do not subscribe to conventional gender distinctions** but identify as neither, both, or fluid between the historically accepted male and female genders. They may describe themselves as nonbinary, genderqueer, or gender nonconforming.
- **Do your own research**. Don't expect the LGBTQ+ person to educate you. It is normal to be curious about people who are different from you in some way, but it is also normal for people to feel upset when the depth of their personality is constantly overshadowed by their gender identity or sexual orientation. The internet provides a wealth of information about LGBTQ+ issues. If you do inquire about their gender(s) or sexuality, be respectful and accept their answer.
- **Take actions that invite trust**. Ask yourself, "What am I doing to make it easier for others to feel safe and valued at work?" (McNaught 2017).
- Keep in mind that everyone has the right to their beliefs, but nobody has the right to impose their beliefs on others. **Speak up** when someone exercises their beliefs at the expense of others, even when the "others" are not present. For example, if someone disparages another person's sexuality or gender, tell them, "That's not acceptable here."
- **Advocate** for workplace equity by communicating with key decision makers in the company, supervisors, and Human Resources. If your employer does not provide diversity training to all employees, or if it does not include LGBTQ+ issues, request such training. If there are no openly lesbian, gay, bisexual, transgender, or queer people in your workplace, your workplace may not be welcoming (McNaught 2017).
- **Build a welcoming culture**. Business executives and managers can develop clear policy on LGBTQ+ workplace rights, roles, expectations, and responsibilities, share this policy with all employees, and consistently stand by it with their words and actions. They can lead by example and create a protocol for supporting gender transition. They can provide trainings to educate staff specifically on LGBTQ+ issues and provide gender-neutral restrooms without an assigned sex on the door. Businesses can standardize leave (e.g., ensure parental leave is equally available regardless of sexual orientation, gender expression, or marital status), encourage internal LGBTQ+ resource groups and

networks, create a strong culture of inclusiveness, and support LGBTQ+ issues in the community (Miller-Merrell 2017). These measures support recruitment and retention of LGBTQ+ people.

Scenario: Someone in the workplace comes out to you as lesbian, gay, bisexual, transgender, or queer. Express your support. Smile and thank them for telling you, ask how you can be supportive (McNaught 2017). Recognize that this is not your story to tell: let them know that you will respect their privacy unless they tell you otherwise.

Additional Actions for LGBTQ+ Wildlifers

- Try not to assume that people with conservative religious beliefs will not welcome you (McNaught 2017).
- Everyone has the right to expect that people will use their correct name and their correct pronouns, and that they will be treated with professionalism and respect in the workplace. It's traumatic to experience harassment at work. As soon as possible, discretely document what happened. You may need to pause to process the injury before reporting it. You may need to reach out to your support network both inside and out of the workplace. If you know others have experienced these kinds of behaviors, consider filing your complaint as a group (J. D. Hollister, personal communication [telephone interview], March 2, 2019).
- Seek out support networks. Find other LGBTQ+ coworkers and wildlife professionals with whom you can confide and find strength. Build a community where none exists or build upon those that are present.

Scenario: your boss is homophobic and transphobic. Don't argue with them about your rights or who you are. Be prepared to report any inappropriate speech or actions to your Human Resources department; document and report, do not let it slide (McNaught 2017).

LGBTQ+ colleagues, especially in group settings and with students, mentees, and supervisees. Go a little out of your way (and possibly your comfort zone) to include illustrative examples of out LGBTQ+ colleagues accomplishing the same tasks as their cisgender and heterosexual counterparts. We can catch the same critters, learn the same methods, and model the same systems as well as anyone else. Our LGBTQ+ status has nothing to do with our ability to complete wildlife-related work tasks. Indeed, our learned experience and hard-earned compassion and appreciation for the innate value of nature's diversity may make us even more effective advocates for the wildlife resources we all dedicate our careers (and often lives) to. The authors of this chapter have each had many people support us in our careers. As a profession, we can do better for LGBTQ+ individuals and anyone who falls below the privilege line. Want to know how? See Box 8.4. Then ask us.

Resources

American Civil Liberties Union. www.aclu.org/issues/lgbt-rights.

Centers for Disease Control and Prevention. "Lesbian, Gay, Bisexual, and Transgender Health." www.cdc.gov/lgbthealth/index.htm.

Diversity Best Practices resource list. www.diversitybestpractices.com.

GLAAD resource list. www.glaad.org/resourcelist.

Human Rights Campaign. www.hrc.org/.

Movement Advancement Project. 2019. How *Employers Can Support LGBT People in Rural Communities*. www.lgbtmap.org/file/rural-lgbt -employers-recommendations.pdf.

Parents and Friends of Lesbians and Gays (PFLAG). https://pflag.org/.

Barres, B., B. Montague-Hellen, and J. Yoder. 2017. "Coming Out: The Experience of LGBT+ People in STEM." *Genome Biology* 18:62.

Booms, T. L. 2019. "I *Am* One of You." *Wildlife Professional* 14(2): 34–35.

McNaught, B. 2017. *Brian McNaught's Guide to LGBTQ Issues in the Workplace*. Independently published. www.brian-mcnaught.com/resources.html.

Nadal, K. L. 2013. *Contemporary Perspectives on Lesbian, Gay, and Bisexual Psychology: That's So Gay! Microaggressions and the Lesbian, Gay, Bisexual, and Transgender Community*. Washington, DC: American Psychological Association.

Patridge, E. V., R. S. Barthelemy, and S. R. Rankin. 2014. "Factors Impacting the Academic Climate for LGBQ STEM Faculty." *Journal of Women and Minorities in Science and Engineering* 20(1): 75–98.

Pettorelli, N. W., J. Barlow, M. W. Cadotte, K. Lucas, E. Newton, M. A. Nuñez, P. A. Stephens. 2019. "Applied Ecologists in a Landscape of Fear." *Journal of Applied Ecology* 56:1034–39.

Roughgarden, J. 2013. *Evolution's Rainbow: Diversity, Gender, and Sexuality in Nature and People*. Berkeley: University of California Press.

Shaw, S., and J. Lee. 2020. *Gendered Voices, Feminist Visions: Classic and Contemporary Readings*. New York: Oxford University Press.

Discussion Questions

1. How many LGBTQ+ colleagues or students do you know at work or school? Given the statistics we've presented and the number of people you know, how many of your acquaintances likely identify as LGBTQ+?

If that number is more than the number of LGBTQ+ acquaintances you have, why might that be the case? What actions might you take to more clearly demonstrate that you are an ally or make the environment in which you study or work more accepting and supportive of closeted LGBTQ+ acquaintances?

2. Using Figure 8.2, where do you fit in the privilege spectrum and what traits do you possess that place you there? Perhaps more importantly, what societal (external) and personal (internal) forces provide you that amount of privilege (or lack thereof)? Can you change any of those forces to increase your amount of privilege? Can you change any of those forces to increase the privilege of those around you? If so, which factors and how can you influence them? If you fall into a place of privilege, do you have a moral obligation to help those who do not?

3. If you have a spouse or partner, think back over the past month of conversations you've had with your coworkers or classmates. How many times did your partner come up in conversation and in what context? Why did they come up? How would your discussions have been different if you withheld the fact that you had a partner? How would that have made you feel? Can you imagine a scenario where you wouldn't feel comfortable divulging the fact that you had a partner? What would that scenario be and why would you feel compelled to hide that information?

Activity: Identity and Pronouns

Learning Objectives: Participants will think about the connection between identity and pronouns, be able to appropriately and comfortably ask for and use individuals' pronouns, practice correctly using the pronouns of others, explain the importance of using the pronouns that an individual identifies with, and understand the importance of how the use of correct pronouns fosters an inclusive environment.

Time: 1–1.5 hr.

Description of Activity

Materials needed: None.

Process:

Part 1: Role Playing (20–30 min.)

The pronouns people use for themselves reflect their gender identity. However, gender identity may not be the same as perceived gender expression. The pronouns people use for themselves are important and meaningful, and it is respectful to give people the dignity of using their correct pronouns. When someone is comfortable sharing their pronouns with you, that is a sign of trust.

Pronoun Examples:

They	Them	Their	Theirs	Themself
She	Her	Her	Hers	Herself
He	Him	His	His	Himself
Ze/Zie	Zim	Zir	Zis	Zieself
Per	Per	Pers	Pers	Perself
Ey	Em	Eir	Eirs	Eirself

Participants will meet in small groups of 2–3 people to work through the role-playing scenarios, which will involve practicing asking for pronouns, using the correct pronouns, dealing with situations of misgendering (when someone uses the wrong pronoun), and addressing groups without using gendered terms.

Scenario 1: You are meeting a new colleague that was just hired to work with you in the same office. How would you introduce yourself and include your pronouns?

Scenario 2: You are a faculty member teaching in an undergraduate wildlife program. How would you ask your students for their pronouns?

Scenario 3: While having lunch with a couple of new friends, you use the wrong pronoun with one of them. What do you do?

Scenario 4: You're leading a meeting with people from across various management districts. How could you encourage attendees to share their pronouns and use others' pronouns?

Scenario 5: You are about to call to order a large group of people to begin a seminar. How would you do this using gender-neutral or inclusive language? Can you think of several options?

Scenario 6: As a supervisor of a field crew, you are meeting with the crew members in the morning to help everyone plan their work day and overhear one person use the incorrect pronouns for another crew member. What do you do?

Part 2 (30 min.): Have everyone form a larger group and ask for examples of responses to the six scenarios.

Part 3: Debriefing and Discussion Questions (30 min.)

1. What did you learn?
2. How did you feel when thinking about your pronouns?
3. Why is it important for people who do not identify as male or female to have descriptive pronouns? Do you think many people identify as "Other," which is a common third option when selecting gender on surveys?
4. How would you feel if someone regularly or repeatedly misgendered you by using the incorrect pronouns?
5. What was most surprising about this exercise?
6. Why is using correct pronouns important?
7. How would you use the information you learned or your experience from this activity in your life as a wildlife professional, student, or citizen?

References and Resources

https://www.glsen.org/article/pronouns-resource-educators
https://www.mypronouns.org

Literature Cited

Allen, N. J., and R. Daneman. 2018. "In Memoriam: Ben Barres." *Journal of Cell Biology* 217(2): 435–38.

Barres, B. A. 2006. "Does Gender Matter?" *Nature* 442:133–36.

Barres, B. A., B. Montague-Hellen, and J. Yoder. 2017. "Coming Out: The Experience of LBGT+ People in STEM." *Genome Biology* 18:62.

Booms, T. 2019. "I *Am* One of You: A Gay Wildlife Biologist's Perspective on Our Profession." *Wildlife Professional* 14(2): 34–35.

Brown, A. 2017. "5 Key Findings about LGBT Americans." Pew Research Center. www.pewresearch.org/fact-tank/2017/06/13/5-key-findings-about-lgbt-americans/.

Bureau of Labor Statistics (BLS). US Department of Labor. N.d. *Occupational Outlook Handbook, Zoologists and Wildlife Biologists.* Accessed April 4, 2019. www.bls.gov/ooh/life-physical-and-social-science/zoologists-and-wildlife-biologists.htm.

California Department of Fish and Wildlife (CDFW). 2008. *Equal Opportunity and Sexual Harassment Prevention Policies.* Director's bulletin 2008/006. https://nrm.dfg.ca.gov/FileHandler.ashx?DocumentID=190591.

Cech, E. A., and M. V. Pham. 2017. "Queer in STEM Organizations: Workplace Disadvantages for LGBT Employees in STEM Related Federal Agencies." *Social Sciences* 6(1): 12.

Commonwealth of Massachusetts. N.d. "Learn about Equal Employment Opportunity/Affirmative Action Programs." Accessed April 4, 2019. www.mass.gov/service-details/learn-about-equal-employment-opportunityaffirmative-action-programs.

Flores, A. R., J. L. Herman, G. J. Gates, and T. N. T. Brown. 2016. *How Many Adults Identify as Transgender in the United States?* Los Angeles: Williams Institute.

Georgia-Pacific. 2019. "Affirmative Action Employment Policy." www.gp.com/legal/careers-notifications/affirmative-action.

Gillig, T. K., E. L. Rosenthal, S. T. Murphy, and K. Langrall Folb. 2018. "More Than a Media Moment: The Influence of Televised Storylines on Viewers' Attitudes toward Transgender People and Policies." *Sex Roles* 78:515–27.

Haldane, J. B. S. 1927. *Possible Worlds and Other Essays*. London: Chatto and Windus. https://jbshaldane.org/books/1927-Possible-Worlds/haldane-1927-possible-worlds.html.

Hughes, B. E. 2018. "Coming Out in STEM: Factors Affecting Retention of Sexual Minority STEM Students." *Science Advances* 4(3): eaao6373.

Human Rights Campaign (HRC). 2019. "A National Epidemic: Fatal Anti-transgender Violence in the United States in 2019." www.hrc.org/resources/a-national-epidemic-fatal-anti-trans-violence-in-the-united-states-in-2019.

Idaho Division of Human Resources (DHR). 2019. "Section 9: Respectful Workplace." In *Executive Branch Statewide Policy*. https://dhr.idaho.gov/wp-content/uploads/RW-Policy.pdf.

James, S. E., J. L. Herman, S. Rankin, M. Keisling, L. Mottet, and M. Anafi. 2016. *The Report of the 2015 US Transgender Survey*. Washington, DC: National Center for Transgender Equality.

Liberty University. 2019. "Non-discrimination Policy." www.liberty.edu/administration/institutionaleffectiveness/index.cfm?PID=30442.

Loftus, J. 2001. "America's Liberalization in Attitudes toward Homosexuality, 1973 to 1998." *American Sociological Review* 66(5): 762–82.

Lopez, G. 2017. "For Years, This Popular Test Measured Anyone's Racial Bias: But It Might Not Work After All." *Vox*. www.vox.com/identities/2017/3/7/14637626/implicit-association-test-racism.

Louisiana Department of Wildlife and Fisheries (LDWF). 2013. "EEO Statement." https://www.ldaf.state.la.us/about/employment/.

Low, T., and K. Tu. 2017. Out at Work. *Nancy* (podcast). October 29. www.wnycstudios.org/podcasts/nancy/episodes/nancy-podcast-out-at-work.

Mallory, C., and B. Sears. 2015. *Evidence of Employment Discrimination Based on Sexual Orientation and Gender Identity: An Analysis of Complaints Filed with State Enforcement Agencies, 2008–2014*. Los Angeles: Williams Institute.

McNaught, B. 2017. *Brian McNaught's Guide to LGBTQ Issues in the Workplace*. Independently published. www.brian-mcnaught.com/resources.html.

Miller-Merrell, J. 2017. "Company Culture: 4 Ways to Create an LGBT-Friendly Workplace." Cornerstone. www.cornerstoneondemand.com/rework/4-ways-create-lgbt-friendly-workplace.

Morgan, K. P. 1996. "Describing the Emperor's New Clothes: Three Myths of Educational (In-)Equity." In *The Gender Question in Education: Theory, Pedagogy, and Politics*, edited by A. Diller, B. Houston, K. P. Morgan, and M. Ayim, 105–22. Boulder, CO: Westview Press.

Movement Advancement Project (MAP). 2019a. Local Non-discrimination Ordinances. Accessed April 4. lgbtmap.org/equality-maps/non_discrimination_ordinances/policies.

Movement Advancement Project (MAP). 2019b. Equality Maps: Religious Exemption Laws. Accessed April 4. http://www.lgbtmap.org/equality-maps/religious_exemption_laws.

Movement Advancement Project (MAP). 2019c. Equality Maps: State Non-discrimination Laws. Accessed April 4. http://www.lgbtmap.org/equality-maps/non_discrimination_laws.

Newport, F. 2018. "In U.S., Estimate of LGBT Population Rises to 4.5%." Gallup. May 22. https://news.gallup.com/poll/234863/estimate-lgbt-population-rises.aspx.

Norton, A. T., and G. M. Herek. 2013. "Heterosexuals' Attitudes toward Transgender People: Findings from a National Probability Sample of U.S. Adults." *Sex Roles* 68:738–53.

Olfenbuttel, C., T. Booms, C. Crow, and K. O'Donnell. 2020. "Out in the Field: A New LGBTQ+ Initiative Takes Shape within the Wildlife Society." *Wildlife Professional* 14(2): 34–37.

Oregon State University (OSU). 2018. "Bias Incident Response Process." https://diversity
    .oregonstate.edu/bias-incident-response-process.
Oregon State University (OSU). 2019a. "Discrimination and Harassment Policies." https://
    eoa.oregonstate.edu/discrimination-and-harassment-policies#protected%20status.
Oregon State University (OSU). 2019b. "OSU Search Advocate Program." https://
    searchadvocate.oregonstate.edu/.
Project Implicit. 2011. Implicit Association Test. https://implicit.harvard.edu/implicit
    /selectatest.html.
Tankard, M. E., and E. L. Paluck. 2017. "The Effect of a Supreme Court Decision Regarding
    Gay Marriage on Social Norms and Personal Attitudes." *Psychological Science*
    28(9):1334–44.
Taylor, J. K., D. C. Lewis, and D. P. Haider-Markel. 2018. *The Remarkable Rise of Transgender
    Rights*. Ann Arbor: University of Michigan Press.
Totenberg, N. 2020. "Supreme Court Delivers Major Victory to LGBTQ Employees." NPR
    News. www.npr.org/2020/06/15/863498848/supreme-court-delivers-major-victory-to
    -lgbtq-employees.
US Department of Justice (DOJ), Federal Bureau of Investigation. 2018. "Hate Crime
    Statistics, 2017." https://ucr.fbi.gov/hate-crime/2017.
US Department of Justice (DOJ), Federal Bureau of Investigation. 2019. "Hate Crime
    Statistics, 2018." https://ucr.fbi.gov/hate-crime/2018.
Wax, A., K. K. Coletti, and J. W. Ogaz. 2018. "The Benefit of Full Disclosure: A Meta-
    analysis of the Implications of Coming Out at Work." *Organizational Psychology Review*
    8(1): 3–30.
The Wildlife Society (TWS). 2012. Bylaws. https://wildlife.org/wp-content/uploads/2014/05
    /bylaws.pdf.
The Wildlife Society (TWS). 2017. Code of Ethics. https://wildlife.org/wp-content/uploads
    /2017/07/Code-of-Ethics-May-2017.pdf.
Yoder, J. B., and A. Mattheis. 2016. "Queer in STEM: Workplace Experiences Reported in a
    National Survey of LGBTQA Individuals in Science, Technology, Engineering, and
    Mathematics Careers." *Journal of Homosexuality* 63(1): 1–27.

Jennifer Malpass, Jamila Blake, Erin Considine

Intergenerational Insight

#MeetTheMillennials

The children now love luxury; they have bad manners, contempt for authority; they show disrespect for elders and love chatter in place of exercise. Children are now tyrants, not the servants of their households. They no longer rise when elders enter the room.
Socrates, 469–399 BCE

#Hashtag

Age is a key form of diversity in contemporary workplaces. In recent decades, the rate of social change has increased, thus shortening generational spans, and now work environments may support as many as five generations at a time (Lyons and LeBlanc 2019). Understanding generational differences can provide context for seemingly inexplicable behaviors and overcome stereotypes that people from a different generation "just don't get it."

The five generations of wildlife professionals working together today have a variety of previous experiences and motivations for pursuing their careers, reflecting the changes we have witnessed since Aldo Leopold offered the first definition of wildlife science in the 1930s. Conservation efforts in the United States often link back to the concerns of hunters to sustain wildlife populations. However, the number of registered hunters in the US has declined in the past 25 years, and by 2016, only about 4% of the population of a given age group held hunting licenses (USFWS 2018). The majority of these hunters were at least 45 years old and 90% were men (USFWS 2018). Rising generations have continued to shift away from such traditional points of entry into the field and may have different inspirations than their predecessors for engaging in the wildlife profession. According to the US Fish and Wildlife Service, approximately one-third of the population participated in "wildlife watching" activities in 2016 (USFWS 2018), including visiting parks and natural areas, wildlife photography, and closely observing wildlife. Members of younger generations that are not as heavily engaged in hunting may have a strong desire to preserve natural spaces for other wildlife-related recreation activities and to tackle issues that transcend species and habitats, like climate change.

All of us writing this chapter are millennials. We all identify as straight, cisgender women, and we grew up in middle-class households. We each have been animal lovers from a young age and explored different career paths before finding our way to the wildlife profession. An interview with a biologist at the San Diego Wildlife Animal Park when chapter author Erin Considine was 12 initially led her toward a degree in zoology, but as her college years approached, she realized this focus meant (too) many classes inside or looking under a microscope. Brief conversations with Dr. Carol Chambers and former dean David Patton while on a precollege visit to the School of Forestry at Northern Arizona University sparked the idea that a career managing wildlife habitat could be hugely impactful and fulfilling. Considine earned both her undergraduate and graduate degrees there, in forestry with an emphasis in wildlife management, preparing her for her current position with the US Forest Service as a district wildlife biologist in western Nebraska focused on habitat enhancement projects that incorporate grazing, recreation, and prescribed fire. Chapter author Jennifer Malpass thought that if you liked school and animals, veterinary medicine was the obvious choice, until she spent a semester abroad in South Africa studying savannah ecology and realized how many more job prospects there were for wildlife biologists than for wildlife veterinarians. She finished her liberal arts degree in biology, then spent a few years working temporary technician positions in the eastern United States and Thailand before returning to The Ohio State University for a PhD in fisheries and wildlife science. She has since worked with the US Fish and Wildlife Service and US Geological Survey on migratory bird projects before moving into her current role with USGS in partner and employee engagement. Chapter author Jamila Blake explored both zoology and veterinary medicine before everything clicked when she took a course in wildlife policy and administration as part of her undergraduate studies at University of Delaware in wildlife conservation and ecology. Blake loved the idea of being able to engage in the wildlife field in a way that supported and empowered others to thrive in their chosen discipline and conduct outreach with the public, legislators, and other organizations. She now works as the professional development manager at The Wildlife Society (TWS) and is concurrently pursuing a master's in public administration with a concentration in nonprofit management.

Millennials as a generation are transformational. Millennials may be viewed positively for our confidence, ambition, collaboration, tech skills, and flexibility. Yet, we also may be accused of being entitled, lazy, and self-obsessed. Millennials are poised to dominate the workforce in coming years based on the size of our population and our progression in our careers, so it is time to understand who we are. Let's start with the hashtag (#). Millennials have transformed the pound sign to a hashtag for use on social media platforms, preceding words or phrases to identify messages related to a specific topic or theme. In this chapter, we tap the hashtag to connect you with the millennial world. We highlight generational differences as well as common threads, including the impacts of technology and college debt, along with the needs that arise with a changing workforce.

#TalkinBoutMyGeneration

What makes a generation? Johnson and Johnson (2010) define it as "*a group of individuals born and living contemporaneously who have common knowledge and experiences that affect their thoughts, attitudes, beliefs, and behaviors.*" First, members of a generation share a common location in history, are about the same age, and are usually born within a decade or two of each other. As such, the generation we are born into is our generation for life: it is not something we grow out of or can change. Second, members of the same generation share experiences and thus may have a common mindset that may help define where break points should fall between generations (Table 9.1). However, it is important to note that characteristics of generations may be generalized across the entire population, and thus there may be exceptions to these characteristics within individuals or groups within a given generation.

Generations are usually defined in 15–20 year periods, in part because people coming of age around the same time experience different generational signposts than those coming of age 15–20 years earlier. A generational signpost is a distinct event or popular phase that is specific to one generation, and it often occurs as members of that generation are coming of age, when the worldview of those individuals is highly malleable (Johnson and Johnson 2010). Dates for each generation are also informed by periods of fluctuating birth rate in the United States (Johnson and Johnson 2010); however, assigning generational dates is not an exact science (Table 9.1). Demographers often disagree about

TABLE 9.1. Comparison of generational signposts within the United States (Johnson and Johnson 2010; Zemke, Raines, and Filipczak 2013).

	Traditionalists	Baby Boomers	Generation X	Millennials	Generation Z
Birth years (approx.)	1928–1944	1945–1964	1965–1980	1981–1995	1996–2012
Defining social involvements, events, and trends	Great Depression New Deal World War II Korean War Rise of labor unions	Vietnam War Hippies Moon landing John F. Kennedy assassination TV Civil Rights Act Sexual revolution	Arab oil embargo Fall of the Berlin Wall Challenger explosion MTV Stonewall rebellion AIDS epidemic Watergate Title IX *Roe v Wade* Pregnancy Discrimination Act	Columbine shooting 9/11 attacks Y2K Social media Reality TV Tea Party Occupy Wall Street Great Recession	Black Lives Matter movement #MeToo Same-sex marriage legalized
Family matters	Women as homemakers "Spare the rod, spoil the child"	Women may have career before kids, then homemakers *Feminine Mystique* published Oral contraceptives introduced Marry early, start families young Nuclear families	Both parents working Latchkey kids Divorce *Mommy Wars* published	*Lean In* published Marry later, kids may come before marriage Unique child names No "one right way" to raise children or define a family Blended families	*To be determined*
Environmental events and trends	Dust Bowl Duck Stamp Act Pittman-Robertson Act Aldo Leopold publishes first textbook in the field of wildlife management	Dingell-Johnson Act Waterfowl Breeding Population Surveys start Clean Air Act IUCN Red List	National Environmental Policy Act First Earth Day Endangered Species Act Recognition of impacts of climate change	Montreal Protocol Rise of wildlife ballot initiatives Information Theoretic Approach / multimodel inference Chytrid fungus discovered	Teaming with Wildlife Restoring America's Wildlife Act introduced White-nose syndrome discovered

TABLE 9.1. *(continued)*

	Traditionalists	Baby Boomers	Generation X	Millennials	Generation Z
	First Wildlife Management Department created by Aldo Leopold The Wildlife Society founded *Journal of Wildlife Management launched*	Wildlife radio telemetry introduced Breeding Bird Survey established Cuyahoga River fire	North American Wetland Conservation Act *Exxon Valdez* oil spill SAS and use of computers to improve data analysis	*An Inconvenient Truth* Deepwater Horizon oil spill MOTUS network launched ICARUS satellite launched	Ivory Crush M-Opinion redefines "take" of migratory birds; subsequently vacated
Milestones for diversity and inclusion in the wildlife profession	Buffalo soldiers are first caretakers of Yosemite and Sequoia National Parks Harriet Hemenway and Minna Hall boycott the plume trade, leading to the formation of National Audubon Society and passage of the Migratory Bird Act	Elizabeth Beard Losey is 1st female member of TWS	Rachel Carson publishes *Silent Spring* Lucille Stickel is 1st female recipient of Aldo Leopold Memorial Award Conservation Sales Tax passes in Missouri	First annual conference of TWS *Human Dimensions of Wildlife* Diana Hallett elected 1st female president of TWS North American Wetland Management Plan incorporates human dimensions	First TWS WOW event Wini Kessler elected 2nd female president of TWS Wini Kessler is 2nd female recipient of TWS Aldo Leopold Memorial Award Carol Chambers elected 3rd female president of TWS Christine Lehnertz serves as first female and openly gay superintendent of Grand Canyon National Park

which specific year is the break point, particularly for younger generations that are still being defined and whose generational signposts are more often related to cultural phenomena instead of a time-specific event (Erickson 2008). For example, those born as World War II ended are easily recognized as baby boomers because of the huge postwar spike in births. In contrast, Generation X, millennials, and Generation Z do not have a similar event marking the beginning of their generations.

Although millennials have distinctive traits that set us apart from other generations, we also possess similarities with previous generations. Like traditionalists, millennials have firm moral convictions, a strong sense of civic duty, and a general can-do attitude (Alsop 2008). Similar to baby boomers, millennials are willing to challenge the status quo, want to know the "why" behind the way things are done, and are team oriented (Alsop 2008). Like members of Generation X, millennials share a propensity to integrate technology into our daily lives to make life easier and bring family and friends closer together (Pew Research Center 2010). Generation X and millennials also have a common desire for greater work-life balance than was available to previous generations, although the motivation may stem from different places: Generation Xers may want more family time to contrast with their own childhood experience as latchkey kids, while millennials may seek to perpetuate close familial and friend relationships established when we were growing up (Alsop 2008).

#KidsTheseDays

Each generation imagines itself to be more intelligent than the one that went before it, and wiser than the one that went after it. —George Orwell

Intergenerational conflict in the workplace affects 75% of workers (Karsh and Templin 2013). Conflicts can stem from differences in communication styles, expectations, and generational perspectives (Myers and Sadaghiani 2010). Multiple generations have always shared the workplace, but the need to negotiate generational differences may be more acute for today's workers for three reasons. First, advances in health care have increased life expectancy. Demographers originally defined four life stages—childhood, adolescence, adulthood, and old age—but have since added odyssey, "*the decade of wandering that frequently occurs between adolescence and adulthood*," as described by *New York Times* columnist David Brooks (2007) and active retirement, given contemporary life expectancy (Erickson 2008). For many, increases in life expectancy can translate to working longer (Vernon 2016). Recently, even those with the means to retire may choose not to for mental stimulation and to maintain social networks in older age (McKie 2018).

The differences in experience for the five generations of wildlifers working together affects us whether we work in academia, government agencies, nonprofit organizations, or consulting firms. For example, members of older generations may have developed outdoor skills early in their childhoods to keep themselves occupied or help around their households. This was the case for coauthor Considine, whose family loved exploring nature and spent many summer nights camping across Arizona and New Mexico. In contrast, some members of younger generations may have been less likely to grow up chopping wood or setting up tents. Coauthor Blake did not grow up in a family that was particularly outdoorsy or had any background in wildlife biology, hunting, or similar fields. She was inspired by the birds, frogs, and lizards she found in her backyard in Tampa, Florida, and started Girl Scouts as a Daisy, choosing badges and projects centered on her areas of interest, usually focused on animals; camping and hiking outdoors; and building community and leadership skills. Similarly, coauthor Malpass grew up flipping rocks in the backyard of her home in a suburb of Chicago, Illinois, and joined the Girl Scouts and later Boy Scouts to take advantage of their camping trips.

A second reason generational differences may be more acute in today's workplaces is that more women are working and are changing long-standing stereotypes of traditional gender roles (Schulte 2014). The growing numbers of women in the workforce can cause discomfort for our male colleagues, but also female colleagues of different generations who have different approaches to the challenges of being a woman in a male-dominated field. The tightrope of appropriate behaviors women wildlifers must walk is epitomized by the struggles we face simply getting dressed for work (Box 9.1). Trailblazing woman of wildlife Elizabeth Beard Losey (née Browne) was the first woman inducted into TWS in 1948 and the first female biologist of the US Fish and Wildlife Service, but with "*mixed emotions*," she left her career upon her marriage in 1953 (Casselman 2006). Today, it is more common for women to continue work after marriage or children, but mixed emotions about appropriate behaviors for married women and mothers persist (Schulte 2014; Williams and Dempsey 2014). In addition, now more women are pursuing graduate education and utilizing those degrees in more technical fields, including natural resources (Lopez and Brown 2011).

A third reason generational considerations may be so important now is that work itself has changed dramatically. Agrarian and manufacturing work of the past has shifted to knowledge-based work, in step with new technologies,

BOX 9.1 #DressForSuccess

In the modern workplace, women are still conscious of the potential repercussions of what they wear. Why is dress so important? What you wear makes a strong first impression, changes how others perceive you, and influences self-confidence. Expectations for how women should dress and behave have changed throughout history, and pants in particular have been a source of contention. Both men and women wore long robes or dresses in the 14th and 15th centuries, but men's robes gradually became shorter. Before long, all that was left to cover men's legs were their "hose," which resembled present-day pants (Brucculieria 2019). In the 1850s, women began to wear "bloomers," trousers loose in most of the leg but gathered at the waist and ankles. Bloomers gave women more freedom in the way they dressed and enabled them to engage in physical activities (Brucculieria 2019). Historically, women have been jailed or even put to death just for "dressing like a man." In the United States, legislation was enacted as far back as the 1840s to prevent people from dressing in conflict with their perceived gender (News Desk 2015).

Wearing pants or a pantsuit became associated with women's empowerment, independence, and control over their bodies starting in the 1960s (Brucculieria 2019). Millennial girls growing up in the 1990s typically had the option to wear pants, shorts, or even skorts, which facilitated play and engagement in the outdoors. In the authors' experience, time spent playing outside during childhood is often what initially sparked an interest in the conservation field as a profession. Skorts help strike a balance between femininity and functionality of clothing, and many outdoor clothing stores now offer hiking skorts for women of all ages.

In the National Park Service, women were not given official uniforms until 1947. Women were dissatisfied with multiple revisions of the uniform, which were impractical, uncomfortable, or ill fitting for work outside (fitted jackets, skirts, and A-line dresses). After much discussion, all employees were authorized in 1977 to wear any uniform configuration, with the additional option for women to wear a skirt (NPS 2017). In preparing this chapter, we and other authors repeatedly heard from women wildlifers of all generations that clothing choices remain a key source of angst today. In our experience, women of earlier generations may choose to err on the side of caution when it comes to selecting clothing for fieldwork to avoid "inviting trouble," while younger women may have stronger objections to this school of thought, especially in light of movements like #MeToo. Wearing pants could be a choice for comfort or practicality but also a way to fit in with men. In fields dominated by men, some women may consider it important to assert their "otherness" through dress, while others take on a more masculine appearance to gain acceptance. The only woman among 64 faculty in her forestry department during the 1990s wore suits or dressed formally to avoid being viewed as a mother, sister, or daughter by her all-male coworkers.

Wildlife professionals may find themselves working in a variety of environments. When working in different regions of the world, cultural considerations may make choices like skorts or clothing that covers shoulders and knees more compelling. In an office environment, women often face a difficult balancing act to determine what clothes they should wear to be comfortable, do their job properly, and avoid unwanted attention. These considerations can limit self-expression through clothing and appearance, as well as lead to stress and anxiety. In addition, dressing for work can lead to decision fatigue because there are too many options and it is difficult to make a meaningful decision that honors the individual but does not cause problems at work (Cohen 2016). Coworkers often offer advice, but recommendations may differ among generations. For example, one wildlifer advised her students, who were younger women, to avoid wearing tank tops when in transit to the field site to avoid catcalls. Even with more options and changing attitudes, figuring out what works for you as individual in your unique workplace can be a frustrating experience. However, because of the ubiquity of concerns related to clothing among women, you will be sure to find others who are willing to discuss their strategies for success.

economic expansion, and transitions from farm to city (Schulte 2014). Wildlife professionals have witnessed a proliferation of roles and responsibilities in the past century. Aldo Leopold's definition of wildlife science has expanded beyond *"the art of making land produce sustained annual crops of wild game for recreational use"* (Leopold 1933). Today's wildlifers work with more species, particularly nongame species, and a wider stakeholder base. The wildlife field has also shifted to incorporate professionals with additional areas of expertise, such as genetics, disease, toxicology, and social sciences. Increased attention on transparency in planning and management activities is now the norm. Such shifts can cause conflicts with older-generation workers who remember when the focus of the profession was narrower.

#MillennialLife

Millennials were part of the second big baby boom in the 21st century, and we experienced a renewed focus on family life, with children as the center of family life. "Baby on Board" car decals popularized in the 1980s and 1990s, declared, *I have a baby in here! The most precious, special, amazing thing in the world, and*

I have one! (Karsh and Templin 2013). The days of "children should be seen but not heard" were over; as children we may have been regularly consulted for input and could have had a voice in family decisions from a young age (Karsh and Templin 2013). In addition, millennial children may not have been denied much of what they wanted (Alsop 2008). We generally were raised to believe we could be anything we wanted to be, and if we wanted something enough, we could get it (Karsh and Templin 2013). If baby boomers were the "me generation," millennials became the "*me me me generation*" (Stein 2013).

Parenting styles also shifted in the 1980s when millennials were young. Making millennial children feel good about ourselves was important, and boosting self-esteem was a crucial parenting goal. Early on, Erin was encouraged by her parents to choose a career that was so well aligned with her passions, she would never have to "work" a day in her life. Even as a youth, she knew this meant finding a career that would allow her to work outside. In contrast to the latchkey kids of Generation X who were often left with little parental supervision, many of us had parents who were able to spend more time at home raising us (Council of Economic Advisers 2014; Karsh and Templin 2013). As a result, many millennials have a strong sense of connection to our family and friends, and value our community. About 50% of millennials say it is important to live near family and friends, compared to about 29% of baby boomers and 40% of Generation X (Council of Economic Advisers 2014). Our parents have even earned their own monikers. The term "helicopter parent" was coined to describe them hovering over us when we had to make decisions or handle difficult situations; others compare parents of millennials to lawn mowers with the propensity to mow down obstacles in front of their kids (Karsh and Templin 2013). Although this parenting style reflected attempts to help and protect us, millennials frequently may be characterized as sheltered with limited capacity to manage challenges on our own (Karsh and Templin 2013).

#ComingOfAge

Generational signposts were revealed as millennials entered adolescence. The World Wide Web was introduced (August 6, 1991), and greater international connection allowed us to keep up with global events and find like-minded people around the world. Creating a Facebook profile was a rite of passage of our generation because it required a college email address before membership was opened to everyone over 13 in September 2006 (AP 2014). We were the generation that experienced school lockdowns during the shooting at Columbine (April 20, 1999) and the attacks on the World Trade Center (September 11,

2001; Table 9.1). Both events profoundly affected Americans of all ages, but to experience inexplicable and previously unimaginable violence as teenagers established the millennial mindset that the world is a dangerous and unpredictable place (Erickson 2008).

Growing up in an era of human-caused terror and natural disasters also exposed millennials to extraordinary examples of how communities come together in crisis, which ingrained in us the importance of civil service, volunteering for a cause, and protecting the planet (Erickson 2008). While earning her undergraduate degree, coauthor Blake worked with a number of local environmental organizations and connected with a community of people who shared her passion. A pivotal moment in shaping her path forward was joining the University of Delaware's student chapter of TWS. Attending meetings, going on field trips, and learning more about the profession made Blake want to become heavily involved, and eventually she served as the vice president and president of the club. Malpass entered her graduate program with the intent of pursuing a career in federal service and found resonance with the US Geological Survey's motto, "*science for a changing world.*" Similarly, throughout her higher education Considine looked for opportunities to involve surrounding communities and youth in landscape restoration projects, which led her to a federal career with the Forest Service, whose motto is, "*caring for the land and serving the people.*"

The financial crash in 2008 and subsequent Great Recession strongly impacted millennials because many of us were entering or preparing to enter the workforce. Those of us with less seniority were the first to be laid off. As those with mid-level positions lost their jobs, competition for entry-level jobs and seasonal positions became fierce. We may not have been as attractive candidates as those who already had experience in the field. Our parents may have been hit hard too, but usually still had the means to invite us back home, reinforcing our strong family relationships. Malpass benefitted from her family's support in early 2009 when the grant funding for her wildlife technician position in Thailand was not renewed. She returned to her family home to regroup before pursuing a volunteer opportunity at a bird banding station that allowed her to gain prerequisite experience for future paid work on bird migration projects and her later dissertation research on urban birds.

As wage growth slowed, many millennials leaned on their parents for financial support or housing (Council of Economic Advisers 2014). Living at home allowed millennials to take lower-paying jobs but came with the associated stigma of not becoming a self-sufficient adult. Financial independence, marriage, children, and home ownership often define the transition to adulthood, and the age at which people reach these milestones has been rising for the past several decades (Erickson 2008). In 1975, 45% of people between the ages of 25 and 34 were employed, lived away from parents, had been or were married, and lived with a child. By 2016, the percentage of 25- to 34-year-olds who had achieved all four of these milestones had dropped to 24% (Vespa 2017). Members of generations who reached these milestones earlier in life sometimes view this as an indicator of maturity or responsibility, which can be a source of conflict in the workplace.

#Technology

Another key area of intergenerational conflict in the workplace is how we handle technology within and outside of work. Millennials lead older generations in technology adoption and use as "*digital pioneers*," because we were adolescents when new technologies like cell phones and the internet first became mainstream (Vogels 2019). Before cell phones with cameras and satellite tracking applications like Google Maps, Considine explored the maze of arroyos around her home in southern Arizona. She would carry a disposable camera to note different wildlife tracks and then had to wait for the camera to be full and get the film developed. Only then could she begin researching her newest discovery at the local library. Now, she takes advantage of advanced cell phone technology in both her work and personal life, utilizing 15+ applications for mapping, camping/hiking, species identification, and community science participation.

In contrast, baby boomers, Generation X, and traditionalists are "*digital immigrants*," having adapted to technology after their teenage years, and Generation Z has had access to mobile technology and internet-enabled devices their entire lives (Karsh and Templin 2013). Today, almost all millennials own smartphones; about 28% exclusively use our smartphones for internet access, and about 80% sleep with our phones next to the bed (Karsh and Templin 2013). Increasing use of technology coincides with decline of memorization; whereas older generations may have depended on rote learning, the ubiquity of search functions allows millennials and Generation Z to focus on skills related to finding, verifying, and applying information (Erickson 2008). Older generations may view our way of retrieving information as a lack of preparation, and complain that today's professionals aren't as ready for the workforce because we don't have scientific names memorized. What older generations may fail to recognize, however, is that the newer generations (millennials, Generation Z, and many Xennials) have honed different skills to match the evolving field of research and management such as geographic information systems (GIS), modeling, remote sensing, and genetics.

Millennials' lives are distinctively interconnected with our online presence and social media profiles (Pew Research Center 2010). Technology influences how we find new jobs, activities, romantic partners, and best friends. Millennials may define themselves by their career choice and technology allows us to share that happiness (or unhappiness) with the world. Our proclivity to share our feelings related to work can be especially true for some wildlifers who are quick to "humble-brag" about our accomplishments and neat critters we study.

#DontQuitYourDaydream

Millennials tend to be more fulfilled by career prospects than previous generations; 47% said they would prefer to work even if it was not a financial necessity, compared to 37% of Generation Xers and 36% of baby boomers (Working Mother Research Institute 2014). Millennials may emphasize finding work that is personally fulfilling, and about 85% of us say it is more important to make a positive difference in the world than it is to receive professional recognition (Meister and Willyerd 2010; Darshan 2011). Compared to baby boomers and Generation X, millennials also tend to value salary earnings, with women driving the emphasis on improving pay (Council of Economic Advisers 2014).

Many millennials may want to incorporate time for recreational activities to achieve work-life balance (Council of Economic Advisers 2014). Baby boomers may have focused on climbing organizational ranks and been willing to put in long hours to be competitive for promotions. In contrast, millennial workers may communicate interest in working flexible hours or exploring multiple career trajectories beyond the traditional "up and out" model because we often prioritize maintaining a certain lifestyle outside of work. A millennial may feel it is equally important to leave work at 5 p.m. to make it to yoga as it is to pick up children from daycare. For millennial wildlifers in particular, not only is outdoor recreation important, but some of us love to broadcast to our peers via social media that hiking, camping, or hunting is how we spend our free time.

The different roles that work plays in our lives may lead baby boomers in leadership positions to distrust millennials' commitment. To compensate, we may over-accommodate the demands of the workplace to demonstrate our dedication to the organization's success. In the wildlife profession, great passion for work may lead some people from any generation to overwork. However, the contemporary attitude shift toward a greater focus on balancing work responsibilities with life outside of work may also contribute to positive organizational change. Ultimately, expanding our definition of "work-life balance" may aid leadership in reevaluating priorities and values for all employees, regardless of each generation's motivation to better balance work and life (Myers and Sadaghiani 2010).

#TeamWorkMakesTheDreamWork

From an early age, millennials may have been pushed toward activities and habits that would result in an impressive resume. It was common for many of us to participate in multiple extracurricular activities both for leisure and to stand out on college applications. Team dynamics were emphasized when we were growing up: in school, on sports teams, and in other clubs or activities. As a result, millennials may be particularly effective working in collaborative environments (Karsh and Templin 2013). In college, team projects acclimated us to sharing responsibility for the final product. We were taught that each person sitting at the table brings a unique perspective to problem solving. Now, as we become managers in wildlife conservation, that "pack mentality" may influence how we address some of the world's most complex wildlife issues, including habitat fragmentation, climate change adaptation, wildlife trafficking, and sea level rise. The interdisciplinary framework is common across federal agencies. In Considine's experience with the Forest Service, an agency that manages 193 million acres, an extensive team environment is commonplace for work on critical landscape-scale issues such as wildfire mitigation and rehabilitation, loss of forested land to urban growth and development, and strengthening existing markets for wood.

Millennials and Generation Z have become the most diverse generations and are more accepting of differences and nontraditional behaviors than previous generations (Pew Research Center 2010). Whereas initial programs to increase workforce diversity focused on tolerance and compliance with equal opportunity laws, some members of younger generations are helping advance this conversation toward embracing diversity and inclusion. Each of the authors has been involved with various diversity and inclusion efforts within the wildlife profession, such as TWS's Inclusion, Diversity, Equity and Awareness Working Group (formerly the Ethnic and Gender Diversity Working Group), the Diversity Joint Venture for Conservation Careers, and Department of Interior's Diversity Change Agents. However, as a Black wildlifer Blake brings a different perspective than Malpass and Considine, who both are white. At times Blake felt a little out of place or lost, especially when she didn't see people like her in the profession or didn't think she fit into any of the career options most graduates of her program chose. She made a point of seeking out and building community to talk about these concerns and find support for herself and others. Blake was fortunate and grateful to form a group of lifelong friends but also threw herself into programs that could help to shape her experience and hopefully help other students from underrepresented communities. She became an ambassador for her college and served as an intern and mentor for a program developed to support underrepresented students and engage in conversations with faculty and staff. In her current role, Blake is the lead member of TWS headquarters staff working on diversity, equity, and inclusion initiatives for the society.

Within the wildlife profession, there are two crucial implications in the shift from simple tolerance to welcoming because of our early lessons in the importance of allowing all members of the group to express opinions. First, millennials may be uniquely positioned to institute the cultural changes necessary to overcome historical underrepresentation in the wildlife profession. Second, millennial wildlifers may be more adept at engaging with people who offer differing perspectives, helping us to address common concerns and benefit from wider knowledge.

#StressedOut

Millennial women in wildlife may face several challenges related to the intersectionality of their generation, age, gender, and profession. Millennials are more likely than any other generation to point to stress as having a significant negative impact on both their physical and mental health, and women report depression and anxiety at higher rates than men (APA 2015). Perfectionism is more widespread among millennials than any previous generation and has been linked to anxiety, depression, and obsessive-compulsive disorder (Frost and Di-Bartolo 2002; Curran and Hill 2019). Women in particular may feel pressured to pursue perfectionism to demonstrate their competence at work (Williams and Dempsey 2014).

Fifty years ago, baby boomers were the first generation to be labeled "workaholics." After World War II, conspicuous consumption may have been used to demonstrate status, and working longer hours and women entering the workforce could translate to more money for families to spend on status purchases like better houses or bigger cars. Today, millennials may be following in their parents' footsteps with workaholic behaviors, but what constitutes "work" and what defines "status" may be different for us. A key measure of status for many millennials is securing a job that reflects well on our parents, is impressive to our peers, and resonates with some of the aspirations we had as children (Petersen 2019). Whereas older generations may have viewed work primarily as a way to pay the bills, work may represent much more for younger generations who feel pressured to find a job we love, engage in long hours of hard work we are passionate about, show everyone how much we love our job via social media, and ultimately link our identity with that of our employer (Griffith 2019).

Unfortunately, working longer hours does not necessarily lead to increased productivity or creativity (Griffith 2019). Relentless self-optimization may have been reinforced for millennials in every facet of our life since we were young. We may experience burnout because we have internalized the idea that we need to be working all the time (Petersen 2019). Burnout goes beyond the state of exhaustion, where your stress and anxiety push you to do more anyway (Petersen 2019). Millennial women are especially driven to overwork because gender bias requires us to repeatedly prove our worth (Williams and Dempsey 2014). As early career wildlifers engage in internships and short-term and entry-level positions, we quite literally have to prove ourselves over and over to gain experience and the respect of our colleagues. When first entering into the workforce, Blake's only interviews were for positions that were three to six months in duration. As is common for wildlifers, when she was hired on as an intern, she was told there were no opportunities for growth into a full-time position with the organization because funding was limited. This lack of job security meant Blake was constantly searching for her next position while working full time, as well as holding a part-time position at a local zoo to supplement her income.

Millennials may have grown up with consistent pressure to do well and get things right the first time. As a result, many of us may fear failure, a fear that is only intensified in fast-paced environments where it may be perceived that there is no recovery time (Alsop 2008). A common solution to burnout is to take a vacation and calm the mind, or at the very least disconnect from technology. However, millennials often remain connected via our smartphones, even when on break, to avoid the anxiety of missing important work or social information (Cohen 2016).

Nearly 20% of millennials who seek mental health resources from their employer do so to address depression, compared to 16% of baby boomers or Generation X (Petronzio 2015). Millennials have made significant contributions to destigmatizing public perceptions of mental health issues, but higher levels of stigma persist within minority communities, even though Black Americans of any age are 20% more likely to experience mental health conditions than members of other races (Allen 2018; Lorusso and Barnes 2019). About 75% of millennials are open to talking with friends and family about mental health issues, and 70% of us say that we would be willing to see a therapist (Hoff 2018). However, millennials also report not seeking treatment because of the associated time and cost expenditures (Lorusso and Barnes 2019). In the wildlife field, early career positions may not provide nonpermanent employees with health care benefits or access to mental health resources. While some young professionals can be covered under their parents' health care plans, this is not always the case. Blake couldn't remain on her mother's insurance plan after she turned 23 and found that securing affordable health care coverage became a top priority after graduation.

Mental health issues may also be exacerbated by daily exposure to social media platforms. Using multiple social media platforms can lead to higher levels of anxiety and depression (Primack et al. 2017). Viewing the lives of others through the filter of social media can make it seem like they have it all together and you are faking it. Almost 70% of millennials feel some sense of imposter syndrome (Kerr 2015): "*high-achieving individuals who, despite their objective successes, fail to internalize their accomplishments and have persistent self-doubt and fear of being exposed as a fraud or impostor*" (Bravata et al 2020). Seeing other people post on social media can make it seem like they live a balanced, satisfied life unaffected by burnout (Petersen 2019).

#NoMoneyMoProblems

Millennials are one of the most educated generations, with a rise in the number of degrees completed after high school, increased numbers of young workers who hold a degree, and higher college enrollment numbers (Council of Economic Advisers 2014). Unfortunately, increased participation in higher education as millennials came of age coincided with rising tuition rates, state government funding cuts, and our parents' inability to offset our college costs (Council of Economic Advisers 2014). About 50% of students had to borrow money during the 2013–2014 school year compared to approximately 30% in the mid-90s, and 63% of millennials have more than $10,000 in student loan debt (Council of Economic Advisers 2014). The extent of student loan debt is particularly bleak for millennial women: 42% owe more than $30,000, compared to 27% of millennial men (Zetlin 2017). Women and millennials are among the groups in the United States who tend to feel the most stress about money. About 75% of millennials report it is somewhat or a significant contributor to their stress levels, and for nearly half the women in the United States, paying for essentials is either somewhat or a very significant source of stress (APA 2015). We are the first generation that does not expect to be financially better off than our parents because of financial instability, student debt, less savings, and the decay of stable, full-time employment (Petersen 2019).

The norm for many wildlifers has been to graduate with a bachelor's degree and spend the next few field seasons job-hopping around the country or world. Working in the field as early-career professionals can be some of the most exciting work we ever do. These positions can even take us around the globe where we gain valuable experience, cultural awareness, and crucial professional connections. Wildlifers from different generations report taking low-paying or unpaid positions to gain professional experience, but in recent years there has been growing concern about how the practice of using new professionals as a source of low-cost labor for full-time positions is unethical and exploitative (Fink 2013; Fournier et al. 2019). Many entry-level positions in the wildlife profession pay less than minimum wage but also preclude taking a second job because of their time commitment or remote location (Whitaker 2003). Use of volunteers or interns in full-time positions, which has been common in the wildlife field for decades, reflects badly on the profession by causing personal and financial hardship, excluding applicants based on economic class, and devaluing natural resources (Whitaker 2003; Fournier and Bond 2015). The willingness of some millennials to work for less than minimum wage to break into the field, perhaps while trying to pick up side hustles to pay the bills, facilitates this continued exploitation (Petersen 2019). While unpaid and underpaid positions have been prevalent in the experiences of previous generations, tradition does not equate to appropriate or sustainable practices (Fournier and Bond 2015).

Efforts to reduce the prevalence of unpaid or underpaid work or develop policies to address the negative impacts of these positions may often be led by younger generations, but progress has been slow and efforts may meet resistance. For example, when asked to support removal of unpaid positions from job boards, members of older generations may respond that they "paid their dues," and so should you. Advocating that our profession demonstrate the value of entry-level positions through appropriate financial compensation should not call into question whether wildlifers from younger generations are willing to work hard. A paradigm shift is needed to disassociate taking on unpaid labor as an indicator of discipline or dedication. The ability, or inability, to take unpaid positions is an indicator of circumstance and access to resources (Fournier et al. 2019). We are cognizant of the fact that organizational funding may be limited, but we need to come up with new solutions if we are serious about sustaining a diverse, equitable, and inclusive conservation workforce. Several groups, like cohorts of TWS Leadership Institute, have proposed the removal of unpaid or "pay to play" positions from job boards as a first step, but to date these efforts have not been successful. Organizations can also explore options like short-term remote work positions or shared-resource partnerships in which two or more organizations pool resources such as housing, equipment, money, and facilities to ensure they can provide paid entry-level positions or internships. Ultimately, intergenerational cooperation is required to fund entry-level positions.

For baby boomer or traditionalist wildlifers, dedication and years of service may have been enough to advance in the profession. For wildlife professionals of younger generations, there is a stronger incentive to have a degree to make ourselves competitive in the tough job market (Council of Economic Advisers 2014). Unfortunately, the likelihood of permanent placement may still be low even with an advanced degree. Most wildlife positions are funded by government agencies and as national budget constraints get tighter, so does the wildlife job market. The outlook for women wildlifers may be even more concerning, because even though the number of women obtaining natural resource–related degrees has exceeded the number of men, men are still more likely to be hired for wildlife jobs (Edge 2016).

Pursuing advanced education can compound our financial challenges. With little in our pocket from subminimum-wage positions and our undergraduate school debt, millennial wildlifers may wonder how to pay for the graduate degree (or two) we believe will help us secure a permanent position. Wildlife graduate programs may provide funding in the form of research or teaching assistantships, but these may not last year-round. Wildlifers may try to fill the gaps with competitive grants that may not cover salary expenses or last throughout a graduate program. Others may get creative by fundraising through platforms like Kickstarter or GoFundMe, the latter of which has crowdfunded more than $100 million for education expenses since 2010 (Lowry 2017). To make ends meet, we may also take out additional federal or private loans. Loan forgiveness programs have been used to recruit or retain wildlifers in federal agencies since the mid-2000s, a testament to its value to attract and secure new hires for rising generations with significant student loan debt (DOI 2014).

#WildlifeMom

Ninety percent of Americans will become parents during their working lives (Cech and Blair-Loy 2019). By 2016, more than half of all millennial women were mothers, and that number has continued to grow by an additional million each year (Walton Family Foundation 2018). Compared to previous generations, millennial mothers tend to be older (26 versus 21 in 1972), especially those with advanced degrees (30; Bui and Miller 2018). Completing graduate school and moving into the workforce often coincides with historical and biological child-rearing years (Erickson 2008). Many women, not only millennials, are confronted with the decision of delaying having children to become more established professionally, in part because they expect their greatest contributions will come at the beginning of their career. Alternatively, mothers may take time off from their careers to raise children. Decisions of when to have children and how much time to take off from work to care for them can be particularly conflicting for women wildlifers whose identities may be closely tied to their work and who feel renewed by their professional roles.

More mothers have participated in the US workforce in the past several decades, and now 45% of two-parent households have both parents working full time, up from 31% in 1970 (Pew Research Center 2015). However, mothers continue to shoulder a disproportionate amount of childcare even as men become more involved in rearing children (Schulte 2014). Forty-one percent of mothers report that being a parent has made it harder for them to advance in their career, whereas only 20% of fathers say the same (Pew Research Center 2015). In science, technology, engineering, and mathematics (STEM) fields, nearly 50% of women leave their full-time job within 4–7 years of becoming a parent (Cech and Blair-Loy 2019). This reduction in the STEM workforce may have detrimental impacts on future generations: there is a positive correlation between girls who are raised by a working mother and education level and choice to work in the future (Working Mother Research Institute 2014).

Mothers who continue to work outside the home may be distressed by the competing needs of their profession versus home life. As women progress in their careers, it can become difficult to find the right balance between professional responsibilities and family commitments. The field of wildlife can present some unique challenges that begin during pregnancy. Occupational hazards common in the wildlife profession, such as exposure to tick-borne or other zoonotic diseases, may be particularly concerning to wildlife moms. For example, there is increased risk of fetal abnormalities for those exposed to Lyme disease during pregnancy (Mylonas 2011). Morning sickness may greatly limit the speed, distance, and number of field surveys a pregnant wildlifer is able to complete. Finding maternity clothes appropriate for fieldwork or completing simple tasks such as lacing up hiking boots or putting on gaiters without assistance may be difficult. One mother recounted her experience working at a wildlife refuge and leading a school group on a winter hike. Forgetting to compensate for baby weight, her usual snowshoes sank a half meter in the snow! It took the teacher and a few students to help get her out, but they finished the hike after exchanging for a larger pair of snowshoes. Other aspects of pregnancy as a wildlife professional that can be both embarrassing and comical include the quantity of food needed to sustain an active pregnant woman and the number of times that she will need to urinate while in the field.

Many new wildlife mothers may struggle with how much maternity leave to take and when to take it. Is there ever a "good" season to take off 6 to 12 weeks? In addition, what each family can afford may dictate how much time a parent can realistically take off from work. Prior to 2020, only 12% of American workers had access to paid leave through their employers, and most of these workers were men (Mugglestone 2015). For people working for public agencies, public or private elementary and secondary schools, and companies with 50+ employees, the Family and Medical Leave Act provides eligible employees with up to 12 weeks of unpaid leave with job protection during any 12-month period (DOL, n.d.).

Once a mother returns from maternity leave, she may again experience some unique challenges. First, many mothers returning to work continue to pump breast milk, which can be awkward in the field. Breast pumps are difficult to fit into a backpack (especially with the food and water required to keep a healthy milk supply), expressed milk must be kept cool, and batteries kept from freezing in cold climates. In addition, finding a private, dry, and warm place to pump multiple times a day is challenging. Second, finding adequate childcare can be a major concern, especially considering many wildlife positions are located in small, sometimes remote, communities and require odd hours that can vary by season. In the past two decades, the cost of childcare has more than doubled, while wages have remained mostly stagnant (DeSilver 2014). The average cost for daycare in 2014 for a 4-year-old varied drastically by location; for example, $4,500/year in Tennessee and $12,300/year in Massachusetts (Mugglestone 2015). In addition, as the wildlife field becomes increasingly more saturated and competitive, new professionals may have to move farther from home and their family support for career advancement, which has a compounding negative effect on wildlife mothers. Many wildlife moms may want the opportunity to bring children into the field to share our passion and help reduce childcare costs. At the time of publication, Considine had two young sons, 5 and 2, who provide endless moments of humor, especially the oldest, who often accompanies her to collect gear in the field or conduct surveys. As much joy as sharing this passion brings, being a wildlife mom has been challenging for Considine in a small rural community where it is difficult to find babysitters who are willing to stay over until the wee hours of the morning or provide last-minute help on the weekend when the weather conditions are just right to conduct a survey. Unfortunately, many employers prohibit bringing children to work because of liability and instead suggest "flexible work schedules" or working from home. Although these accommodations may work for wildlifers whose main responsibilities are computer based, they may be inadequate or unfeasible for those whose primary work is in the field. Moreover, field-based positions often are lower-graded and lower-paid jobs, making childcare costs even more burdensome.

Many of the struggles of balancing life as an employee and a parent may not be limited to millennial wildlife mothers. Working moms may be judged for spending too much time focused on work and not giving adequate attention to their children or, alternately, as taking too much time for family so insufficiently committed to their work (Williams and Dempsey 2014). Unfortunately for millennial mothers, two additional pressures may be increasingly present. First, the majority of millennials' time in the workforce has coincided with rapid adoption of technology to increase communication. With the increase in work laptops and mobile phones, as well as a proclivity to remain "always connected" to others using social media, many millennials may feel an expectation to respond during nonwork hours. The idea of "flexible work hours" still leaves mothers feeling judged when we leave the office at 5 p.m., knowing we still have a few hours of work after bedtime or in the early morning. In contrast, many employers expected baby boomer or Generation X mothers to separate work and home life (Ferrante 2018). Second, millennial mothers may face pressure to celebrate every accomplishment of our children's lives through posts to social media. Competition between mothers to show their parenting success on platforms like Pinterest, Facebook, or Instagram has led to a host of negative behavioral and mental health outcomes (Coyne et al. 2017). Millennial wildlife moms may find it particularly difficult to meet some of these arbitrary expectations with variable work hours and remote research sites. As described in *The Mommy Myth* (Douglas and Michaels 2005), today's mothers are subjected to an onslaught of beatific imagery, romantic fantasies, self-righteous sermons, psychological warnings, terrifying movies about losing their children, and even more terrifying news stories about abducted and abused children. In addition, they must cope with totally unrealistic advice about how to be the most perfect and revered mom in the neighborhood, or even in the whole country. These ideas are predominantly fueled by the media (and now broadcast across social media platforms), and the book coined the term "*new momism—the insistence that no woman is truly complete or fulfilled unless she has kids, that women remain the best primary caretakers for children, and that to be a remotely decent mother, a woman has to devote her entire physical, psychological, emotional, and intellectual being 24/7, to her children. A highly romanticized and yet demanding view of motherhood in which the standards for success are impossible to meet.*"

However, small battles are being won each year to help working mothers. The Affordable Care Act, passed in 2010, requires more insurance plans to cover the cost of a breast pump, and employers to provide a private place that is not a bathroom and adequate break time for an employee to nurse or pump (DOL, WHD n.d.). As of 2018, every state in the country has some form of legal protection for breastfeeding in public (NCSL 2021). Breast pump designs are also being modernized to include hands- and tube-free designs that give more flexibility to working mothers. The passage of the National Defense Authorization Act in December 2019 included the Federal Employee Paid Leave Act, which approved up to 12 weeks of paid parental leave for federal employees beginning in October 2020. A federal travel regulation issued in October 2021 formally recognizes that nursing mothers on official travel have a special need and allows them to bring a caregiver for the child between the employee's reasonable break periods to breastfeed or to pump, store, and ship breast milk home to the child at the government's expense (GSA 2021). Lastly, many agencies and organizations, including TWS, are implementing their own policies to support working mothers. Starting in 2017, TWS began implementing a Quiet Room at the annual conference for participants needing a peaceful setting, including those with small children and nursing mothers.

#MeetMeInTheMiddle

"What's missing from this talk about 'having it all' is the recognition that if it's left for women to work out for themselves, like it is now, they can't have it all. . . . [W]omen have made all of the changes unilaterally that they really can" (Schulte 2014: 35). Just as it may be hard to be a millennial, it also may be challenging to work with us. However, to effectively leverage the assets of rising generations, both new and established professionals will need to adapt. Employers may report several common issues when working with millennials. First, they may view us as entitled. For older generations, demonstrating duty was an important characteristic, and it was expected to "pay one's dues" with years of service in an entry-level position before receiving benefits like better pay or titles. Blake had experiences with short-term positions or internships that had the potential to become full-time positions, but these provided no employee benefits, like paid leave or accrual of time toward becoming vested, before being hired on full-time. For some millennials, fewer summer or after-school jobs, often in favor of resume-building activities like camp or sports, may have fostered a distaste for "menial" work, leading us to seek or expect advancement on faster

timelines (Karsh and Templin 2013). In addition, the rise of young business icons, like Facebook founder and CEO Mark Zuckerberg, reinforces the idea that bosses need not have decades of experience (Karsh and Templin 2013).

Second, employers and colleagues may feel millennials lack skills needed to succeed in the workplace. This criticism may be more related to our age versus characteristics that are unique to millennials as a generation: as we age and move through our first few jobs, we gain experience. While some members of older generations, especially independent Generation Xers, may feel empowered to take on a project with little direction or guidance, millennials may crave structure and be driven by the need to get it done right the first time. We may also be viewed as lacking skills for dealing with difficult people. As millennials gain more experience as professionals, we will likely also gain more tools for dealing with difficult situations. In addition, for some of us, being accepting of all people may make us hyperaware of controversial issues, political correctness, and sensitivity in the workplace.

Third, generations may prefer different communication styles, and it can be challenging for managers to accommodate these differences. Most millennials will follow rules, but we may want to know the reasoning behind decisions because this provides a broader context to see our individual impact within our organization, and thus find more meaning in our tasks no matter our current role in the organization (Karsh and Templin 2013). We may thrive on sincere, specific and frequent feedback. While some may belabor millennial's desires for more frequent feedback as coddling, we may view feedback as a way to course correct. In terms of workplace performance, about 41% of millennials will do what their managers tell them to do, a higher percentage compared to previous generations (Pew Research Center 2010). Millennials may tend to place more faith in authority than Generation X because of our heightened level of parental supervision (Center for Women and Business 2017).

#OnwardAndUpward

We millennials are making our own way in the wildlife profession and building on the momentum created by trailblazing women of the generations that came before us. In this chapter, we have shared how world events when we are coming of age influence each generation's mindset. For millennials, the effects of changing parenting styles, the Great Recession of 2008, and technological advances in our teenage years had a profound impact on how we approach challenges and relate to others. We have shown how our unique perspective may provide new opportunities for connecting with wildlife stakeholders, but also contributes to the stresses we may feel when trying to "make it" in this field. As some of us are having children of our own, we have highlighted the barriers working mothers may continue to face today alongside what positive progress has been made. To solve some of the most pressing problems for millennial women in the wildlife profession, it will take the energy and vigor of all generations. Understanding the dynamics of each generation that must work together is integral in developing a productive dialogue and ensuring success in the wildlife profession. We conclude our chapter with ways to support millennial women of wildlife.

#HowToBeAnAlly

- **Embrace generational diversity.** Viewing our workplace through an intergenerational lens can help reduce conflict, but remember to seek to understand colleagues as individuals without falling prey to generational stereotypes. In addition, check your language for slights that may promote intergenerational conflict. Referring to onboarding a younger colleague as "babysitting" or commenting, "I have a daughter (or niece) your age," when meeting a new team member is not appropriate in the workplace. Just think: Would you like us to say, "I have a parent your age," when we are introduced?
- **Define and communicate expectations for employee availability outside of work hours.** While the intent of flexible work policies is to provide a higher level of self-determination and work-life balance, flexibility may imply that employees need to be always available. Since many in the millennial generation are always plugged in, there may be an element of anxiety when emails arrive on our phones at late hours. Leadership plays an important role in setting the tone for this flexible work environment: if leaders are always available and send/receive emails at any hour of the day, everyone else may feel obliged to do the same (Correia 2016).
- **Create mentorship opportunities and encourage women to serve as mentors.** Women tend to be empowered by other women in leadership and by strong mentoring relationships (Correia 2016). Employers and colleagues can create spaces where women feel empowered and valued to do things our way and help women find healthy ways to quiet feelings of anxiety and self-doubt in the workplace (Ledbetter and Kinsman 2019). Organizations can also support formal efforts for mentoring and career guidance, for example, facilitating regularly scheduled check-ins help to alleviate feelings of otherness and the anxiety that comes with proactively reaching out (Ledbetter and Kinsman 2019). Also find ways

to support women as mentors: when women are exposed to powerful women, we are more likely to endorse the notion that women are well suited for leadership roles (Ledbetter and Kinsman 2019).

- **Consider criticism carefully.** Constructive criticism can be hard for millennials to take, especially since we may have grown up thinking we were special and then hustled to hone ourselves to perfection. Millennials may respond more positively to criticism that focuses on observable behaviors, reaffirms faith in us as people, and defines specific actions for improvement. These techniques highlight that it is the behavior, not the person, that is the issue and empower us to take steps to change our behavior (Zemke et al. 2018).

- **Support transparent, collaborative, and cooperative organizations and leaders that help to create that environment.** Generally, millennials value leaders who are transparent about what they are doing and why they are doing it and they help increase our perceived level of competence in our work. Employers can offer town hall meetings to create opportunities for leadership to connect directly with employees. Millennials are empowered by signals of respect from leadership and typically have high expectations to be respected in the workplace (Correia 2016).

Resources

Douglas, Susan, and Meredith Michaels. 2005. *The Mommy Myth: The Idealization of Motherhood and How It Has Undermined All Women.*

Karsh, Brad, and Courtney Templin. 2013. "Manager 3.0: A Millennial's Guide to Rewriting the Rules of Management." New York: AMACOM.

Petersen, Anne. 2019. "How Millennials Became the Burnout Generation." *BuzzFeed News*, January 5. www.buzzfeednews.com/article /annehelenpetersen/millennials-burnout-generation-debt-work.

Schulte, Brigid. 2014. *Overwhelmed: Work, Love, and Play When No One Has the Time.* New York: Picador.

Slaughter, Anne-Marie. "Why Women Still Can't Have It All." *Atlantic*, July/ August 2012. https://www.theatlantic.com/magazine/archive/2012/07 /why-women-still-cant-have-it-all/309020/.

Williams, Joan, and Rachel Dempsey. 2014. *What Works for Women at Work.* New York: New York University Press.

Discussion Questions

1. What generational signpost(s) do you most attribute to your development? What about those events changed the way you see the world and the people you interact with?
2. Aldo Leopold described the axe, cow, plow, fire, and gun as the five most crucial tools for wildlife management in 1937. Describe the five tools you think are most important in wildlife management today. Why did you choose them and how do they compare to Leopold's?
3. How would you describe imposter syndrome specific to the field of wildlife? How do you think impostor syndrome differs by gender, generation, and socioeconomic status? If you experience impostor syndrome personally, how do you address it?
4. How can we be better allies to wildlifers of different generations? How do other aspects of diversity (e.g., gender, sexual orientation, educational background, socioeconomic status) influence strategies to support wildlifers of different generations?

Activity: Generational Continuums

Learning Objectives: Participants will understand that generational differences are not necessarily reflective of age but more about shared characteristics that result from shared experiences among a group of people. Understanding generational differences, both our own and those of others, can lead to better supporting individual strengths and improving everyone's experience in a group. Generations are defined as the traditionalists (born 1928–1944), baby boomers (born 1945–1964), Generation X (born 1965–1980), millennials (born 1981–1995), and Generation Z (born 1996–2012). For this activity, only the first four generations (traditionalists–millennials) will be highlighted; Generation Z has not yet been in the workforce long enough to accurately describe their characteristics.

Time: 30–40 min.

Description of Activity

Materials needed: The Generational Continuum Worksheet and a writing utensil.

Process:

Part 1: Activity (5 min.)

Provide the group with time to complete the Generational Continuum Worksheet.

Generational Continuum Worksheet

For each of the bolded work values below, draw an X on the line symbolizing a continuum between two characteristics or descriptions of that value. There is no right or wrong answer.

Work Ethic. *Is your dedication to your job or employer your top priority in life (ambitious/driven)? Do you enjoy working, understand the need to work, but also value your life outside of work (balanced)?*

Ambitious/Driven————————————————————————Balanced

Relationships. *Are you fiercely devoted to your friends/family/romantic partners (loyal)? Do you appreciate and enjoy the people in your life but prefer to keep options open for your time?*

Loyal————————————————————————Reluctant to commit

View of Authority. *Do you interact with people in positions of power and/or leadership like other people in your life (relaxed/unimpressed)? Or do you interact with people in these positions more formally (respectful)?*

Relaxed/unimpressed————————————————————— Respectful

Project Style. *When considering your preferences to working on a project, do you prefer to work alone (self) or are you more likely to prefer working in a group (team)?*

Self————————————————————————————Team

Technology. *Are you able to quickly learn new software, programs, apps, or adopt new hardware (completely integrated)? Or are you able to figure things out if necessary (adapting)?*

Completely integrated—————————————————————Adapting

Career Route. *Do you expect to work in a job until another option or a better option comes along (change)? Do you plan to work with the same company or organization for the majority of your career (promotion)?*

Change————————————————————————Promotion

Civic Perspective. *Do you prefer to dedicate your time outside of working hours to focusing on yourself (self)? Are you involved in your community (your town, religion, political party, hobbies, education, career, gender, culture, etc.) and working with groups to make meaningful contributions (team)?*

Self————————————————————————————Team

Communication. *Do you prefer to communicate via a written, electronic message like in a text/email if you need to ask a question or solve a problem with a team (written)? Or do you prefer to speak with people face to face or over the phone (oral)?*

Written——————————————————————— Oral

Activity Direction. *Are you most likely to be working on several projects or activities at once (multi-active)? Do you prefer tackling and completing a task before moving on to the next one (linear-active)?*

Multi————————————————————————— Linear

Work Time. *Do you primarily consider work an important aspect of life constrained to 40 hours a week and what you do between the weekends (time)? Are you project oriented and commit the necessary time to get the job done (project)?*

Time————————————————————————Project

Focus. *When thinking about your job, do you focus more on how much you produce and the time you work (productivity) or do you focus more on the contributions you are making (contribution)?*

Productivity——————————————————— Contribution

Part 2: Table Comparison (15 min.)
After everyone has completed the worksheet, have participants work in
    pairs or small groups (no larger than 4) to compare their responses to
    Table 9.2.

TABLE 9.2. Generational differences chart (modified from www.wmfc.org).

	Traditionalists	Baby Boomers	Generation X	Millennials
Birth Years	1928–1944	1945–1964	1965–1980	1981–1995
Attributes	Committed to company Confident Conservative Dedication Doing more with less Fiscally prudent Hard-working Loyal to organization/ employers (duty, honor, country) Organized Patriotic Rules of conduct Sacrifice Task oriented Thrifty; abhor waste Trust hierarchy and authority	Ability to handle a crisis Ambitious Challenge authority Competitive Consensus leadership Consumerist Good communication skills Idealism Live to work Loyal to careers and employers Multitaskers Optimistic Political correctness Strong work ethic Willing to take on responsibility	Adaptable Anti-establishment mentality Confident Focus on results Free agents High degree of brand loyalty Independent Loyal to manager Pragmatic Results driven Skeptical of institutions Willing to put in the extra time to get the job done Work to live	Aspire to contribute At ease in teams Attached to gadgets and parents Best educated Diversity focused Eager to spend money Focus on change using technology Individualistic yet group oriented Innovative—think outside the box Loyal to peers Optimistic Respect competency not title Seek responsibility early Sociable—make workplace friends Think globally
View of education	A dream	A birthright	A way to get there	An incredible expense
Dealing with money	Put it away Pay cash Save, save, save	Buy now, pay later	Cautious Conservative Save, save, save	Earn to spend
Entitlement	Seniority	Experience	Merit	Contribution
Respect for workplace authority	Based on seniority and tenure	Originally skeptical but are becoming similar to traditionalists	Skeptical of authority figures Will test authority repeatedly	Will test authority but often seek out authority figures for guidance
Workplace view on skill building	Training happens on the job New skills benefit the company, not the individual	Skills are not as important as work ethic and "face time"	Amassed skills will lead to the next job The more you know, the better Work ethic is not as important as skills	Training is important and new skills will ease stressful situations Learning motivated Want immediate results
Work ethic and values	Adhere to rules Duty before fun Expect others to honor commitments and behave responsibly Linear work style Work is their bond Value due process and fair play	Dislike conformity and rules Heavy focus on work as an anchor in their lives Loyal to the team Strive to do the very best Value collaboration Value equality	Care less about advancement than work-life balance Expect to influence the terms and conditions of the job Like a casual work environment Move easily between jobs Outcome oriented	Decrease in ambition in favor of more family time and less personal pressure Expect to influence the terms and conditions of the job Obsessed with personal development Tolerant Value mentorship

TABLE 9.2 *(continued)*

	Traditionalists	Baby Boomers	Generation X	Millennials
	Value attendance Value compliance Value good attitude Value practical knowledge	Value personal growth and gratification Want respect from younger workers Want a flexible route into retirement Work efficiently	Output focused Prefer diversity, technology, informality, and fun Get the work done, and on to the next thing	Want long-term relationship with employers, but on their own terms Want to be evaluated on work product, not when, how or where they got it done
Preferred work environment	Conservative Hierarchical Clear chain of command Top-down management	"Flat" organizational hierarchy Democratic Equal Opportunity	Functional, positive, fun Fast paced and flexible Informal Access to leadership and information	Collaborative Achievement oriented Highly creative Diverse Fun, flexible, positive
What they are looking for in a job	Recognition and respect for experience Value placed on history and traditions Job security and stability Company reputation and ethics Clearly defined rules and policies Do what needs to be done	Ability to "shine" or "be a star" Company represents a good cause Fit in with company vision and mission Need clear and concise job expectations and will get it done Like to achieve work through teams	Dynamic young leaders Cutting-edge systems and technology Forward-thinking company Flexibility in scheduling Input evaluated on merit, not age or seniority Will question tasks if reasoning is unclear Will seek another position when unengaged	Want to be challenged Expect to work with positive people and company that can fulfill dreams Treated with respect in spite of age Expect to learn new knowledge and skills Friendly environments Expect to make a difference Social network Strong, ethical leaders and mentors
Keys to working with	Work is not supposed to be fun Follow rules well but want to know procedures	Want to hear that their ideas matter Before they do anything, they need to know why it matters, how it fits into the big picture and what impacts it will have on whom	Want independence in the workplace Desire informality Give time to pursue other interests Allow for fun at work	Like a team-oriented workplace Learn about personal goals Feel disrespected by any requirement to do things just because it has always been that way or to pay one's dues

(continued)

TABLE 9.2 *(continued)*

	Traditionalists	Baby Boomers	Generation X	Millennials
	Tend to be frustrated by what they see as lack of discipline, respect, logic and structure, especially if the workplace is more relaxed or spontaneous Like the personal touch	Do well in teams Are motivated by their responsibilities to others Respond well to attention and recognition Need flexibility and freedom	Provide latest technology	Provide engaging experiences that develop transferable skills Provide rationale and value for assigned tasks Provide variety Offer structured, supportive work environment
View of authority	Respectful	Impressed	Unimpressed	Relaxed
Feedback and rewards	No news is good news Satisfaction is a job well done Feedback on performance as they listen Want subtle, private recognition on an individual level without fanfare	Feel rewarded by money Will often display awards and recognition for public view Enjoy public recognition Appreciate awards for their hard work and the long hours they work	Want to be rewarded with time off Not enamored by public recognition Freedom is the best reward Need constructive feedback Are self-sufficient: give them structure, some coaching, but implement a hands-off supervisory style	Like frequent, specific feedback Need clear goals and expectations Communicate regularly Provide supervision and structure Want recognition from bosses and family
Mentorship	Support long-term commitment Show support for stability, security and community Provide actions with focus on standards and norms Allow the employee to set the "rules of engagement" Respect their experience Emphasize similarity of current situation to the past	Appreciate they paid their dues under hierarchical rules Help them find balance: work, family, finances, etc. Show them how you can help them use their time wisely Demonstrate importance of a strong team and their role Pre-assess their comfort level with new technology before new projects	Offer a casual work environment and lighten up Encourage creativity Allow flexibility They work with you, not for you Offer variety and stimulation Present yourself as an information provider, not boss Use their peers as testimonials Follow up and meet your commitments They are eager to improve and expect you to follow through with information	Raise the bar on self as they have high expectations Set goals in steps and actions Honor their optimism Welcome and nurture them Be flexible Offer customization Offer peer-level examples Provide information and guidance

Part 3: Debriefing and Discussion Questions (10–20 min.)

1. How did you feel completing the worksheet and comparing your answers to Table 9.2?
2. What did you learn?
3. Were you surprised about the characteristics you tend to share with different generations?
4. Were you most similar to the generation to which you belong based on your birthdate?
5. What trait or characteristics was most notable for you?
6. Was there a trait or character included that you never thought of before as describing a generation?
7. How can having members of different generations within a group or an organization influence your group's success?
8. Can you envision conflict that might arise between members of different generations based on these characteristics? How would you approach solving these conflicts?
9. How would you use the information you learned or your experience from this activity in your life as a wildlife professional, student, or citizen?

Literature Cited

Allen, Maya. 2018. "Black Therapists Explain the Stigma of Mental Health in Minority Communities." *The/Thirty*, July 12. https://thethirty.whowhatwear.com/mental-health-in-minority-communities--5aa98297a77b1.

Alsop, Ron. 2008. *The Trophy Kids Grow Up: How the Millennial Generation Is Shaking Up the Workplace*. San Francisco: Jossey-Bass.

American Psychological Association (APA). 2015. *Stress in America: Paying with Our Health*. February 4. www.apa.org/news/press/releases/stress/2014/stress-report.pdf.

Associated Press (AP). 2014. "Timeline: Key Dates in Facebook's 10-Year History." February 4. https://phys.org/news/2014-02-timeline-key-dates-facebook-year.html.

Bravata, Dean M., Sharon A. Watts, Autumn L. Keefer, Divya K. Madhusudhan, Katie T. Taylor, Dani M. Clark, Ross S. Nelson, Kevin O. Cokley, and K. Heather. 2020. "Prevalence, Predictors, and Treatment of Impostor Syndrome: A Systematic Review." *Journal of General Internal Medicine* 35(4): 1252–75. doi: 10.1007/s11606-019-05364-1

Brooks, David. 2007. "The Odyssey Years." *New York Times*. October 9. www.nytimes.com/2007/10/09/opinion/09brooks.html

Brucculieria, Julia. 2019. "Women and Pants: A Timeline of Fashion Liberation." *HuffPost*. March 10. www.huffpost.com/entry/women-and-pants-fashion-liberation_l_5c7ec7f7e4b0e62f69e729ec.

Bui, Quoctrung, and Claire Cain Miller. 2018. "The Age That Women Have Babies: How a Gap Divides America." *New York Times*. August 4. www.nytimes.com/interactive/2018/08/04/upshot/up-birth-age-gap.html.

Casselman, Tracy. 2006. "Elizabeth Beard Losey, First Female Member of The Wildlife Society." *Wildlife Society Bulletin* 34(2): 558. doi: 10.2193/0091-7648(2006)34[558:EBLFFM]2.0.CO;2.

Cech, Erin, and Mary Blair-Loy. 2019. "The Changing Career Trajectories of New Parents in STEM." *Proceedings of the National Academy of Sciences of the United States of America* 116(10): 4182–87. https://www.pnas.org/content/116/10/4182.

Center for Women and Business at Bentley University. 2017. *Multi-generational Impacts on the Workplace.* https://www.bentley.edu/centers/center-for-women-and-business/multigenerational-impacts-research-report-request#DownloadReport.

Cohen, Josh. 2016. "Minds Turned to Ash." *Economist 1843 Magazine*, August/September. https://www.1843magazine.com/features/minds-turned-to-ash.

Correia, Tricia. 2016. "Women, Millennials, and the Future Workplace: Empowering All Employees." Georgetown Institute for Women, Peace, and Security. https://giwps.georgetown.edu/wp-content/uploads/2017/08/Women-Millennials-and-the-Future-Workplace.pdf.

Council of Economic Advisers, Executive Office of the President of the United States. 2014. *15 Economic Facts about Millennials.* https://obamawhitehouse.archives.gov/sites/default/files/docs/millennials_report.pdf.

Coyne, Sarah M., Brandon T. McDaniel, and Laura A. Stockdale. 2017. "Do You Dare to Compare? Associations between Maternal Social Comparisons on Social Networking Sites and Parenting, Mental Health, and Romantic Relationship Outcomes." *Computers in Human Behavior* 70:335–40. doi: 10.1016/j.chb.2016.12.081.

Curran, Thomas, and Andrew Hill. 2019. "Perfectionism Is Increasing over Time: A Meta-analysis of Birth Cohort Differences from 1989 to 2016." *American Psychological Association* 145 (4): 410–29. https://www.apa.org/pubs/journals/releases/bul-bul0000138.pdf.

Darshan, Goux. 2011. "Millennials in the Workplace." *Bentley University Center for Women & Business.* https://www.scribd.com/doc/158258672/CWB-Millennial-Report?secret_password=2191s8a7d6j7shshcctt.

DeSilver, Drew. 2014. "Rising Cost of Child Care May Help Explain Recent Increase in Stay-at-Home Moms." Pew Research Center, April 8. www.pewresearch.org/fact-tank/2014/04/08/rising-cost-of-child-care-may-help-explain-increase-in-stay-at-home-moms/.

Douglas, Susan, and Meredith Michaels. 2005. *The Mommy Myth: The Idealization of Motherhood and How It Has Undermined All Women.* New York: Free Press.

Edge, Daniel W. 2016. "Learning for the Future: Educating Career Fisheries and Wildlife Professionals." *Mammal Study* 41(2): 61–69. doi: 10.3106/041.041.0203.

Erickson, Tamara J. 2008. *Plugged In: The Generation Y Guide to Thriving at Work.* Massachusetts: Harvard Business Press.

Ferrante, Mary Beth. 2018. "The Pressure Is Real for Working Mothers." *Forbes.* August 27. www.forbes.com/sites/marybethferrante/2018/08/27/the-pressure-is-real-for-working-mothers/#5523c74f2b8f.

Fink, Eric M. 2013. "No Money, Mo' Problems: Why Unpaid Law Firm Internships are Illegal and Unethical." *University of San Francisco Law Review* (47):1–25.

Fournier, Auriel M., and Alexander L. Bond. 2015. "Volunteer Field Technicians Are Bad for Wildlife Ecology." *Wildlife Society Bulletin* 39(4): 819–21. doi: 10.1002/wsb.603.

Fournier, Auriel M. V., Angus J. Holford, Alexander L. Bond, and Margaret A. Leighton. 2019. "Unpaid Work and Access to Science Professions." *PloS One* 14(6): e0217032. doi: 10.1371/journal.pone.0217032.

Frost, Randy O., and Patricia Marten DiBartolo. 2002. "Perfectionism, Anxiety, and Obsessive-Compulsive Disorder." In *Perfectionism: Theory, Research, and Treatment*, edited by Gordon Flett and Paul Hewitt, 341–71. Washington, DC: American Psychological Association.

General Services Administration (GSA). 2021. "Federal Travel Regulation (FTR); Applicability of the Federal Travel Regulation Part 301-13 to Employees Who Are Nursing." *Federal Register* 86 (205, October 27): 59391 www.federalregister.gov/documents/2021/10/27/2021-23397/federal-travel-regulation-ftr-applicability-of-the-federal-travel-regulation-part-301-13-to.

Griffith, Erin. 2019. "Why Are Young People Pretending to Love Work?" *New York Times.* January 26. www.nytimes.com/2019/01/26/business/against-hustle-culture-rise-and -grind-tgim.html.

Hoff, Victoria. 2018. "Our State of Mind: 13 Important Statistics about Millennials and Mental Health." *The/Thirty.* May 30. https://thethirty.whowhatwear.com/millennial -mental-health-statistics/slide2.

Johnson, Megan, and Larry Johnson. 2010. *Generations, Inc.: From Boomers to Linksters— Managing the Friction between Generations at Work.* New York: AMACOM.

Karsh, Brad, and Courtney Templin. 2013. *Manager 3.0: A Millennial's Guide to Rewriting the Rules of Management.* New York: AMACOM.

Kerr, Breena. 2015. "Why 70% of Millennials Have Impostor Syndrome." *Hustle,* November. https://thehustle.co/why-70-of-millennials-have-impostor-syndrome/.

Ledbetter, Bernice, and Michael Kinsman. 2019. "Ensuring #MeToo Movement Advances Diversity in Leadership." *Workforce.* February 28. https://workforce.com/news /ensuring-metoo-movement-advances-diversity-in-leadership/

Leopold, Aldo. 1933. *Game Management.* New York: Scribner's Sons.

Lopez, Roel, and Columbus H. Brown. 2011. "Why Diversity Matters: Broadening Our Reach Will Sustain Natural Resources." *Wildlife Professional* 5(2): 20–27. http://wildlife .org/dbadmin/twp_archive/Summer%202011.pdf.

Lorusso, Marissa, and Sophia Barnes. 2019. "Survey: Stress, Stigma, and Access Loom Large for millennials." *Matters of the Mind.* www.themillennialminds.com/survey/.

Lowry, Erin. 2017. "Crowdfunding College Costs: Does It Work?" *Discover,* April 13. www .discover.com/student-loans/college-planning/how-to-pay/costs/crowdfunding-college -tuition.

Lyons, Sean T., and Joshua E. LeBlanc. 2019. "Generational Identity in the Workplace: Toward Understanding and Empathy." In *Creating Psychologically Healthy Workplaces,* edited by Ronald J. Burke and Astrid M. Richardsen, 270–91. Cheltenham: Edward Elgar.

McKie, Robin. 2018. "Want to Enjoy a Longer, Happier Life? Just Keep on Working." *Observer,* November 18. www.theguardian.com/science/2018/nov/18/life-expectancy -ageing-population-working-older-retirement.

Meister, Jeanne, and Karie Willyerd. 2010. "Mentoring Millennials." *Harvard Business Review,* May. https://hbr.org/2010/05/mentoring-millennials.

Mugglestone, Konrad. 2015. "Finding Time: Millennial Parents, Poverty, and Rising Cost." *Young Invincibles.* April 28. www.luminafoundation.org/files/resources/finding-time.pdf.

Myers, Karen, and Kamyab Sadaghiani. 2010. "Millennials in the Workplace: A Communi- cation Perspective on Millennials' Organizational Relationships and Performance." *Journal of Business and Psychology* 25(2): 225–38. www.ncbi.nlm.nih.gov/pmc/articles /PMC2868990/.

Mylonas, Ioannis. 2011. "Borreliosis during Pregnancy: A Risk for the Unborn Child?" *Vector-Borne and Zoonotic Diseases* 11(7): 891–98. doi: 10.1089/vbz.2010.0102.

National Conference of State Legislatures (NCSL). 2021. "Breastfeeding State Laws." August 26. www.ncsl.org/research/health/breastfeeding-state-laws.aspx.

National Park Service (NPS). 2017. "Breeches and Blouses." August 14. www.nps.gov /articles/breeches-blouses.htm.

News Desk. 2015. "Arresting Dress: A Timeline of Anti-cross-dressing Laws in the United States." *PBS News Hour,* May 31. www.pbs.org/newshour/nation/arresting-dress -timeline-anti-cross-dressing-laws-u-s.

Petersen, Anne Helen. 2019. "How Millennials Became the Burnout Generation." *BuzzFeed News,* January 5. www.buzzfeednews.com/article/annehelenpetersen/millennials -burnout-generation-debt-work.

Petronzio, Matt. 2015. "Millennials Experience Depression at Work More Than Any Other Generation, Study Finds." *Mashable,* May 21. https://mashable.com/2015/05/21 /millennials-depression-work/.

Pew Research Center. 2010. "Millennials: A Portrait of Generation Next, Confident. Connected. Open to Change." February 24. www.pewresearch.org/wp-content/uploads /sites/3/2010/10/millennials-confident-connected-open-to-change.pdf.

Pew Research Center. 2015. "Millennial Moms Give Themselves High Marks." *Parenting in America*. December 14. www.pewsocialtrends.org/2015/12/17/parenting-in-america/st _2015-12-17_parenting-04/.

Primack, Brian A., Ariel Shensa, Cesar G. Escobar-Viera, Erica L. Barrett, Jaime E. Sidani, Jason B. Colditz, and A. Everette James. 2017. "Use of Multiple Social Media Platforms and Symptoms of Depression and Anxiety: A Nationally-Representative Study among U.S. Young Adults." *Computers in Human Behavior* (69):1–9. doi: 10.1016/j. chb.2016.11.013.

Schulte, Brigid. 2014. *Overwhelmed: Work, Love, and Play When No One Has the Time*. New York: Picador.

Stein, Joel. 2013. "Millennials: The Me Me Me Generation." *Time*, May 20. http://time.com /247/millennials-the-me-me-me-generation/.

US Department of the Interior (DOI). 2014. *Personnel Bulletin No. 14-02: Department Policy on Student Loan Repayment Benefit Plan*. March 14. www.doi.gov/sites/doi.gov/files/uploads /departmental_policy_on_the_student_loan_repayment_benefit_plan_mar_14_2014 .pdf.

US Department of Labor (DOL). N.d. "Family and Medical Leave (FMLA)." Accessed February 5, 2020. www.dol.gov/general/topic/benefits-leave/fmla.

US Department of Labor, Wage and Hour Division (DOL, WHD). N.d. "Section 7(r) of the Fair Labor Standards Act—Break Time for Nursing Mothers Provision." Accessed February 7, 2020. www.dol.gov/agencies/whd/nursing-mothers/law.

US Fish and Wildlife Service (USFWS). 2018. *2016 National Survey of Fishing, Hunting, and Wildlife-Associated Recreation*. April. www.fws.gov/wsfrprograms/subpages /nationalsurvey/nat_survey2016.pdf.

Vernon, Steve. 2016. "Americans Are Living Longer—and Working Longer." *CBS News*. June 16. www.cbsnews.com/news/americans-are-living-longer-and-working-longer/.

Vespa, Jonathan. 2017. "The Changing Economics and Demographics of Young Adulthood: 1975–2016." *Current Population Reports*, P20-579. Washington, DC: US Census Bureau. www.lemcapital.com/wp-content/uploads/The-Changing-Economics-and -Demographics-of-Young-Adulthood-1975-2016-US-Census-Bureau.pdf.

Vogels, Emily A. 2019. "Millennials Stand Out for Their Technology Use, but Older Generations Also Embrace Digital Life." Pew Research Center. September 9. www .pewresearch.org/fact-tank/2019/09/09/us-generations-technology-use/.

Walton Family Foundation. 2018. *Millennial Parents and Education*. Walton Family Foundation. Echelon Insights. December 6. https://8ce82b94a8c4fdc3ea6d-b1d233e3bc3cb108 58bea65ff05e18f2.ssl.cf2.rackcdn.com/1f/b9/c41cff2942ff9e9fba4f54b06794/millennial -parents-12.2018%20FINAL.pdf.

Whitaker, Darroch. 2003. "The Use of Full-Time Volunteers and Interns by Natural-Resource Professionals." *Conservation Biology* 17(1). doi: 10.1046/j.1523-1739.2003.01503.x.

Williams, Joan, and Rachel Dempsey. 2014. *What Works for Women at Work*. New York: New York University Press.

Working Mother Research Institute. 2014. *Mothers and Daughters: The Working Mothers Generations Report*. http://wmmsurveys.com/wp-content/uploads/2017/10/10 _Generations_Report.pdf.

Zemke, Ron, Claire Raines, and Bob Filipczak. 2018. *Generations at Work: Managing the Clash of Boomers, Gen Xers, and Gen Yers in the Workplace*. 2nd ed. New York: AMACOM.

Zetlin, Minda. 2017. "63 Percent of Millennials Have More Than $10,000 in Student Debt. They'll Be Paying for Decades." *Inc.* January 11. www.inc.com/minda-zetlin/63-percent -of-millennials-have-more-than-10000-in-student-debt-theyll-be-paying.html.

Jessica A. Homyack, Tabitha A. Graves, and Stephanie S. Romañach

10

Multi-scale Approaches to Increase Inclusion

I was told to work at home as a way to resolve a hostile work environment.

I was told to focus on male candidates for a position I supervise because "there's already too much estrogen in the group."

—Natural resources professionals quoted in Kriese et al., 2020

Shortly after my first child was born, my spouse and I (Jessica Homyack) dressed her in a lavender outfit, packed the diaper bag, fumbled our way through buckling her in the car seat, and headed to a gathering organized by the dean for graduate education at Virginia Tech. This represented a big first—taking her out in public (oh, new parents!) to a graduate student event. My husband and I were both PhD candidates, and we wondered how we were going to pay for an unexpected C-section and medical insurance for our healthy newborn. But we did not have to worry about an unpaid leave of absence from my teaching and research duties because of a new Work-Life Grant program to financially support graduate students through pregnancy and childbirth. I was among the first students to use the grant program, which was enacted based on recommendations from the university-wide National Science Foundation (NSF) AdvanceVT program (Box 10.1). The program explicitly recognized the tug-of-war between personal and professional life that is exacerbated for graduate students who are often young, overworked, underpaid, and excluded from parental leave benefits. Since its inception, the grant program has been incorporated into a broader commitment to a university campus climate that values community and diversity.

BOX 10.1 An Example That Works: The National Science Foundation ADVANCE Program

Higher education has long been viewed as an "ivory tower," distancing intellectual pursuits from everyday life. Because education was for the privileged, powerful, and usually white male, there is a long-standing thread of elitism in academic life against those who deviate from that historical standard. In science and natural resources disciplines, gaps still exist in reaching demographic and resource equity based on gender in colleges and universities, particularly for faculty and administrative roles. Only about one-quarter of college presidents are women, despite balanced gender ratios in enrollment (Cook 2012). Among faculty, women and underrepresented groups are still much more likely to have precarious part-time or non-tenure-track positions (Finkelstein et al. 2016). Here, we provide an example of a comprehensive program that has removed barriers to greater equity for women and underrepresented groups in academic STEM programs by supporting local approaches ranging from policy changes to improvements in practices, such as training (e.g., Smith et al. 2015).

In the United States, the NSF sets topical and cultural directions for scientific disciplines as a leading funding agency. Receiving an NSF grant can be a ticket to tenure and promotion for those in academic positions. The NSF recognizes that the systemic gaps and challenges for recruitment and retention of a diverse STEM workforce require a multifaceted approach to increasing participation of underrepresented groups. Since 2001, the NSF has invested more than $270 million in its ADVANCE program, which was established to increase the participation of women in STEM fields in academia (NSF n.d.; Cheruvelil et al. 2014), in part because greater innovation results from a more diverse workforce (Knapp et al. 2015). Because most STEM professionals begin their educational and professional journey at a college or university, academia serves a key role in socializing gender norms in science. The ADVANCE program has granted funds to 177 institutions of higher education, including 40 minority-serving colleges or universities. Its broad reach has altered the policy, organizational, and personal barriers to inclusion and diversity in academic STEM. These grants, often over $1 million per institution, provide visible opportunities and legitimacy to women and others engaged in diversity efforts at their workplace and can help garner support across the campus and from people in positions of authority. This funding provides resources to implement change long after the grant money has been spent. For example, many colleges and universities have retained their ADVANCE offices and programs for more than a decade past the receipt of funding. It was established to be nimble and allows for individuality in the scope and scale of how ADVANCE responds to the culture and needs of the receiving institution.

Virginia Tech, an institution with a well-known wildlife degree program, focused on improving policies and benefits related to work-life balance for faculty such as additional flexibility for family leave and options to extend the length of time an applicant can take to meet tenure requirements. Montana State University, which also has a well-known wildlife program, focused on improving faculty searches through implementing structural changes, training the hiring committee about implicit bias, and providing a family advocate (separate from the hiring committee) to be a resource to interviewees (Smith et al. 2015). Other institutions have focused workshops and lectures for faculty and future faculty to hone skills critical to many academic jobs like grant writing, networking, negotiation, and interviewing. Most colleges and universities that are ADVANCE recipients also individually recognize women in STEM through institutional grants, fellowships, or entire programs designed to enhance scholarly activities and productivity. Holistically, the program strives to transform the institutional culture with policies and practices that are equitable to women rather than asking women to change to fit the system. ADVANCE's funding, dedicated staff, and network of local experiments linked to a nationwide program have all contributed to its success at organizational cultural shifts.

How far has ADVANCE moved the needle on equitable representation and climate for women in academia? In other words, does the institution still act as an ivory tower with an elitist and privileged culture? Some scholars suggest that the program falls short of full inclusion by favoring white women (Bumpus 2020). In other words, without also considering sexual orientation, race, and socioeconomic status, the system is still geared toward a narrow view of diversity (Morimoto and Zajicek 2014). However, the surge of resources to nearly 200 US institutions has resulted in a flow of knowledge and information about gender disparity in academia that has likely led to ripples of change across higher education.

Creating a shift toward an organizational culture that values ideas and skills brought by a diverse workforce is necessary to reach equity, but it is neither simple nor rapid to implement. Past efforts to improve the workplace for women in wildlife have often relied on technical fixes, such as employer-required trainings on the legal definitions of sexual harassment (Brown et al. 2010) or playing the numbers game, where increasing counts of staff from underrepresented groups were reported as progress toward poorly defined goals (Hirsh and Tomaskovic-Devey 2020). However, trainings and compositional diversity reports often accomplish little more than checking the box that "diversity is valued." These activities can serve as a symbol of change rather than a true culture shift. In fact, research indicates that legal training on issues such as sexual harassment does not prevent future harassment (NASEM 2018). The US Forest Service responded to broad concerns about a lack of diversity and discrimination by attempting organizational changes during the 1990s. The agency increased women hires, placed more women in managerial positions, and conducted multiple trainings on sensitivity and harassment. However, subsequent surveys indicated that the agency had not created a welcoming environment (Brown et al. 2010). Recent reports of harassment, assault, and retaliation against women in the US Forest Service exemplify how difficult cultural change is, particularly in sexist workplace cultures, for example, among wildland firefighters (OIG 2019).

As the US Forest Service exemplifies, focusing on hiring more women without examining the workplace environment leads down a path without sincere change. To create dynamic, supportive workplaces where women are empowered to fully contribute, organizations must critically view their internal culture and transform it with intentional efforts that value inclusion and welcome contributions from every member of the workforce. A recent report reviewing more than 800 companies described four levels of maturity with regard to diversity, equity, and inclusion (DEI; Bersin 2021). Maturity ranges from a focus on risk mitigation and compliance to a focus on accountability and outcomes, where diversity is intertwined with the daily activities of the organization. Mature companies have higher levels of innovation, financial performance, and retention. This aligns with research suggesting that higher diversity leads to higher-impact research (Freeman and Huang 2015): an inclusive workplace climate supports organizational missions as well as scientific excellence. In this chapter, we describe the roles of policies, organizational structures, and individuals in developing a diverse and inclusive workplace. Although we focus on approaches determined to counteract gender bias, our goal is to describe approaches to attain positive cultural shifts toward a wholly welcoming culture that weaves diversity, equity, and inclusion through the institutional fabric. We start by describing the underlying problem of implicit bias and successful approaches that treat biased behavior as a bad habit that can be changed. We then share case studies that take a comprehensive approach to creating a culture of inclusiveness. Finally, we explore more specific tools from the perspectives of employers, professional societies, and individuals.

Unconscious Bias

As described in Chapter 4, implicit biases underlie many patterns of unequal opportunities and inequity that result in cultures favoring members of the dominant group. Implicit bias can increase stereotypes and lead to more harassment for underrepresented groups. Such harassment increases the likelihood of departure of diverse employees from STEM (science, technology, engineering and mathematics) fields (NASEM 2018). Frequently, bias exists and can conflict with a person's values.

For example, as a woman, I (Tabitha Graves) value giving my employees equal opportunities; however, when I took the Implicit Association Test (IAT; Greenwald et al. 1998), I felt growing horror as I responded more slowly when associating female with career and male with family. My results correspondingly suggested I had very high levels of bias. Although some research finds IATs may not accurately measure an individual's behavior from implicit bias (Uhlmann et al. 2012), I identified areas where I could reduce the effects of implicit bias on my own behaviors. At the time, I was hiring a data assistant. While this was not my first hire, I hadn't received training on equitable hiring practices, let alone implicit bias. I did not know that both men and women were more likely to hire men than women. During the interviews a male student with experience in restaurants who described how those skills would translate to the position impressed me. I was less excited by a female student with a high-pitched voice. However, she had direct experience in database management, and I hired her based on my rubric. She excelled as the data assistant, and I felt good about my decision, the decision process, and learning about my biases.

This dichotomy between values and behaviors is common across individuals, and substantial research illustrates that women and men exhibit sexist behavior equally (e.g., Moss-Racusin et al. 2012). Learning about the influence of society on my behaviors helped me understand that I had adopted discriminatory behavior. This awareness shed light on previous observations of female supervisors unintentionally favoring male employees and showed me ways I could balance my supervisory style. Perhaps most importantly, it changed my primary emotions from anger and hurt to humility when speaking about the topic. Instead of feeling like I am fighting colleagues exhibiting discrimination, I see myself coaching a teammate on a skill set that will help all wildlife professionals. If you value women in the wildlife profession or you want the positive outcomes of a diverse workforce, wherever you are on the gender spectrum, you can create change at an individual level by addressing your own implicit bias.

Policies that promote awareness of implicit bias and approaches for reducing it offer a way forward for organizations and individuals. Awareness can be developed through activities such as inviting biologists to read one of the hundreds of studies about implicit bias (e.g., Moss-Racusin et al. 2012) and following up with a discussion or other interactive activities (e.g., see Chapters 1 and 12, Fitzgerald et al. 2019). Raising awareness has been shown to be particularly effective when the benefits of addressing discrimination are discussed and bias is framed as a habit versus taking a blaming approach or considering bias to be a character trait (Cox and Devine 2019).

A series of experiments have tested the effectiveness of trainings that use a habit-breaking empowerment framework (Cox and Devine 2019). These trainings first raised awareness of one's own vulnerability to implicit bias and to the consequences of stereotype-based bias (Carnes et al. 2015). They then used habit-breaking approaches to promote participants' motivation to reduce their own bias and set expectations that, like changing habits, changing bias requires sustained effort. Cox and Devine (2019) shared a toolbox of bias reduction strategies, including imagining what it would feel like to be a member of a different group and making positive and personal contact with members of stereotyped groups. These trainings led to greater perceptions of fit, feeling valued, and confidence to raise issues not only for those who had attended the training but also for their colleagues who had not been trained, with effects lasting more than two years (Forscher et al. 2017). The proportion of women hired in departments who had the training increased by 18% over other departments (Devine et al. 2017).

Other approaches to address implicit bias reduce the decision maker's knowledge of the gender of a potential employee or award recipient. A classic example of the success of these structural approaches is in the orchestra world, where moving auditions behind a screen and using a carpet to prevent cues like hearing high heels increased the likelihood a woman would be hired in the final round by 30% (Goldin and Rouse 2000). This, along with advertising for positions publicly (versus allowing conductors to handpick applicants) resulted in a shift from 6% to 21% women in the orchestra between 1970 and 1993. We are not aware of published research on blind approaches in the wildlife field. However, examples relevant for wildlife biology include blind journal reviews, abstract submissions, and award committees (Budden et al. 2008; Darling 2015). While people may still be identifiable in some cases, this approach could still reduce bias. Alternatively, intentional selection from underrepresented groups has successfully increased recognition in outgroups (Débarre et al. 2018).

Supporting an Inclusive Culture

Clear, prominent, and consistent policies supporting diversity can set the stage for an inclusive culture. However, as culture includes the values, practices, systems, traditions, and behaviors within an organization, creating cultural change may require addressing both visible elements of culture, such as administrative structures, and invisible elements, such as unspoken priorities and subconscious attitudes of community members. Policies can be used to define structures, prioritize diversity and inclusion, and design training to enhance inclusion and ally skills to create institutional excellence (e.g., Williams 2007).

Although comparing the relative effectiveness of policies can be difficult, using multiple methods across scales more effectively increases inclusivity (Bezrukova et al. 2016). One example of a successful multiscale approach, the National Science Foundation ADVANCE program at Montana State University, tested a three-step faculty search intervention using a randomized controlled study (Smith et al. 2015). The intervention included (1) a faculty search tool kit with concrete strategies for conducting a broad applicant search, which enhanced the competence of the search committee; (2) a 30-minute oral presentation by a faculty member on the role of implicit bias in skewing the hiring processes, which improved the ability of the search committee to limit their own biases; and (3) support via (a) the search committee and a peer faculty member and (b) a family advocate for job finalists for a completely confidential 15-minute conversation, which both addressed concerns of finalists and demonstrated organizational commitment to work-life balance. The intervention increased the number of women interviewed by phone (40.5% versus 14.2%), women finalists interviewed on campus (40.3% versus 18.2%), and offered positions (nine women versus two in the standard search process), with women being 5.8% more likely to accept the offer in the intervention approach (Smith et al. 2015). As another example, the National Aeronautics and Space Administration (NASA) consistently earns the highest ratings in the "support for diversity" category for large federal agencies. NASA has a clear-cut and absolute zero tolerance stance toward harassment, with accessible policies presented in easy-to-read formats. The harassment brochure clearly communicates, "*It is NASA's longstanding policy that harassment in the workplace is prohibited and will not be tolerated.*" This wording sets a strong tone of organizational support. NASA has also developed a strategic plan to integrate inclusion into decision-making to enhance their mission. Its framework targeted (1) principles of demonstrated leadership commitment, (2) employee engagement and effective communication, (3) ongoing diversity and inclusion education and skills development, (4) commitment to community partnerships, and (5) shared accountability and responsibility for diversity and inclusion. NASA's policies can serve as a model for other agencies to create a culture of inclusiveness.

Additional support for a comprehensive approach comes from a recent assessment of sexual harassment programs in 805 companies over 32 years, which found complex patterns of success in increasing the number of women managers (Dobbin and Kalev 2019). Around half of the ~80,000 discrimination and harassment complaints filed annually with the US Equal Employment Opportunity Commission result in retaliation. Dobbin and Kalev (2020 and citations within) noted that grievance procedures carry confidentiality clauses that permit serial abusers to carry on; accused harassers are more likely to be struck by lightning than to be transferred or lose their job. To avoid these pitfalls, employers can adopt alternatives for addressing discrimination such as employee assistance plans, ombuds programs, and dispute resolution systems. A long-term study of the US Postal Service's comprehensive approach showed that 90% of participants were satisfied, supervisors reported it improved their conflict management and listening skills, and 30% of complainants received an apology, a rarity in formal grievance systems (Bingham and Scherer 2001).

Organizations with an Equal Employment Opportunity Office can provide support and understanding of diversity and equity issues within a workplace. For instance, in the US federal government, which employs numerous wildlife professionals, many agencies have employee resource groups to create community among people who share characteristics such as gender. These organized groups have some legal protections and can provide formal input to programs, internal support, and mentoring opportunities. Organizational support for these groups signals a clear investment in workplace culture as one part of a comprehensive program (Table 10.1).

Making Change

Why do some organizations have a welcoming culture that values the backgrounds and opinions of their employees? Organizations with a diverse workforce tend to have less gender bias and harassment (NASEM 2018), but organizations can proactively institute actions, policies, and procedures that promote inclusion regardless of their composition. These structural changes can be made to the formal systems in place to support diversity and inclusion (Holvino et al. 2004). Organizations may have the greatest ability to create

TABLE 10.1 Clues for evaluating the commitment of a prospective employer, professional organization, college, or university to a welcoming, inclusive culture.

Component	Type	Source of information
Inclusive climate	Legal and policy	Anti-harassment policy in place and diversity training for all employees
	Structure	DEI council in place
	Legal and policy	Policy and statement of support for DEI is prominently displayed on webpages and documents
	Structure	Outgroups or Employee Resource Groups exist to create community
	Legal and policy	Organization-wide implicit bias training and discussions
	Structure	Organization offers clothing and equipment for multiple genders and body types
	Legal and policy	Codes of conduct in place for conferences or fieldwork with ombudspersons for reporting
Hiring practices	Legal and policy	Dual career statement, policy, and officer in place to support partner hires
	Structure	Interviews are scheduled to avoid national and religious holidays, and candidates receive accommodations for special dietary or physical needs
	Legal and policy	Outreach includes funding/approaches to maximize diversity of applicant pool
	Structure	Hiring committees are diverse and best practices used to minimize bias in decisions
Awards and recognition	Structure	Rubrics are developed and used to rank candidates
	Structure	Diversity metrics for awardees are available and improving
	Legal and Policy	Outreach includes approaches to maximize diversity of applicants
	Structure	Keynote and plenary speakers are from diverse backgrounds
Salary and Promotion	Legal and Policy	Equal pay policies supported
	Structure	Transparent salary information is readily available
	Structure	Transparent process for promotions or lateral assignments
	Legal and policy	Promotions are considered at a structural level to maintain equity
	Structure	Collaboration, diversity and inclusion work, and mentoring are explicitly valued in performance reviews and promotions
Family-friendly policies	Legal and policy	Family leave policy supported and used by all genders
	Structure	Range of supportive family friendly employer benefits, such as child or eldercare assistance, shipping of breast milk while traveling, lactation rooms
	Structure	Conferences include accommodations for childcare and lactation support
Individual support for well-being	Structure	Wellness benefits provided for all employees including fitness, mental health, and employee assistance programs
	Individual	Formal or informal mentoring networks developed

positive change because they serve as a link between legal requirements and individual actions by rewarding those contributing to inclusivity.

And what does a truly inclusive organization look, feel, and sound like for an employee? Organizations that develop, describe, and communicate their aspiration to be welcoming and inclusive illustrate a commitment to creating a better future. One of the first steps to improving culture is defining the current climate and describing the beliefs and values the organization aspires to embody. For example, an organization that only passively commits to inclusion and equity for women with a diversity statement but does not commit to specific changes to support women may be in a transitional stage of compliance (Holvino et al. 2004). Contrast that transitional stage to an employer who actively incorporates an intersectionality of styles and perspectives and approaches diversity with an attitude of continual change and improvement. Individuals can observe the cultural norms of their organization and use clues to evaluate the cultural climate of a potential employer using a diversity and inclusion scorecard (Kohl et al. 2017; Smith et al. 2015).

Workplaces that have mostly male employees, particularly at the highest levels of the organization, are more likely to have a culture that tolerates harassment (NASEM 2018). Research on group dynamics indicates that when a minority group exceeds about 15% of the total, the group's chances of survival and continued existence grows (Etzkowitz et al. 1994). Feelings of isolation, tokenism, and pressure to represent an entire gender, race, or other identity diminish for individuals of that underrepresented group. This further emphasizes that an inclusive culture goes beyond reaching compositional diversity targets to "tick the diversity box."

Best Hiring and Promotion Practices

Often, organizations want to increase the diversity of their workforce but are unclear how to increase diversity in their applicant pool. The number of women and other minority applicants for wildlife positions can be influenced by unintentional biases within the job announcement. Whereas men apply for positions if they meet 60% of the qualifications, women tend to only apply for positions if they meet nearly 100% of the desired characteristics (Mohr 2014). Further, women view job descriptions with masculine language (e.g., dominant, assertive, competitive) as less appealing, and they may avoid applying. To address unintentional biases, organizations can develop job ads that only include their most important qualifications and avoid gender-coded language.

Gaucher, Friesen, and Kay (2011) provide a list of common gendered words to avoid and online gender decoders are available (e.g., http://gender-decoder .katmatfield.com/). Creating specific rubrics for examining applications or conducting interviews also helps reduce unconscious bias (Moss-Racusin et al. 2012). Hiring committees can ensure that interviews are welcoming by avoiding national and religious holidays during scheduling and working with their human resources department to ensure needs are met for applicants with dietary restrictions and special travel or physical needs.

In addition, a tradition of unpaid internships and volunteer positions in wild-life biology creates an equity divide. Women and minorities may avoid applying for internships or seasonal technician positions if they cannot afford to work for low pay or be separated from family. Evaluating all aspects of the hiring processes, developing and updating procedures to increase inclusion through formalized best hiring practices, and considering the consequences to diversity of long-held traditions that might be barriers to entry to the field can promote DEI (e.g., Smith et al. 2015).

Developing transparent processes including sharing information about salary, use of and improvements in benefits, and the pathway to promotions and lateral assignments can reduce biases that negatively affect underrepresented groups. The gender wage gap, or differences between salary for women and men in the same position, decreased after legislative change to regulate pay transparency in Denmark (Bennedsen et al. 2019). The California Equal Pay Act has been amended and precludes unfair pay gaps among employees conducting similar work who differ by race and gender (California Labor Code Section 1197.5). Employers reap benefits from transparency, with increased productivity and worker loyalty (Huet-Vaughn 2013). On average, white women receive only 79% of wages that white men make, and the percentages drop precipitously for women of color (remember intersectionality? Black: 63%, Native American: 57%, Hispanic: 54%; AAUW 2017). Over a career, pay discrepancies cascade into women earning hundreds of thousands of dollars less than men, significantly impacting retirement. However, negotiating a fair salary can be difficult when you are unaware of what your peers are paid. Many agencies, universities, and nonprofits publish salary information, but access to these data is not publicized, and pay transparency in the private sector is rare. Several internet sites anonymously publish salaries, but these sites are seldom specific to the wildlife profession. Employers who fear backlash from their employees against publicly available salary information could reduce obfuscation by including salary ranges in job announcements and stated metrics for evaluating candidates for positions. Increased transparency of salaries and promotions sends a clear message that employees are reviewed based on skills and merit, not gender or other aspects of identity.

Transparency extends to actions against sexual harassment or other inappropriate behaviors. Organizations that do not leave whispered stories of harassment for water cooler conversations can instead discuss harassment publicly to shift their culture toward increased civility. Regular work-climate surveys with questions about bias and being valued as an individual can provide information on how the culture is changing and whether new initiatives are successful (Hirsh and Tomaskovic-Devey 2020). Making results of these climate surveys accessible, with the message that the organization cares about implementing a transformative shift in climate, can support deeper change. Anonymous incident reporting systems can also provide critical data for evaluating movement toward inclusivity and equity.

Family

Significant life events occur at all career stages, highlighting the need to implement flexible gender-neutral benefits as employees take on personal challenges such as elder care. Organizations that develop and actively encourage use of family friendly benefits for all employees offer structural solutions to remove perceived barriers from women and invisible barriers to men. Women in the workplace are affected by family responsibilities more than men are, in part because many organizations continue to use an outdated stereotype of family, where the husband works and his wife stays home with the children (Eagly and Carli 2007). Many publications describe the leaky pipeline in the sciences, where nearly twice as many women as men drop out or switch to part time after having a child. The causes of the leaks are complex, but the overlap among child-bearing years, graduate education, and early-career positions are especially corrosive (Homyack et al. 2014). Many people view this period of their lives as dichotomous: they can be an involved parent or a productive employee. Broadening the range of benefits available to employees signals a supportive work environment. Beyond teleworking, altering the promotion timeline for new parents, and providing childcare subsidies, employers could offer childcare resources, pay to ship breast milk during overnight travel, and allow infants in the workplace. Some cultures may have greater challenges related to job mobility, such as expectations that women support extended family through their lives (see Chapter 5). Benefits that reduce the "mental load" and help with family responsibilities are critical for women who wish to step into conference travel, short-term assignments, or remote field assignments, gain varied experiences, and build professional networks. For example, several medical schools allocate resources to hire "extra hands" (e.g., technicians) to provide women time to succeed professionally. This equity-based assistance is critical because despite gains in gender parity, women still spend more time on family responsibilities outside of work than men (Eagly and Carli 2007). Creating an environment that fully supports a personal and family life outside of work hours translates into a welcoming multicultural workplace and improves individual productivity and mental health. Another structural change for employers and professional organizations such as The Wildlife Society (TWS) is the creation and support of committees and staff positions dedicated to diversity, equity, and inclusion. Diversity councils are a visible commitment to improving work climate but can only succeed when managers and those with authority support employees with time and resources (e.g., US Geological Survey 2016). Like other groups within an organization, DEI councils should start with developing a mission that is tied to the broader organization. For example, how would reaching an aspirational state of diversity and inclusion better poise the agency or professional society to meet its mission? The mission and objectives should be clearly laid out in a charter document with specific, measurable goals, and the council should check in on progress and adapt when encountering roadblocks. The US Geological Survey (USGS) provides an example of diversity charters with strategic goals and implementation plans, and Holvino et al. (2004) provided a useful overview of developing and implementing a strategic plan for organizational diversity.

The Power of a Society

Professional societies are in a powerful position to affect change for the wildlife field because they can serve as a catalyst for individual actions to connect to a broader common purpose. Currently, TWS has a position statement on workforce diversity and several working groups that have made important strides in educating the wildlife community about the need and challenges to increasing diversity and inclusion in our field (TWS 2020). Further, a diversity vision describing the TWS's aspirational state was recently published and actionable steps to improve inclusion are in development by TWS leaders (TWS n.d.). TWS has also taken one of the most common first steps toward changing culture, connecting those within outgroups. The Women of Wildlife networking group began with socials and a Facebook page organized by several women and a few men who recognized the inequity in the wildlife biology profession. Similarly, the Out in the Field initiative was developed to increase the visibility and develop a supportive community of LGBTQ+ wildlife professionals (see Chapter 8). Both forums connect people who are facing challenging situations and extend a professional support network. The momentum of these outside groups has led to symposia, workshops, and panel discussions at the TWS's annual conference on topics related to diversity in the wildlife profession such as microaggressions (2016) and "Beyond #MeToo" (2018), which are often standing-room-only events. Training on inclusion and diversity, both by professional societies and in the workplace, can be effective educational tools when they are part of comprehensive cultural change that links policy, structure, and individual actions (Bezrukova et al. 2016; Moss-Racusin et al. 2018). For example, privilege walks, where people are provided opportunities to reflect on how society favors or disfavors individuals, can be integrated into meetings and conferences. We've participated in privilege walks in groups of 15–105 people (see Privilege Walk activity at the end of the chapter). It may take courage to lead such activities, but the rewards are great (Cox and Devine 2019 and citations within).

Professional societies that organize conferences, publish technical information, and administer awards can provide great opportunities to underrepresented groups. A task group of the Inclusion, Diversity, Equity, and Awareness Working Group of The Wildlife Society scoured past annual conference programs and found that from 1994 to 2017, only 20% of 83 plenary speakers in opening sessions were women. Research shows that it takes a conscious effort to change this pattern. Including women on committees that select distinguished speakers increases the probability of having women selected (Farr et al. 2017; Débarre et al. 2018). Policies to ensure gender equity of conference speakers and providing financial resources for speakers also reduces bias against women. Internet resources like 500 Women Scientists are valuable for identifying skilled women to invite (500 Women Scientists n.d.). As described in Chapter 2, women have been critical contributors to wildlife biology since the beginning of the profession, but even those who persist through significant challenges are rarely recognized for their accomplishments (Miriti et al. 2020). Few recipients of the Aldo Leopold Memorial Award and Fellows administered by The Wildlife Society are women (by 2021, <3% and 18%, respectively). Award panels may shift this pattern if they review their procedures and consider broader contributions instead of longevity or short periods of extreme productivity. Panels can increase inclusion by making efforts to diversify their pool and explicitly rank candidates by predefined criteria. Otherwise, awards may be devalued when candidates that the panelists know personally through social networks are ranked the highest. Finally, award criteria should include ethics clauses to avoid rewarding people for their scientific contributions if they do not serve as role models for equity and inclusion.

Codes of conduct for conferences, societies, or organizational units can address inclusion (Favaro et al. 2016). By requiring that participants read and sign codes of conduct, entities demonstrate a clear commitment to diversity and inclusion, reassure attendees concerned about unfair or harassing treatment, and set a clear target for inclusive and respectful behavior. Foxx et al. (2019) found that only 24% of biology conferences had a code of conduct at the time of their study. Of them, 46% did not mention sexual misconduct, and 26% did not specify how to report a violation. They also highlighted other challenges, such as ensuring that trained people, such as ombudspersons, are available to respond to reports with care and concern for anonymity and reducing fear of retaliation. More comprehensive national conference codes of conduct could be used as templates for regional meetings to adopt (e.g., ESA 2018). Often codes of conduct are paired with other initiatives managed by an equity or diversity committee (Tulloch 2020). For example, the Citizen Science Association conference in 2019 included not only a code of conduct but also the option of including pronouns on name badges, diverse and accomplished keynote speakers from a wide range of perspectives, and family-oriented activities. Conference organizers also coordinated substantial outreach and funding that brought in people from the environmental justice movement who are disproportionally nonwhite and scheduled a networking session that disrupted attendees' normal habits of talking to people they know (and who look like them, which is known as affinity bias). This meeting enacted many of the components known to improve inclusivity for meetings and provided a positive case study for achieving broad representation (Pendergrass et al. 2019; Tulloch 2020). Many strategies that succeed at increasing the inclusivity at conferences transfer to other situations as well.

For example, in the wildlife profession, employees often live in field environments with limited access to communication for weeks. In these situations, organizational codes of conduct are critical to protect employees from harassment and unsafe conditions. Surveys of professions that conduct fieldwork indicate that harassment is common. In one study, 64% of field scientists had personally experienced harassment (Clancy et al. 2014). In most instances, policies and rules regarding harassment in field conditions are either not present or present but not enforced, with significant consequences for mental health and career development (Nelson et al. 2017). For the field environment, codes of conduct should also include measures for personal safety, access to transportation and communication devices, and multiple pathways to contact outside help.

Increasing Inclusivity in Research

Women face bias in publishing and in obtaining grants that negatively impacts their career advancement in a complex series of interactions that perpetuate attrition from the profession (Cameron et al. 2013). Without equitable access to resources and publishing, women, particularly those in research positions, are less likely to meet metrics for receiving tenure or promotions (Ceci and Williams 2011). Moreover, with fewer grants and peer-review publications on their curriculum vitae (and pervasive fear of rejection), women's self-confidence may decline, and they may compete less well in job applications, which adds into the negative feedback loop of lower publication rates. Granting and publication bodies may offset these trends with blind submission and by using rubrics that acknowledge publication impact along with quantity. More grant-awarding bodies are improving access to funding by using external requests for proposals (versus internal unadvertised paths only) and requesting that diversity goals be incorporated into research proposals. Traditional metrics of publication quality place significant value on single-authored manuscripts as a signal of research independence and maturity. However, women and others who manage multiauthored publications may have collaboration skills and emotional intelligence critical to progress on complex conservation challenges (Goring et al. 2014). Reviewers and promotion committees could acknowledge that managing professional relationships requires time and skills that could result in less collaborative research and fewer publications. For women submitting grants, curriculum vitae, tenure, or review packages, highlighting activities such as coordination, contributions to diversity and inclusion, mentorship, fund raising, and idea generation can better communicate one's role and accomplishments. Finally, recognizing that reviewers of all genders have implicit bias and incorporating training into ranking proposals and manuscripts can help to reduce bias (Smith et al. 2015b).

Individual Actions and Change

As Frances Moore Lappé wrote, "*Every aspect of our lives is, in a sense, a vote for the kind of world we want to live in*" (Lappé 2011). To begin to effect change starting with ourselves, we need to understand our own associations. Implicit cognition, or unconscious learning, is based in experience and exposure

(Greenwald and Banaji 1995), but it does not mean that an individual desires to hold unconscious biases (Karpinski and Hilton 2001), highlighting the importance of follow-up conversations to create change. Organizationally, implicit association or implicit bias tests are often taken without follow-up or discussion. As individuals, we can make use of resources such as using self-reflection templates to understand biases: keeping a log of behaviors related to the bias, changing or finding new experiences and exposure, developing strategies to counteract underlying biases from previous exposure, and joining discussion forums. The key is to recognize that these biases are based on experience and that attitudes and behaviors can change, particularly where the social context or office culture actively works at it (Cox and Devine 2019; Vuletich and Payne 2019).

The Role of Allies

Considering the pervasive discrimination and harassment faced by women in the workplace, forming allies is a powerful means to reaching career goals and sometimes to simply survive at work. The idea behind building allies is not to insulate yourself but rather to find support. A boss can be an ally, but, importantly, people outside of your organization may offer alternative perspectives. Some components to finding and becoming an ally are to:

- communicate openly to find potential allies;
- determine who has shared values through conversations and observation, by setting up lunches or coffee breaks to talk through career issues and challenges; and
- give of yourself and your time with your allies. Allyship is a two-way street.

The return on investment of building your own network can help career advancement in ways that may be challenging without this support (Madsen et al. 2020). Plus, these relationships are typically mutually beneficial and promote a more diverse workplace. Where men were engaged in improving gender diversity, companies made much greater progress compared to companies where men were not engaged (Krentz et al. 2017).

Acting as a Sponsor and Mentor

Sponsors and mentors offer useful and sometimes necessary additional workplace relationships. A mentor provides advice, but a sponsor serves as an advocate for career advancement. A mentor can help craft career goals, provide a sounding board for ideas, model work-life balance, and assist with retirement goals. A mentor can be anyone who has been around long enough to help navigate through organizational and career advancement challenges. Although mentors can provide great advantages, they often do not translate into career advancement (Carter and Silva 2010). In the white, male-dominated sciences, finding a sponsor in a higher position in the organization can be critical for career advancement. The gender of the sponsor is unimportant, so long as they are willing to put their name and reputation on the line in support of your career advancement (Hewlett et al. 2010). In the corporate arena, 46% more men than women have a sponsor advocating for them. Not only can a sponsor promote the protégé and help make connections to senior management, they can also provide opportunities that expand current skills ("stretch" opportunities) for professional growth. Finding the right sponsor initially requires narrowing in on a position, department, or project and then looking within that group. The key is in cultivating a relationship so that your sponsor believes in your potential and chooses to act on your behalf to provide opportunities for further exposure and new assignments that might be difficult to access otherwise. Performing well on the job is the first step in commanding their attention.

Finding Allies

Psychologists have shown that exposing women to female leaders has a powerful, positive impact on their unconscious beliefs about their suitability for leadership roles (Dasgupta and Asgari 2004). However, people from all segments of society have struggled with seeing themselves as an intellectual fake or impostor who were mistakenly given their jobs (Matthews and Clance 1985). Imposter syndrome is particularly pervasive among high-achieving women (Clance and Imes 1978). One approach to this prevalent phenomenon is to engage in a group setting with allies, for example, other women who might be experiencing similar struggles. These groups can provide a safe space for camaraderie and first steps toward changing their mindset, for example, through keeping a record of positive feedback for future reflection (Clance and Imes 1978).

Finding others who share values and can help in DEI work can be challenging in some organizational settings, but those connections have great rewards. One step to is to spend time thinking through your own values and the kinds of connections you're seeking. Ultimately, personal connections can help lessen the mental exhaustion and split the daunting work needed to achieve diversity and inclusion.

Asking for help can be hard. Some may consider asking for help a sign of weakness or poor problem-solving ability. However, if work demands are not sustainable, people run the risk of burning out, feeling overwhelmed, and losing motivation (Maslach and Leiter 2008). These impacts might be even quicker to surface in the emotionally charged work of reaching DEI goals, where traction for change often feels slow (Atkinson 1990).

Being Asked to Represent an Entire Group

Often, the burden of representation can fall on one or a few women or people of color. Connecting with a network can turn a single voice into a powerful force. Groups like the 500 Women Scientists can provide unity and strength and command public attention on issues of discrimination and bias in the sciences.

The US public has debated the burden of representation in advertisements, on television, and in movies, but the question is just as relevant in any professional science organization. Should one Black woman be asked to represent all minorities in the organization? Should she be pushed into the spotlight? Or can she be free to interact as an individual with her own personal motivations and experiences? Scientists recognize the limitations of small sample sizes. We should embrace the same precision and avoid extrapolating from one or two individuals to the whole. In the context of DEI goals, people have their own personal motivations; therefore, when forming diversity committees, one woman or one Hispanic man should not be expected to bear the burden of representation. Being asked to be "the representative" can create a difficulty for those from underrepresented groups that is not faced by a white person. Some people of color might not identify with their heritage. Some women might not have experienced significant effects of bias. In both cases, these individuals might feel like frauds.

Mental Health Awareness

With busy days of meetings, deadlines to meet, and field surveys to complete, it may feel as though little time remains for anything else, including self-care. Heathcare professionals are taught to practice self-care to more effectively help others (Shapiro et al. 2007; Dyess and Sherman 2009), a lesson from which all professions can benefit. Mindfulness, the state of being present and aware of our thoughts and feelings, has been shown to have many physical health benefits (Kabat-Zinn 2003; Grossman et al. 2004) and to reduce cortisol and stress levels in the body (Turakitwanakan et al.). Practices such as meditation, physical activity, lunch breaks, and generally taking time for oneself can help decrease stress. Some organizations, like Ben & Jerry's ice cream company, have even installed nap rooms to allow employees to recharge on the job (Mishra 2009). Engaging in activities to reset and recharge our minds can put professionals in a much better position for success, provide endurance, and make our work more focused and effective.

Balancing Achievements with Time Spent on DEI

Scheduling tasks, prioritizing them, asking for help when needed, and making sure to take breaks are essential for working efficiently. Taking on diversity and inclusion is no different, and because the topic is so personal and emotive, it can spill into our personal time, putting dedicated employees at risk for burn out. We could ask whether we are trying to push too many ideas at once and would benefit from focusing on only one or two, or whether we need to add more self-care to our weekly routines. Do we need help or advice from a colleague or mentor who cares about us, our well-being, and our successes and is not involved in the challenge directly?

A few years ago, Google created a program called Diversity Core, which empowers employees to spend 20% of their workday on innovating around diversity and inclusion. This program builds DEI into the workplace culture instead of leaving it as something extra that employees must take on. The idea to give employees time to spend on diversity came from Google's business practice of giving employees 20% free time more broadly. This free time has yielded incredible innovation for the company, such as the development of Gmail (Guynn 2015), and hopes are high that providing time for diversity will achieve amazing breakthroughs for DEI in the workplace. One method for freeing work time for innovation is to structure more effective meetings and use those hours for deep thinking and problem solving (Perlow et al. 2017). For example, disseminate information via email if an interaction is not required, create 10-minute mini-meetings, or designate Fridays as no-meeting days.

Implementing Approaches That Work

The high-profile attention given to gender disparity and harassment in the United States has illuminated problems and increased opportunities for more inclusion in the profession of wildlife science. The question now is, can the wildlife profession build upon the foundation laid before us and create meaningful progress toward our aspirational state? Perhaps now more than ever, the answer is yes. External factors play substantial roles in accelerating cultural change, and we hope that the combination of societal fervor and scientific knowledge about how to improve workplace equity are producing a climate for rapid growth. New research on best practices continues to emerge and will improve scientific achievement and meritocracy (Moss-Racusin et al. 2014; Devine et al. 2017).

Our chapter summarized the change that can occur across scales from organizations to individual people, from universities to government agencies, and from field work to conferences to research. Much as species interact with each other from the subcellular to the ecosystem level, the most progress toward a welcoming and inclusive environment will be made through strategic actions at multiple scales. As in other sciences, more research will continue to improve understanding about successful approaches. We've been asked, "How will we know when we've made it to our aspirational state?" When we are part of a truly inclusive profession, policy and legal changes will support organizational structures, and individuals will be energized and rewarded for their work on DEI. In short, when diversity and inclusion are woven into the institutional fabric, we will acknowledge and appreciate the phenomenal breadth of human diversity and the multitude of identities that each person brings to their workplace each day. We will make decisions on hiring, promotions, and resources based on skills and competencies, not appearance or perceived identity.

How to Be an Ally

- Consider how you can address your own habit of implicit bias.
- Invite diverse applicants for jobs, speaking engagements, awards, and other opportunities.
- Share some of these resources with your organization to consider a multipronged approach to initiate cultural change.

Discussion Questions

1. Have you ever participated in an organization (e.g., club, society, or workplace) that was inclusive and welcoming of all people? What policy, organizational, or personal structures supported this culture? If not, what supportive structures were missing?
2. Reflect on the role of self in developing and maintaining an inclusive organizational culture. How do individual actions, behaviors, and

values interact with policies and organizational structures to advance or hinder change toward inclusivity?

3. Are you aware of any new methods to improve inclusivity supported by research?

Activity: Privilege Walk: Who Are the Unusual Voices?

Learning Objectives: Participants will gain awareness of their personal privileges and the privileges of other people and will recognize the impact of privilege on diversity and inclusion.

Time: 20–45 min.

Description of Activity

Materials needed: For the physical part of the activity, no additional materials are needed. For the seated version, the facilitator will need a single sticky note for each participant.

The activity also can be created within software like Quizziz to be accomplished individually, with results that can be summarized and visualized for the group in real time. Quizziz also allows for analysis by question (including time spent answering each question) and data summaries, useful if assessment is a goal.

Part 1 (10 min.): There are two forms for this activity, depending on your access to space and the desire for visibility of participants' privilege within the group. In the less anonymous format, participants stand in a line with space ahead and behind them, and they step forward or backward based on their answer to each statement below. This format demonstrates physically the ways in which society favors some participants over others, which ultimately can shape our lived experience. Alternatively, participants can write a +1 or −1 on a sheet of paper in response to each question and tally their final score on a sticky note. The facilitator then collects the sticky notes and places them on the wall or a board in order of the scores to visualize the privilege in the group. The latter method maintains some anonymity in responses.

Introduction (facilitator reads to group): Your identity influences "privilege," the power you have in this world. Race, ethnicity, gender, age, and socioeconomic background are all intertwined and linked to power and influence. Sometimes these privileges are visible and obvious. Sometimes they aren't. There are no right or wrong answers. Everyone should feel free to respond to each question however they feel most comfortable.

1. If English is your first language, take one step forward.
2. If either of your parents graduated from college, take one step forward.
3. If you have been divorced or impacted by divorce, take one step back.
4. If you skipped a meal because there was no food in your house, take one step back.
5. If you have visible or invisible disabilities, take one step back.
6. If your parents and family encouraged you to attend college, take one step forward.
7. If your family has health insurance, take one step forward.
8. If you studied the culture or history of your ancestors in elementary school, take one step forward.
9. If you have been bullied or made fun of based on something you cannot change (e.g., your gender, ethnicity, age, sexual orientation), take one step back.
10. If you have ever been offered a job because of your association with a friend or family member, take one step forward.
11. If you have ever been stopped or questioned by the police because they found you suspicious, take one step back.
12. If you or your family have ever inherited money or property, take one step forward.
13. If you have ever been uncomfortable about a joke or statement you overheard about your race, gender, appearance, or sexual orientation but felt unsafe to confront it, take one step back.
14. If you took out loans for your education, take one step back.
15. If there were more than 50 books in your home growing up, take one step forward.
16. If you are a white male, take one step forward.
17. If you have worried about assault when walking alone at night, take one step back.
18. If the Band-Aids at mainstream stores blend in with or match your skin tone, take one step forward.
19. If you work with people you consider like yourself, take one step forward.
20. If your household employs help for gardening, lawn mowing, house cleaning, child care, etc., take one step forward.
21. If you would ever think twice about calling the police when trouble occurs, take one step back.

22. If you can show affection for your romantic partner in public without fear of ridicule or violence, take one step forward.

23. If you were ever discouraged from an activity because of race, class, ethnicity, gender, disability, or sexual orientation, take one step back.

24. If you have ever tried to change your appearance, mannerisms, or behavior to fit in more, take one step back.

25. If you have found someone in your career or education who served as a mentor, take one step forward.

26. If there has ever been substance abuse in your household, take one step back.

27. If you live in an area with crime and drug activity, take one step back.

28. If someone in your household suffers from or has suffered from mental illness, take one step back.

29. If you have been a victim of sexual harassment, take one step back.

30. If you have you ever been asked to speak on behalf of a group of people who share an identity with you, take one step back.

31. If you can make mistakes and people attribute your behavior to flaws in your racial or gender group, take one step back.

32. If you attended graduate school, take one step forward.

33. If your work or school holidays never overlap with your religious holidays, take one step back.

After finishing the questions, give participants a minute to calculate totals on their sticky note and arrange on the wall if using the sticky-note version. Ask participants to look around at where people are physically located in the space or at the wall. Then have everyone move to sit in a circle for discussion or into small groups for discussions.

Part 2: Debriefing and Discussion Questions (20–30 min.)

1. How did you feel during this activity?

2. What did you learn?

3. What was most surprising about the experience?

4. What privilege stuck out most to you?

5. Was there a privilege included that you never thought of before as privilege?

6. What are factors that influence privilege that you may never have considered before?

7. How can privilege shape the diversity within a group or an organization?

8. Were there any statements or a privilege that were not included that you think should be included?

9. How would you use the information you learned or your experience from this activity in your life as a wildlife professional, student, or citizen?

Teaching Notes

This activity can be emotionally heavy for some participants, especially those who have a lower privilege score. Group members should have already developed trust for each other, and this exercise should only be conducted in a safe environment. Facilitators should be prepared for emotional responses. It may also be useful to evaluate whether people understand and value the difference between equality and equity given concerns about backlash from some people (Cox and Devine 2019).

Literature Cited

American Association of University Women (AAUW). 2017. "The Simple Truth about the Gender Pay Gap." www.aauw.org/resources/research/simple-truth/.

Atkinson, Philip E. 1990. *Creating Culture Change: The Key to Successful Total Quality.* Bedford, UK: IFS.

Bennedsen, Morten, Elena Simintzi, Margarita Tsoutsoura, and Daniel Wolfenzon. 2019. "Do Firms Respond to Gender Pay Gap Disclosure?" National Bureau of Economic Research. doi: 10.3386/w25435.

Bersin, Josh. 2021. *Elevating Equity: The Real Story of Diversity and Inclusion.* joshbersin.com.

Bezrukova, Katerina, Chester S. Spell, Jamie L. Perry, and Karen A. Jehn. 2016. "A Meta-analytical Integration of Over 40 Years of Research on Diversity Training Evaluation." *Psychological Bulletin* 142(11): 1227–74.

Bingham, Shereen G., and Lisa L. Scherer. 2001. "The Unexpected Effects of a Sexual Harassment Educational Program." *Journal of Applied Behavioral Science* 37(2): 125–53. https://journals.sagepub.com/doi/abs/10.1177/0021886301372001.

Brown, Greg, Chuck Harris, and Trevor Squirrell. 2010. "Gender Diversification in the U.S. Forest Service: Does It Still Matter?" *Review of Public Personnel Administration* 30(3): 268–300. doi: 10.1177/0734371x10368219.

Budden, Amber E., Tom Tregenza, Lonnie W. Aarssen, Julia Koricheva, Roosa Leimu, and Christopher J. Lortie. 2008. "Double-Blind Review Favours Increased Representation of Female Authors." *Trends in Ecology and Evolution* 23(1): 1. doi: 10.1016/j.tree.2007.07.008

Bumpus, Namandjé. 2020. "Too Many Senior White Academics Still Resist Recognizing Racism." *Nature* 583(661): 1. doi: 10.1038/d41586-020-02203-w.

Cameron, Elissa Z., Meeghan E. Gray, and Angela M. White. 2013. "Is Publication Rate an Equal Opportunity Metric?" *Trends in Ecology and Evolution* 28(1): 7–8. doi: 10.1016/j.tree.2012.10.014.

Carnes, Molly, et al. 2015. "Effect of an Intervention to Break the Gender Bias Habit for Faculty at One Institution: A Cluster Randomized, Controlled Trial." *Academic Medicine* 90(2): 221–30. doi: 10.1097/ACM.0000000000000552.

Carter, Nancy M., and Christine Silva. 2010. "Pipeline's Broken Promise." February 1. www.catalyst.org/research/pipelines-broken-promise/.

Ceci, Stephen J., and Wendy M. Williams. 2011. "Understanding Current Causes of Women's Underrepresentation in Science." *Proceedings of the National Academy of Sciences* 108(8): 3157–62. doi: 10.1073/pnas.1014871108.

Cheruvelil, Kendra S., et al. 2014. "Creating and Maintaining High-Performing Collaborative Research Teams: The Importance of Diversity and Interpersonal Skills." *Frontiers in Ecology and the Environment* 12(1): 31–38. doi: 10.1890/130001.

Clance, Pauline Rose, and Suzanne Ament Imes. 1978. "The Imposter Phenomenon in High Achieving Women: Dynamics and Therapeutic Intervention." *Psychotherapy: Theory, Research & Practice* 15(3): 241.

Clancy, Katherine B., Robin G. Nelson, Julienne N. Rutherford, and Katie Hinde. 2014. "Survey of Academic Field Experiences (SAFE): Trainees Report Harassment and Assault." *PLoS One* 9(7): e102172. doi: 10.1371/journal.pone.0102172.

Cook, John G. 2012. "The American College President Study: Key Findings and Takeaways." *Presidency*. Spring Supplement.

Cox, William T. L., and Patricia Devine. 2019. "The Prejudice Habit-Breaking Intervention: An Empowerment-Based Confrontation Approach." In *Confronting Prejudice and Discrimination: The Science of Changing Minds and Behaviors*, edited by Robyn K. Mallett and Margo J. Monteith, 249–73. San Diego, CA: Academic Press.

Darling, Emily S. 2015. "Use of Double-Blind Peer Review to Increase Author Diversity." *Conservation Biology* 29(1): 297–99. doi: 10.1111/cobi.12333.

Dasgupta, Nilanjana, and Shaki Asgari. 2004. "Seeing Is Believing: Exposure to Counter Stereotypic Women Leaders and Its Effect on the Malleability of Automatic Gender Stereotyping." *Journal of Experimental Social Psychology* 40(5): 642–58.

Débarre, Florence, Nicolas O. Rode, and Line V. Ugelvig. 2018. "Gender Equity at Scientific Events." *Evolution Letters* 2(3): 148–158.

Devine, Patricia G., et al. 2017. "A Gender Bias Habit-Breaking Intervention Led to Increased Hiring of Female Faculty in STEMM Departments." *Journal of Experimental Social Psychology* 73:211–15.

Dobbin, Frank, and Alexandra Kalev. 2019. "The Promise and Peril of Sexual Harassment Programs." *Proceedings of the National Academy of Sciences* 116(25): 12255–260.

Dobbin, Frank, and Alexandra Kalev. 2020. "Making Discrimination and Harassment Complaint Systems Better." In *What Works? Evidence-Based Ideas to Increase Diversity, Equity, and Inclusion in the Workplace*. Amherst: University of Massachusetts, Amherst Center for Employment Equity. /www.umass.edu/employmentequity/what-works -evidence-based-ideas-increase-diversity-equity-and-inclusion-workplace.

Dyess, Susan M., and Rose O. Sherman. 2009. "The First Year of Practice: New Graduate Nurses' Transition and Learning Needs." *Journal of Continuing Education in Nursing* 40(9): 403–10.

Eagly, Alice, and Linda L. Carli. 2007. "Women and the Labyrinth of Leadership." *Harvard Business Review*, September.

Ecological Society of America (ESA). 2018. "Code of Conduct for ESA Events." https://www .esa.org/events/code-of-conduct-for-esa-events/.

Etzkowitz, Henry, Carol Kemelgor, Michael Neuschatz, Brian Uzzi, and Joseph Alonzo. 1994. "The Paradox of Critical Mass for Women in Science." *Science* 266:51–54.

Farr, Cooper M., et al. 2017. "Addressing the Gender Gap in Distinguished Speakers at Professional Ecology Conferences." *BioScience* 67(5): 464–68. doi: 10.1093/biosci/ bix013.

Favaro, Brett, et al. 2016. "Your Science Conference Should Have a Code of Conduct." *Frontiers in Marine Science* 3:1–4. doi: 10.3389/fmars.2016.00103.

Finkelstein, Martin J., Valerie Martin Conley, and Jack H. Schuster. 2016. "Taking the Measure of Faculty Diversity." In *Advancing Higher Education*. New Work: TIAA Institute.

Fitzgerald, Chloë, Angela Martin, Delphine Berner, and Samia Hurst. 2019. "Interventions Designed to Reduce Implicit Prejudices and Implicit Stereotypes in Real World Contexts: A Systematic Review." *BMC Psychology* 7(1): 29. doi: 10.1186/s40359-019-0299-7.

500 Women Scientists. N.d. "Request a Woman in STEMM." Accessed September 23, 2020. https://500womenscientists.org/.

Forscher, Patrick S., Chelsea Mitamura, Emily L. Dix, William T. L. Cox, and Patricia G. Devine. 2017. "Breaking the Prejudice Habit: Mechanisms, Timecourse, and Longevity." *Journal of Experimental Social Psychology* 72:133–46.

Foxx, Alicia J., et al. 2019. "Evaluating the Prevalence and Quality of Conference Codes of Conduct." *Proceedings of the National Academy of Sciences* 116(30): 14931–136. doi: 10.1073/pnas.1819409116.

Freeman, Richard, and Wei Huang. 2015. "Collaborating with People Like Me: Ethnic Coauthorship within the United States." *Journal of Labor Economics* 33: 289–318.

Gaucher, Danielle, Justin Friesen, and Aaron C. Kay. 2011. "Evidence That Gendered Wording in Job Advertisements Exists and Sustains Gender Inequality." *Journal of Personality and Social Psychology* 101(1): 1–20.

Goldin, Claudia, and Cecilia Rouse. 2000. "Orchestrating Impartiality: The Impact of "Blind" Auditions on Female Musicians." *American Economic Review* 90(4): 715–41.

Goring, Simon J., et al. 2014. "Improving the Culture of Interdisciplinary Collaboration in Ecology by Expanding Measures of Success." *Frontiers in Ecology and the Environment* 12(1): 39–47. doi: 10.1890/120370.

Greenwald, Anthony G., and Mahzarin R. Banaji. 1995. "Implicit Social Cognition: Attitudes, Self-Esteem, and Stereotypes." *Psychological Review* 102(1): 4–27.

Greenwald, Anthony G., Debbie E. McGhee, and Jordan L. K. Schwartz. 1998. "Measuring Individual Differences in Implicit Cognition: The Implicit Association Test." *Journal of Personality and Social Psychology* 74(6): 1464–80.

Grossman, Paul, Ludger Niemann, Stefan Schmidt, and Harald Walach. 2004. "Mindfulness-Based Stress Reduction and Health Benefits: A Meta-analysis." *Journal of Psychosomatic Research* 57(1): 35–43.

Guynn, Jessica. 2015. "Google Gives Employees 20% Time to Work on Diversity." *USA Today*, May 13. www.usatoday.com/story/tech/2015/05/13/google-twenty-percent-time -diversity/27208475/.

Hewlett, Sylvia Ann, Kerrie Peraino, Laura Sherbin, and Karen Sumberg. 2010. "The Sponsor Effect: Breaking through the Last Glass Ceiling." *Harvard Business Review*, December. https://www.wearethecity.com/wp-content/uploads/2014/10/The-Sponsor -Effect.pdf.

Hirsh, Elizabeth, and Donald Tomaskovic-Devey. 2020. "Metrics, Accountability, and Transparency: A Simple Recipe to Increase Diversity and Reduce Bias." In *What Works? Evidence-Based Ideas to Increase Diversity, Equity, and Inclusion in the Workplace*. Amherst: University of Massachusetts, Amherst Center for Employment Equity. www .umass.edu/employmentequity/what-works-evidence-based-ideas-increase-diversity -equity-and-inclusion-workplace.

Holvino, Evangelina, Bernardo M. Ferdman, and Deborah Merrill-Sand. 2004. "Creating and Sustaining Diversity and Inclusion in Organizations: Strategies and Approaches." In *The Psychology and Management of Workplace Diversity*, edited by Margaret S. Stockdale and Faye J. Crosby, 245–76. Malden, MA: Blackwell.

Homyack, Jessica A., Sara Schweitzer, and Tabitha Graves. 2014. "Glass Ceilings and Institutional Biases: A Closer Look at Barriers Facing Women in Science and Technical Fields." *Wildlife Professional* (Fall): 48–52.

Huet-Vaughn, Emiliano. 2013. "Striving for Status: A Field Experiment on Relative Earnings and Labor Supply." PhD diss., University of California, Berkeley.

Kabat-Zinn, Jon. 2003. "Mindfulness-Based Stress Reduction (MBSR)." *Constructivism in the Human Sciences* 8(2): 73–107.

Karpinski, Andrew, and James L. Hilton. 2001. "Attitudes and the Implicit Association Test." *Journal of Personality and Social Psychology* 81(5): 774–88.

Knapp, Bernhard et al. 2015. "Ten Simple Rules for a Successful Cross-Disciplinary Collaboration." *PLoS Computational Biology* 11(4): e1004214. doi: 10.1371/journal.pcbi.1004214.

Kohl, Michel, Serra Hoagland, Ashley Gramza, and Jessica A. Homyack. 2017. "Professional Diversity: The Key to Conserving Biological Diversity." In *Becoming a Wildlife Professional*, edited by Scott Henke and Paul Krausman, 188–95. Baltimore: Johns Hopkins University Press.

Krentz, Matt, Olivier Wierzba, Katie Abouzahr, Jenn Garcia-Alonso, and Frances Brooks Taplett. 2017. "Five Ways Men Can Improve Gender Diversity at Work." Boston: Boston Consulting Group.

Kriese, Ken, Sarah Reif, and John Davis. 2020. "Where There Is Heat, There Is Fire: Leaning into Hard Truths about the Conservation Culture." Paper presented at 85th North American Wildlife and Natural Resources Conference, Omaha, Nebraska, March 8–13.

Lappé, Frances Moore. 2011. *Diet for a Small Planet: The Book That Started a Revolution in the Way Americans Eat*. New York: Ballantine Books.

Madsen, Susan R., April Townsend, and Robbyn T. Scribner. 2020. "Strategies That Male Allies Use to Advance Women in the Workplace." *Journal of Men's Studies* 28(3): 239–59.

Maslach, Christina, and Michael P. Leiter. 2008. "Early Predictors of Job Burnout and Engagement." *Journal of Applied Psychology* 93(3): 498–512.

Matthews, Gail, and Pauline Rose Clance. 1985. "Treatment of the Impostor Phenomenon in Psychotherapy Clients." *Psychotherapy in Private Practice* 3(1): 71–81.

Miriti, Maria N., Karen Bailey, Samniqueka J. Halsey, and Nyeema C. Harris. 2020. "Hidden Figures in Ecology and Evolution." *Nature Ecology & Evolution* 4: 1282. https://doi.org/10.1038/s41559-020-1270-y.

Mishra, Jitendra M., 2009. "A Case for Naps in the Workplace." *Seidman Business Review* 15(1): 9.

Mohr, Tara Sophia. 2014. "Why Women Don't Apply for Jobs Unless They're 100% Qualified." *Harvard Business Review*, August 25.

Morimoto, Shauna A., and Anna Zajicek. 2014. "Dismantling the 'Master's House': Feminist Reflections on Institutional Transformation." *Critical Sociology* 40(1): 135–50.

Moss-Racusin, Corinne A., John F. Dovidio, Victoria L. Brescol, Mark J. Graham, and Jo Handelsman. 2012. "Science Faculty's Subtle Gender Biases Favor Male Students." *Proceedings of the National Academy of Sciences* 109(41): 16474–79.

Moss-Racusin, Corinne A., Christina Sanzari, Nava Caluori, and Helena Rabasco. 2018. "Gender Bias Produces Gender Gaps in STEM Engagement." *Sex Roles* 79: 651–70. doi: 10.1007/s11199-018-0902-z.

Moss-Racusin, Corinne A., et al. 2014. "Scientific Diversity Interventions." *Science* 343(6171): 615–16.

National Academies of Sciences, Engineering, and Medicine (NASEM). 2018. *Sexual Harassment of Women: Climate, Culture, and Consequences in Academic Sciences, Engineering, and Medicine*. Washington, DC: National Academies Press.

National Science Foundation (NSF). N.d. "ADVANCE at a glance." Accessed July 29, 2020. www.nsf.gov/crssprgm/advance/.

Nelson, Robin G., Julienne N. Rutherford, Katie Hinde, and Kathryn B. H. Clancy. 2017. "Signaling Safety: Characterizing Fieldwork Experiences and Their Implications for Career Trajectories." *American Anthropologist* 119(4): 710–22. doi: 10.1111/aman.12929.

Office of the Inspector General (OIG). 2019. *Forest Service Initiatives to Address Workplace Misconduct.* Audit report 08601-0008-41. www.usda.gov/oig/audit-reports/forest -service-initiatives-address-workplace-misconduct.

Pendergrass, Angie, et al. 2019. "Inclusive Scientific Meetings: Where to Start." 500 Women Scientists. https://500womenscientists.org/inclusive-scientific-meetings.

Perlow, Leslie A., Constance Noonan Hadley, and Eunice Eun. 2017. "Stop the Meeting Madness." *Harvard Business Review*, July/August.

Shapiro, Shauna L., Kirk Warren Brown, and Gina M. Biegel. 2007. "Teaching Self-Care TO Caregivers: Effects of Mindfulness-Based Stress Reduction on the Mental Health of Therapists in Training." *Training and Education in Professional Psychology* 1(2): 105.

Smith, Jessi L., Ian M. Handley, Alexander V. Zale, Sara Rushing, and Martha Potvin. 2015. "Now Hiring! Empirically Testing a Three-Step Intervention to Increase Faculty Gender Diversity in STEM." *Bioscience* 65(11): 1084–87.

Smith, Kristin A., Paola Arlotta, Fiona M. Watt, Initiative on Women in Science and Engineering Working Group, and Susan L. Solomon. 2015. "Seven Actionable Strategies for Advancing Women in Science, Engineering, and Medicine." *Cell Stem Cell* 16(3): 221–24.

Tulloch, Ayesha I. 2020. "Improving Sex and Gender Identity Equity and Inclusion at Conservation and Ecology Conferences." *Nature Ecology & Evolution* 4: 1311–20.

Turakitwanakan, Wanpen, Chantana Mekseepralard, and Panaree Busarakumtragul. 2013. "Effects of Mindfulness Meditation on Serum Cortisol of Medical Students." *Journal of the Medical Association of Thailand / Chotmaihet thangphaet* 96(suppl. 1): S90–S95.

Uhlmann, Eric Luis, T. Andrew Poehlman, and Brian A. Nosek. 2012. "Automatic Associations: Personal Attitudes or Cultural Knowledge?" In *Series in Political Psychology: Ideology, Psychology, and Law*, edited by J. Hanson, 228–60. New York: Oxford University Press. doi: 10.1093/acprof:oso/9780199737512.003.0009.

US Geological Service (USGS). 2016. "Diversity Council." www.usgs.gov/about/organization /science-support/human-capital/diversity-council.

Vuletich, Heidi A. and B. Keith Payne. 2019. "Stability and Change in Implicit Bias." *Psychological Science* 30(6): 854–62.

The Wildlife Society (TWS). N.d. "Diversity, Equity, and Inclusion." Accessed December 4, 2021. https://wildlife.org/dei/.

The Wildlife Society (TWS). 2020. "TWS Standing Position: Workforce Diversity within the Wildlife Profession." October 13. https://wildlife.org/tws-standing-position-workforce -diversity-within-the-wildlife-profession/.

Williams, Damon A. 2007. "Achieving Inclusive Excellence: Strategies for Creating Real and Sustainable Change in Quality and Diversity." *About Campus* 12(1): 8–14.

Gordon R. Batcheller, Paige Schmidt, Dale Caveny,
and John L. Koprowski

11

The Responsibilities of Men in Supporting Women

Empathy is seeing with the eyes of another, listening with the ears of another and feeling with the heart of another. —*Attributed to Alfred Adler*

Why Men Matter in Moving the Wildlife Profession Forward

In the wildlife field, one of the most popular and exciting jobs that an eager undergraduate seeks is as a field technician. For wildlife research, fieldwork typically entails living far from civilization for an extended period, often camping in tents, no running water or electricity, and little to no outside communication. When Elaina,* an undergraduate, was hired for her first field assistant position working for Mark, a graduate student, she was delighted. Before heading to the field, Elaina discovered that Mark was engaged to be married. A few nights after arrival at their camp, Mark tried to kiss Elaina. Elaina was shocked and pulled away but didn't say anything more. After this incident, Mark's behavior changed. He treated Elaina rudely and was hostile toward her. She worked alone for most of the field excursions. Elaina felt trapped with no way to communicate or remove herself from the situation; she had to wait for Mark's advisor to pick her up six weeks later. When the advisor arrived, Elaina described

* Throughout this chapter, all names are fictional. The anecdotes are not.

her experience. The advisor said, "Well, when I dropped you off, I figured you would either hate each other or become lovers." The advisor took no action. Elaina was traumatized by this experience.

Fifteen years later, Mark, now a faculty member at a university, crossed paths with Elaina, now a seasoned professional working in a conservation area. Mark requested funding to conduct research in the conservation area, and Elaina put aside her earlier, negative experiences and decided to support the project. It was not until several years later, as Mark was evaluated for promotion and tenure, that Elaina became aware that Mark's sexual misconduct had continued. While working on a project in the conservation area, Mark had an affair with his graduate student Amy. When Amy's boyfriend Jeff, a graduate student hired to work on the project, confronted Mark about his actions, Mark threatened Jeff with retaliation. Jeff reported the threat to Robert, the conservation area biologist. Robert encouraged Jeff to complete his project and thesis. However, at the time of writing (and more than five years after these events occurred), neither Robert nor Elaina could locate Jeff or determine whether he completed his thesis. Mark received tenure the year these events occurred.

Although this situation may seem unlikely, it is not a unique story and raises several questions. Why would a faculty member leave two students alone in the field for such a long time without checking on them, knowing Elaina had no way to leave or communicate if the situation became unsafe or hostile? How could a young faculty member threaten a student and get away with it? Several things went wrong. Mark was in a position of authority and abused it. When Susan reported the circumstances to her supervisor, he dismissed her concerns. Jeff was encouraged by a person in a position of authority to ignore Mark's misbehavior for the sake of his career advancement. Three students, first Elaina then Jeff and Amy, were not protected by the university system, and two faculty members advanced in their careers with no repercussions for their behavior. These and other themes will be explored in this chapter.

Nearly all state and federal agencies address sexual harassment in their personnel policies, and training requirements are enforced. By its nature, sexual harassment training typically addresses the behavior of men, while recognizing that this training benefits women as well. Although the focus of most training programs of this nature is on sexual harassment, its character, causes, and prevention, more comprehensive training on diversity, equity, and inclusion has been delivered in recent years. For example, an Executive Order on Diversity, Equity, Inclusion, and Accessibility in the Federal Workforce was issued on June 25, 2021:

As the Nation's largest employer, the Federal Government must be a model for diversity, equity, inclusion, and accessibility, where all employees are treated with dignity and respect. Accordingly, the Federal Government must strengthen its ability to recruit, hire, develop, promote, and retain our Nation's talent and remove barriers to equal opportunity. It must also provide resources and opportunities to strengthen and advance diversity, equity, inclusion, and accessibility across the Federal Government. The Federal Government should have a workforce that reflects the diversity of the American people. A growing body of evidence demonstrates that diverse, equitable, inclusive, and accessible workplaces yield higher-performing organizations. (Exec. Order No. 14035)

Saucedo and Hentze (2020) report on growing awareness of the need for more comprehensive training within the state employment sector.

Work, whether in an office, a laboratory, or the field, is the most likely place to encounter sexual harassment (Clancy et al. 2014; NASEM 2018; Pew Research Center 2018). Fieldwork in the sciences can pose unique challenges (Clancy et al. 2014) that require unique solutions. Individuals who identify as women account for over 50% of undergraduate students enrolled in wildlife and fisheries programs, numbers that have never been greater (see Chapter 3; Sharik et al. 2015). Inclusion of Black, Indigenous, and women of color remains a major concern in wildlife and fisheries, with only about 10% of undergraduates representing minority groups, irrespective of gender identity (see Chapter 3; Sharik et al. 2015). We must do better, but to do better, we must acknowledge the problem, then act.

To understand the current environment for women, we examined their real-life experiences in the wildlife profession. We heard stories that illustrated the subtlety of gender-biased discrimination and the need for men to be responsible for correcting and eliminating discriminatory behaviors. We talked with 12 women employed in the wildlife profession from a wide spectrum of experiences: academic, federal agency, state agency, nongovernmental; and with a wide span of service, ranging from recent hires to mid- or late career. We asked, "What do men need to be aware of, and what do men need to do to help women achieve their professional aspirations?" This simple question became a bridge to wide-ranging and complex discussions about the role of men in fostering success for women in the wildlife profession.

As we conducted these interviews, it became clear that discrimination was often subtle and nuanced. Often, the behavior doesn't violate personnel policies, and the person doing it might be completely unaware. Yet all our interviewees had experienced debilitating sexual discrimination, and the accumulation of ostensibly minor events, over time, is like the biological magnification of contaminants. Eventually, microaggressions (subtle behaviors that alone may be seemingly inconsequential or innocent but accumulate over time) become debilitating and toxic.

For example, men might assume that women are unable to complete certain physical tasks that may require strength or agility. This is manifested in field scenarios where lifting, pulling, or maneuvering a heavy object may be required (e.g., carrying field gear to a remote site). If a colleague or supervisor who is a man assumes that women cannot handle the load, he may automatically do it himself or ask another man to. They may not even consciously analyze the decision as a gendered choice. However, women are strong, and men and women should assume that everyone is physically capable unless they ask for assistance (Figure 11.1).

We must do better to be inclusive in our workplaces so that we can maximize our efficacy in taking on the substantial challenges that natural resource management and conservation require. There is abundant research that supports the benefits of a diverse workforce to natural resource management. Managerial performance (Dezsö and Ross 2012), profitability (Litz and Folker 2002), and employee health and quality of life (Melero 2011) positively correlated with gender diversity in an organization. Companies with women leaders and board

FIGURE 11.1 The physical tasks associated with wildlife fieldwork often require strength and strong technical knowledge. Men should always assume that women are capable of performing these tasks, unless demonstrated otherwise, and with the woman's assent. Photo courtesy of Derek Reich.

members adopted strategies more favorable to the environment and sustainability (Glass et al. 2015). Conservation and natural resource management organizations were more collaborative, practiced reciprocity, and emphasized conflict resolution when women participated (Westermann et al. 2005).

In this chapter, we assess the unique aspects of the wildlife field in the context of other professions. We follow by examining real-life experiences of women wildlife professionals. We provide case studies that illustrate both the subtle and overt nature of gender-based discrimination, with a focus on the responsibilities of men to identify these subtleties, to establish awareness, to assertively correct discriminatory behaviors, and to change the culture of the profession. We use the responses of our interviewees and our experiences to develop suggestions for men to support women in the workplace and serve as allies as we collectively labor to end gender-based discrimination. Ultimately, our objective is to ensure that the work environment experienced by wildlife professionals is a supportive and uplifting place where each person attains professional fulfillment, free from gender-based harassment, discrimination, and crime. Such a work culture is essential if the wildlife profession is to succeed in conserving wildlife and natural resources for future generations. Men have a responsibility to contribute to creating and maintaining a diverse, inclusive, and safe workplace for all. Men who are in positions in management and administration have not only the responsibility but also the authority to address discrimination in the workplace. Those in a position of leadership must lead through their vision, words, actions, and most importantly, by example.

In reading this chapter, remember that we must begin with empathy, the awareness among men that women experience discrimination. This was very clear as we interviewed the women who contributed to this chapter. Men must assertively, proactively, and deliberately behave in a manner that respects skills and abilities without consideration of gender.

Sexual Harassment of Women in the Wildlife Profession

Many choose the wildlife profession because it provides exhilarating workplaces outside a traditional office or laboratory setting and enables us to make a difference in the future of our planet. The opportunity to work under the cloak of an old-growth forest, in the unending sea of grasslands of a prairie, the starkness of the Grand Tetons jutting from the rolling plains, the tranquility of a coastal marsh, or the respite of an urban woodlot are the idyllic settings that attract many to our profession. The nature of fieldwork and the conservation field itself pose unique challenges for our science, but also for our scientists, educators, students, and conservation professionals. In turn, the conservation workplace also presents unique challenges to prevent, address, and end sexual harassment.

The workplace is where the greatest amount of sexual harassment occurs (Pew Research Center 2018). The statistics in any workplace are alarming: 69% percent of women and 61% of men reported experiencing sexual harassment at work (Pew Research Center 2018) despite laws and policies in place to address this issue. The workplace is therefore the most dangerous setting for those facing sexual harassment and requires a change in culture.

In the wildlife profession, field sites are workplaces often in remote locations with small groups of people who share a camp or cabin. The statistics on sexual harassment at these field sites mirror those of the workplace. In a recent survey, 64% of respondents reported experiencing sexual harassment, and 21% experienced sexual assault (Clancy et al. 2014). Similar rates of harassment were found for field placements in archeology (Meyers et al. 2015), geosciences (Sexton et al. 2016), and social work (Moylan and Wood 2016). These surveys clearly demonstrate the failure of supervisors and the organizations they represent to protect their employees and volunteers. Employees continue to find themselves in situations with poor tools, training, support, or resources, as was the case for Elaina during her first field position. Our agencies and universities most often require training in sexual harassment in the office; unfortunately, few if any require explicit training for fieldwork environments. Although Title VII of the Civil Rights Act of 1964 (Pub. L. 88-352) and Title IX of the Education Amendments of 1972 (Pub. L. 92-318) have resulted in policies and training on sexual harassment, the emphasis often seems to be symbolic compliance to avoid liability with lesser effort to end sexual harassment (NASEM 2018). Our next example illustrates institutional failures to protect women in wildlife from harassment and discrimination.

Karen, in her early twenties, started her first year of graduate school with a first-year advisor of graduate students, Dr. Smith, a tenured professor and man whose lab was staffed entirely by young women. Karen hired a technician, Julie, to assist her at a remote study site. Karen and Dr. Smith traveled together to the conservation area to meet and train Julie. This was a successful trip and work relationship. Two months later, Dr. Smith traveled to the field site to train Julie on surveys unrelated to Karen's project. During this training, Dr. Smith and Julie were alone at the field site.

Immediately after Dr. Smith left the field site, Julie wanted to speak with Karen. During this conversation, Julie asked Karen whether she or any of the other women in the lab had experienced inappropriate behavior from Dr. Smith.

Karen acknowledged a few instances when she thought his behavior was questionable. Julie then told Karen that she and Dr. Smith were alone at the field site, drinking, and he encouraged her to drink more, massaged her legs and thighs, complimented her, and then leaned in for a kiss. Julie turned away and he kissed her head. She then took his hands off her and walked away.

In retrospect, Karen realized that Dr. Smith had groomed her with seemingly harmless, flattering comments, and she had nearly become another victim. Dr. Smith clearly sought to make her feel special, and his actions were subtly seductive: complimenting her appearance, sending selfies, inviting her over for a drink when his wife was out of town, running together, even suggesting they camp together outside of work. Dr. Smith made Karen feel special, with great potential as a scientist. He sought to establish a peer-to-peer relationship, where a consensual sexual encounter could be within normal bounds. He failed to honor the behavioral norms expected in an advisor (powerful) and advisee (vulnerable) relationship.

Reporting inappropriate behavior is a serious matter, so Karen sought guidance from a trusted former advisor. She asked whether she was required to disclose the information, even when the victim did not wish to file a complaint.

Convinced that the incident must be reported, Karen disclosed the facts to the university Affirmative Action (AA) Office. The AA representative was empathetic and took the story seriously but conveyed that for the incident to be addressed in a meaningful way, Karen would have to identify the individuals involved, and Julie would need to speak with AA. As a result, Karen could not maintain her own anonymity, and Dr. Smith knew she had reported his behavior.

Later that field season, Karen traveled to the field site to meet with Julie for the first time since the incident. Karen told Julie to inform her if she was uncomfortable since Karen, Julie, and Dr. Smith would be sharing living quarters for the next week (the field station had separate corridors for men and women). Julie seemed completely comfortable with Dr. Smith's presence.

Once back at the university, Karen met with the AA office on the status of her filing. The AA office provided vague answers: they "took action," they "took the case seriously," Dr. Smith was counseled. The only observed change in his behavior was that he was no longer allowed to be alone with students. He was still employed by the university, was still in a tenure track position, and still taught and advised graduate and undergraduate students. However, Karen had to alter her graduate research to accompany her advisor, so he was not alone with Julie, which kept her from conducting some aspects of her project, essentially making the safety and comfort of fellow subordinates Karen's responsibility.

Six months after she left the university, Karen received a message from someone identifying herself as a graduate student in journalism seeking an interview on any experiences with sexual harassment Karen might have had as a woman in the sciences. Through the interview process and the disclosure of additional instances of inappropriate behavior, it became clear that Dr. Smith previously engaged in sexual harassment with undergraduate laboratory assistants at a previous university.

Through additional interviews and the simple process of "connecting the dots," it became clear that Dr. Smith had a history of abusing women. He used his power as an advisor to manipulate women, including a consensual but inappropriate affair with an undergraduate. It was also clear the university failed to protect women.

Mansfield et al. (2019) detailed a powerful narrative of sexual harassment of women in academia, documenting potentially dangerous relationships between students and established, tenured, and highly respected faculty men. Indeed, they opined that sexual harassment may *benefit* the perpetrator: *"The relationship between these serial harassers and scholarly influence is no coincidence: prestige and ability to advance is at least partly predicated on the damage they have done to women and others with feminized bodies and non-heteronormative identities. Abuse of power contributes to their professional success in multiple ways"* (Mansfield et al. 2019, 82–83). In essence, Mansfield et al. (2019) asserted that harassers dominate their victims to such an extent that the accomplishments and potential of women is eroded or eliminated. They cited an insidious culture of acceptance of inappropriate behavior by highly accomplished academics as just another form of the abuse of power punctuated by the victimization of vulnerable students and faculty, especially women. They made a strong case that men in academia have a responsibility to stop such abuses. *"We must make destructive power visible and refuse to celebrate abusive individuals. We can refuse to honor them with nominations and invitations, and we can institute new review requirements for disciplinary honors and other forms of institutional recognition"* (Mansfield et al. 2019, 85). The recent revocation of a prestigious award presented to Richard Vogt, known within the herpetologist community for inappropriate behavior, exemplifies this need (Diep 2018). Clearly, men must accept responsibility for changing the culture as a step in eliminating discrimination and harassment experienced by women in wildlife careers. It is not acceptable to look the other way, thereby enabling the continuation of abuse and inequity. Further, men in academia have the power and responsibility to report and stop the abuse of women.

Improvements in the field are necessary to provide safe work conditions for all. Not surprisingly, sexism and sexual harassment in the workplace increase the likelihood that women will drop out of STEM (science, technology, engineering, and mathematics) careers (Sexton et al. 2016; Nelson et al. 2017; St. John et al. 2018). Nelson et al. (2017), Clancy et al. (2014), and the National Academies of Sciences, Engineering, and Medicine (NASEM 2018) provide important suggestions to enhance protection for all participants during fieldwork, focusing on transparency and direct communication. Most of the suggestions centered on two components of the field experiences: (1) the lack of clarity about what constitutes appropriate professional behavior at field sites, and (2) the need to provide clear access to professional resources to benefit victims and to be firm in punishing misbehavior.

Significant institutional reforms are required, as NASEM (2018, 3) summarizes: "*Thus, the cumulative effect of sexual harassment is a significant and costly loss of talent in academic science, engineering, and medicine, which has consequences for advancing the nation's economic and social well-being and its overall public health.*" To that end, we now provide specific examples germane to the wildlife profession with the intention of addressing areas for awareness and reform.

The Advisor-Student Relationship

The advisor-student relationship in academia is often one of the most important relationships in the life of a wildlife professional (Levin and Hussey 2007). An effective advisor serves as an introduction to the profession, professional behavior, and a lifetime of experiences that permit wildlifers to work in exciting locations and make long-lasting differences in our natural world. However, NASEM (2018) also notes that the academic environment has a poor history of addressing sexual discrimination, thereby limiting career advancement of women in the sciences. They attribute this to (1) the dependence on advisors and mentors for career advancement, (2) declines in productivity and morale resultant from sexual harassment, (3) the "macho" culture or "toxic masculinity" in some fields of study, and (4) the relatively small social network in specialized fields through which rumors and accusations are spread. Both our earlier stories clearly demonstrate these effects of sexual harassment and discrimination. Effective advisors must be aware of these challenges and address them.

Academic professionals have an important opportunity to influence the lives of their advisees and an obligation to create equal opportunities for all students. Advising students on the quality of their science and career choices is only part of the job. Modeling and actively cultivating appropriate behaviors and environments that provide everyone a safe opportunity to flourish should be a goal of all academics. Stopping and reporting misconduct must be a requirement.

Academic mentors should provide a strong foundation in both knowledge and skills required for success. Knowledge might come from classes that the student takes or discussions during lab meetings. However, important skills are often related to communication, decision making, team building, and collaboration (Blickley et al. 2013). An effective academic advisor communicates the need for such skills and helps evaluate the strengths and weaknesses of their students. The faculty member should also go beyond treating students equally, since equality does not account for need. An advisor who treats students equitably, providing different resources to each student depending on need, will be more successful in encouraging women or those from underrepresented groups. Mentors who seek opportunities to encourage their mentees can have a profound impact early in their career. Women are disproportionately underrepresented in indicators of career progression and productivity including corresponding author on publications, grants, and leadership of conferences (reviewed in Hinsley et al. 2017). Proactively including women and members of other underrepresented groups can begin to correct these imbalances.

Women and men also differ in their behavior in professional settings. For example, women tend to ask fewer questions. If unaware of gender differences, women may be incorrectly perceived as having less investment and enthusiasm in the profession (Hinsley et al. 2017). An effective advisor and mentor recognizes and provides opportunities for improvement and success. Nominating and encouraging woman to participate in leadership roles and mentoring them beyond academia is the mark of an excellent advisor, mentor, supervisor, or friend. The mentor also shares the benefits of mentorship (Wells et al. 2005). In contrast, an advisor or mentor who only reports misconduct against women is meeting the lowest bar.

Supervisor-Supervisee Relationship

Following graduation from an academic institution or upon entering the conservation workforce, another important relationship is that of the supervisor/supervisee, particularly for early-career professionals. The power differential in the student/advisor relationship is present upon entry to paid employment. It can be exploited by supervisors to the detriment of women. Further, while the numbers of women studying wildlife-related fields continues to grow (see Chapter 3), the numbers of women actually entering and progressing in the profession continues to lag. The lack of other women in the wildlife profession exacerbates the power differential between employees and supervisors as women are more isolated and often lack suitable mentors, supervisors, or advocates to assist them in avoiding and addressing issues within the workplace.

Of the five academic institutions that grant the most degrees in wildlife biology, 57% of graduates are women (DataUSA n.d.). Of the women graduating with bachelor's degrees in wildlife biology, most are white. Notwithstanding these data, Rouleau et al. (2017) found that "*women are hesitant to enroll in a forestry and related natural resources program due to concerns about their gender, work locations, and work conditions.*" These findings illustrate two key points: (1) women are entering the wildlife profession in growing numbers, but (2) significant obstacles to their career satisfaction and growth remain. To illuminate those barriers, we now turn to real stories from the women we interviewed.

In Their Words: Stories of Women Working in Wildlife

These stories are not hypothetical; they are real-life experiences along the career path of women in the wildlife profession. They illustrate the subtle and at-times ignored aspects of sexual discrimination. Each event, in and of itself, may not constitute a debilitating event, but each illustrates the potential role of men, especially supervisors, colleagues, and leaders, in supporting women by stamping out discriminatory practices in the workplace. We use these case studies to build several key principles in defining how to be an ally.

The Committee

By its nature, the work of wildlife professionals often requires collaboration in a team or committee setting. We all attend meetings at some point or another, and it is obvious that skilled leaders or facilitators can ensure that the meeting runs effectively. Beyond the traditional skill set associated with effective facilitation (e.g., maintaining rules like "one speaker at a time"), it is less obvious that women face discrimination in a committee setting. Men should be aware of at least two common types of discrimination. The first is an assumption that a woman is the most appropriate person to serve as secretary and take notes for the meeting. At the beginning of a breakout session at a recent major professional conference, the chair looked to the closest woman and asked, "Would you be willing to take notes?" It would be obviously awkward for the woman to either say no or to ask why a man wasn't asked. Few people wish to knowingly embarrass a colleague, especially when it is reasonable to conclude that no harm was intended. But this is discrimination. Empathetic men can preempt both discrimination and any potential awkwardness. They could step up

BOX 11.1 How the Toss of a Paper Ball Turned into a Call to Action

It all started with the simple action of a woman standing to read a personal account of sexual harassment. Her words carried weight in part because she was one of the more soft-spoken members of the class. But the real power of her words came from the heartfelt way they were spoken.

After she shared her experience, she tore a page from her journal, wadded it into a tight ball and tossed in on the floor in the middle of the classroom. "I read one story," she said in a measured tone, "but that story," pointing to the ball of paper, "will go unread unless it is joined by others." She sat and over the next minutes, one after another, other balls of paper, each with a personal account of sexual harassment, assault, or gender discrimination were tossed onto the floor.

Then the stories were read.

"The guys in my work unit took bets on who could have sex with the new girl first."

"At a post-commission meeting social, a commissioner made multiple sexual advances even after I asked him to stop. My HR person told me to keep it to myself."

"A male colleague often calls me 'the big-boobed chick on the 2nd floor.' "

"My supervisor describes the weekend as rutting season."

"I was date raped when I was 16."

"My job was threatened if I reveal the sexual nature of our office 'banter.' "

"I was told that I needed a good f*ck to straighten me out."

"I was sexually assaulted by multiple men at a wedding. The more who groped me, the more joined in—like a pack of wild animals tearing me apart."

"I was groped by a colleague and when I reported it, was laughed at and told 'Boys will be boys.' "

"Our work group's truck first aid kit contains condoms."

There were 20 women in the room, all from the conservation field, with enough stories to fill four pages of text. It was generally accepted that these stories were only the beginning.

Tom Vania watched this event unfold with a sense of bewilderment, shock, and disbelief. "I was blind to what my colleagues have experienced, and in many cases, continue to experience," he later shared.

Perhaps an even more difficult question that Vania began wrestling with was what to do with this revelation? Once back home in Alaska, he determined to address the issue of harassment in the workplace.

From his platform as regional supervisor of the Alaska Department of Fish and Game's Division of Sport Fish, Vania set out to bring greater awareness to harassment in the workplace with a focus on seasonal field employees. These employees are often college students or recent college graduates. They work in remote locations, with limited communications and in living conditions with little privacy (wall tents, cabins). These employees generally work in small teams of two to four. They are often sent out with little training or orientation on what constitutes acceptable workplace behaviors. Compounding the problem, they don't receive information on how to report improper behavior.

Using his own experience of enlightenment, Vania began speaking to the supervisory staff that selects, trains, and oversees these seasonal field positions. He has given his presentation eight times to more than 125 field supervisors. He plans to follow up with the supervisors with the question that continues to haunt him: "Now that you know what is taking place, what do you plan to do?"

A woman from Vania's cohort at the National Conservation Leadership Institute recently asked him a challenging question: "With the steps you are taking, are you making an impact or difference in your agency's culture? Can you count simply having this conversation as a measure of success?" It's an appropriate question with no clear answer.

It's often difficult to measure impact when discussing cultural change. However, one thing is certain, for Tom Vania the seemingly innocuous tossing of a paper ball created a jarring splash that is now rippling across his agency.

Small interventions can have a remarkable effect.

---

and say, "I would be happy to take notes today." At a different breakout session at the same conference, a man did just that. The second common example of discrimination in a meeting is the failure to credit a woman's input. A group of biologists were discussing information about a wildlife species and the appropriate management response. A woman made a comment germane to the problem but received neither response nor acknowledgment. The discussion continued later in the meeting, and a man made a very similar comment and most of the group concurred. This is discrimination. An empathetic chair should have immediately acknowledged the input of the woman in the first instance. Absent that, an empathetic colleague could offer a gentle rebuke: "That is a good idea, and it was also a good idea when Shazia brought it up earlier today."

The Field Excursion

Depending on the age of a breastfeeding child, a nursing mother may need to pump up to every two hours. At the time of let-down the mother's breasts will become engorged and painful. If she doesn't express the milk in time, they will leak. Failure to nurse or pump can lead to clogged ducts and mastitis. The nature of fieldwork can make it a challenge to find adequate time and a place for lactating mothers to take care of these needs. For example, for a remote location, the woman would have to pack in a portable, battery-powered breast pump, a cooler with ice for storage, and necessary cleaning and sanitation supplies. This should not be seen as a liability in hiring but rather an indication of her dedication, perseverance, and tenacity. A new mother early in a new job was taking a break to pump breast milk during a site visit. A colleague commented, "I remember when my wife breastfed our kids. Good for you." Another colleague voiced his support. This was both surprising and helpful to the young mother.

The Banquet

In the nongovernmental wildlife conservation community, fund-raising banquets are a common event. These are typically held on a Friday or Saturday evening and are largely attended by the organization's members, volunteers, and officers. They are often hunters, trappers, or anglers who support the cause of conservation by attending the annual banquet, a major undertaking for the organization. While most members are men, spouses, partners, and in some cases, children attend the banquets. Professional wildlife biologists who work for these nongovernmental organizations (NGOs) are expected to attend as well. These banquets are enjoyable for some attendees, but for women, they may be misogynistic. The fund-raising model typically includes both "silent auctions" and live auctions. Alcohol is served, and at times, in large quantities. Members and volunteers attend with the anticipation of fun and levity. Unfortunately, however, sexual overtones and banter may occur. For example, a woman may be asked to hold an auction item in the air so that everyone can see it. Not accidentally, the volunteer is a young, pretty woman, and dressed attractively. In some cases, *professional* models are hired to perform this function. In most cases, these women dress in revealing clothing, augmenting the blatant sexual atmosphere of the events. Ticket sales and auction prices get a boost from this misogynistic behavior. For a professional woman, it can be an uncomfortable environment, and she may find herself the victim of blatant sexual discrimination or harassment. For one woman, the atmosphere at one such banquet grew decidedly uncomfortable, and she left early, even though her employers (and the volunteers) expected the professional staff to be on hand for networking and banter. The remedy? Leaders of these organizations need to ensure that fundraising banquets are managed with professionalism and respect, places where you would be comfortable bringing family members, including young children. Sex should not be used to sell conservation. PERIOD. The employers of women, often men, should have frank discussions with local leaders to ensure they recognize their members have, in the past, engaged in inappropriate behavior, and that such behavior will not be tolerated. Organizations that use sex to sell conservation and perpetuate discrimination and misogyny should not be supported. Men in leadership roles in these organization must demand these practices be discontinued immediately. In the absence of corrective measures, professional women should not be expected to attend them. This is a clear example where men in positions of authority can establish standards for behavior among their conservation partners.

The Public Meeting

Wildlife biologists frequently must attend meetings for a variety of reasons including outreach, information, and participatory decision making (Figure 11.2). Public meetings are often dominated by men. For a woman biologist, this can lead to uncomfortable encounters. At one public meeting, a biologist started her presentation, and following her introduction, a man said, "It's about time you sent someone good looking to these meetings." Raucous laughter followed. This is both embarrassing and belittling and constitutes sexual harass-

FIGURE 11.2 Wildlife biologists frequently interact with the public and may encounter inappropriate behaviors, including sexual harassment or discrimination. Supervisors should ensure that training includes methods for women to respond to such behaviors, and men who witness such behaviors should step up and take corrective measures appropriate to the setting.
Photo courtesy of Derek Reich.

ment. Men at these meetings should preempt such behavior whenever possible. However, spontaneous comments of this nature may not be anticipated. It is appropriate, therefore, for men to talk through scenarios that may be encountered and ask their women colleagues how they would like them to respond. In this case, a professionally delivered but public rebuke may be appropriate. It might be initiated by a woman and then supported by a man, or it may come from a supervisor. In the case of a moderated meeting, it should be the moderator. If an organization demonstrates the inappropriate behavior repeatedly, a direct dialogue with its leaders should be initiated. Leaders should insist that all professionals be treated with dignity and respect.

The Multiagency Meeting

Often, multiple agencies, NGOs, and academic institutions will collaborate to address common resource issues. Within these multiagency meetings, attendees will often range in age and professional stature from interns or seasonal employees to heads of departments and agencies. Members of the public may be invited to participate. If not open to the public, laypeople are often still encountered during these events. The following example demonstrates inappropriate behavior by a mid-level participant from a natural resources agency. A multiagency meeting was being held at the state agency's environmental education center and was not open to the public. However, a large group of high school students were participating in an unrelated outdoor environmental education event. Throughout the meeting, students were coming to the doors of the building trying to get in to use the restroom. It was disruptive but ignored. Toward the end of the meeting, a young woman who was visibly pregnant came to the door and knocked for access. Most attendees who saw wanted to let her in to use the restroom. The mid-level employee from the state agency remarked, "If she had kept her legs closed, she wouldn't have that problem." A man from the same agency quickly responded, "That isn't appropriate." Another person got up to let the young woman in. We don't need to belabor that the first man's comments were both rude, inappropriate, and judgmental, but we do want to affirm that the intervention was a great way to step up and shut down inappropriate and harassing behaviors. A woman witnessing the entire episode expressed her gratitude that she did not have to be the one to assert herself in a room full of men.

These stories detail examples of harassment, inappropriate behavior, or difficulties women in wildlife may experience, regardless of the setting or source. Importantly, several demonstrate how men either made a proactive positive difference to lessen a difficulty (i.e., support of breastfeeding mothers, asking men to take notes at a meeting) or reactively corrected inappropriate behavior of other men. Lastly, they show significant room for improvement.

How Men in the Wildlife Profession Can Be Allies

The suggestions in Boxes 11.2–11.5 outline how to be an ally, and they apply to men (and in some cases, women) in the wildlife profession and guide behaviors and attitudes focused on fostering diversity. They are specific, tangible, and based on our interviews and experiences. Our recommendations are practical but generally promote Jones and Solomon's (2019) suggestion that effective supervisors: (1) provide opportunities, (2) share their network and champion women, (3) share feedback and guidance, (4) learn the needs of women, and (5) demonstrate confidence in women. We address work scenarios in the field and office, and principles that apply to all levels of employment (i.e., academic, governmental, and non-governmental). Recommendations for the office and meetings (Box 11.2) are applicable to situations and circumstances involving office and meeting behaviors, including those encountered during professional conferences. Field recommendations (Box 11.3) target circumstances encountered in the unique field settings that occur within the wildlife profession. Academic recommendations (Box 11.4) focus specifically on the advisor-student relationship. In many cases, recommendations in one category may be entirely applicable to other categories. General (Box 11.5) recommendations are applicable to nearly all circumstances.

North Carolina Wildlife Resources Commission colonel Jon Evans received his indoctrination on the value of diversity in the army. *"At Fort Benning we were a group of individuals from every conceivable background and ethnicity. . . . The army instills a mission focus. We may have looked and sounded different, but our collective success required teamwork in spite of any differences."*

Col. Evans was later hired by the Raleigh Police Department. Those early lessons from Fort Benning were foundational at the police academy. Now, in addition to racial and ethnic diversity, gender was added to the mix. *"We studied together, ate together, and trained together. There was no distinction between race or gender,"* he recalls. As if to emphasize the diversity point, Evans was assigned a woman field training officer.

He describes this experience as transformative. *"We stopped cars together, we chased suspects on foot, and conducted building searches. Raleigh had approximately 800 officers on the force. While she taught me how to be a cop, more importantly she demonstrated how to be part of a large team."*

BOX 11.2 Office and Meeting

- When organizing a conference, ensure that women serve as keynote speakers, plenary speakers, moderators, and discussion leaders. These opportunities provide professional and career development.
- "Hospitality suites" hosted in conjunction with some conferences may not be safe spaces for women. They are held late at night, alcohol is served, there may be many men, and there is not a lot of space. When organizing a hospitality suite, ensure that as many women are invited as men. Alternatively, hold these social gatherings in a more public space, such as a hotel meeting room, instead of a hotel guest suite. Making such events alcohol-free may create a more welcoming and safer environment. Rules and expectations for appropriate behavior should be presented ahead of these gatherings and should include sober monitors to hold everyone accountable to the standards of safe and appropriate behavior.
- During a meeting, men should volunteer upfront to serve as the notetaker or facilitator if one is used, or a rotating schedule should be established to avoid bias. Note taking and facilitation can preclude active participation in discussions.
- The person chairing a meeting should ensure that all participants have an equal opportunity to contribute and all are immediately recognized for their thoughts.
- When members of the public make comments that are either sexist or sexual in nature, an appropriate rebuke needs to be made. Men must speak up.
- Agencies, institutions, businesses, and nongovernmental organizations should adopt rules and guidelines for member and volunteer behavior at all events, especially fund-raising events (e.g., annual banquets) to ensure that the event is comfortable for women. Using sex to sell conservation should be condemned and policies adopted to preclude such abuses.
- Agencies, institutions, businesses, and nongovernmental organizations should also ensure that robust sexual harassment policies are adopted, and that members and volunteers are aware of and follow these policies, rules, and applicable laws. Women must have recourse to file formal complaints of sexual harassment that will be investigated and ensure there are consequences.
- Because wildlife professionals regularly engage with volunteers and other members of the public, employers should establish clear rules for appropriate behavior, as is done with any other official training within the organization. Leaders must communicate those rules to everyone, including volunteers.

- Some men associated with the wildlife profession have gained a reputation as notoriously inappropriate. Many women know who they are. Supervisors should bring disciplinary action against these men. How can a supervisor know this? Ask others, gain evidence, and work with Human Resources and Affirmative Action.
- Funding agencies should hold recipients accountable to all applicable policies on harassment and discrimination.

In 1996, Evans became a North Carolina wildlife officer. Compared to his military experience and time in Raleigh, he found that the agency was composed of largely white men. There were only two women among the state's 208 wildlife officers. He had worked many hours with one of those women and recognized the value her perspective brought to their conservation efforts. He realized that white men didn't look like North Carolina's population. He knew that law enforcement effectiveness is enhanced when the force reflects the demographics of the general population.

When Evans became the division's training director in 2011, he assumed responsibility for recruiting, hiring, and training the agency's wildlife officers. The first year, he hired two women recruits, effectively doubling the number of woman officers. In his second class in 2013, he hired two more.

As North Carolina's new law enforcement chief, his first step was to address his own limited knowledge of laws and policies related to sexual harassment, so he attended the Personnel Issues for Law Enforcement Managers training at the state's justice academy. This gave him a clearer understanding of the laws concerning equal employment opportunities (EEO), helped him define harassment, and clarified actions that create hostile work environments. He found the class so valuable that he sent the majors of field operations and administrative functions, and the new training director. After this training, it was obvious to all of them that the division's culture must change to achieve greater diversity and inclusion.

- Col. Evans personally conducted the annual EEO training for all division staff. While he taught the law concerning harassment, bullying, and other illegal behaviors, he also established a baseline for acceptable and unacceptable behavior for all officers and staff.
- He suggested that interns be chosen from underrepresented groups. The goal was to expose nontraditional applicants to the agency and division. This created opportunities for someone without a strong

BOX 11.3 **Field**

- Ensure that women have field gear that is both comfortable and functional. Field gear (e.g., waders, helmets, gloves, safety equipment, jump suits, fire-retardant clothing) are often sized for men and are uncomfortable or unsafe for women. Properly fitted gear not only provides for comfort but improves functionality and performance and can ensure personal protective equipment (PPE) provides the requisite level of safety. Supervisors should make deliberate efforts to acquire the appropriate field gear for women and should be encouraged to confer with them in advance of doing so.
- Make certain that tampons and pads are available in rest rooms and in field kits for everyone because transmen may need them as well as cisgender women.
- Always assume that a woman is capable of doing all physical field activities that a man may be asked to do, but it should be clear to all involved that anyone should ask for assistance if they want it (Figure 11.3).
- In a field setting (e.g., following a day of work), alcohol should be served with caution. In general, it should be avoided, especially in small groups, regardless of gender identity. A field venue is a work setting, even after the fieldwork is completed for the day, and alcohol is never appropriate in the work environment. If alcohol is consumed, a sober monitor should be designated.
- Men need to be sensitive to special circumstances women encounter while working in the field. For example, women with nursing children will need access to a clean and comfortable place to express and store breast milk. Women may feel uncomfortable addressing their needs during menstruation in a field setting or may request additional time or privacy. Receiving support from men goes a long way to ensuring that a woman is comfortable, and a man should not admonish a woman for addressing these tasks. Men and women both need to have their privacy respected when attending to bodily functions. They should give one another sufficient space by moving to a location out of view of others or at least looking away. Procedures should be established in advance.
- Conservation organizations should prominently feature women, of all ages, participating in field activities as wildlife professionals (e.g., web content, publications).

FIGURE 11.3 Women should have equal access to acquiring field skills requiring intense training, such as technical climbing to monitor or capture wildlife. Male supervisors should ensure that training opportunities are provided to both men and women.
Photo courtesy of the North Carolina Wildlife Resources Commission.

FIGURE 11.4 Handling wildlife, in this case a nuisance beaver, often requires fieldwork in remote locations. Supervisors must be sensitive to the behaviors of their field crews, especially when men and women are paired or when camping is required. Training programs should include specific discussions on sexual harassment and other forms of discrimination. Photo courtesy of the North Carolina Wildlife Resources Commission.

- Supervisors need to recognize that certain men, because of their reputations or behavior or both, make women uncomfortable. Supervisors must document inappropriate behavior, take necessary disciplinary actions, and ensure these behaviors are stopped. Even if the behaviors are addressed and stopped, women may still be uncomfortable around offending men and should not have to work with them.
- Training should occur prior to start of field work. All personnel should be educated about which behaviors are inappropriate and what disciplinary actions will be taken should they occur. This conversation needs to be held in a trusting and open atmosphere. Everyone, especially women, need to feel comfortable expressing any concerns with confidence that no adverse action will be taken for reporting inappropriate behavior, based on clear and strong policies (Figure 11.4).

wildlife background to develop the baseline skills and knowledge needed to become a strong officer applicant.

- He emphasized that each officer is responsible for diversity outreach. Instead of a couple of brief paragraphs for his annual EEO outreach report, he now receives dozens of pages of truly impactful connections.
- The recruit academy training staff now includes women as instructors. Evans believes strongly that instruction by women breaks down gender stereotypes before they become ingrained in young officers.
- Evans continually challenges supervisors' terminology. He corrects them when they refer to "my guys" and encourages them to use "my officers." He changed the commendations document from the "Atta Boy" to the "Way to Go" list.

The division now has nine women employed as wildlife officers with four more in the current recruit class with a goal of gender parity. "We have experienced this growth without lowering our standards," reports Evans. "While the numbers are important, it is the cultural change that excites me." He pauses before adding, "We have accomplished much and are making great strides, but we still have a long way to go. We are dedicated to continued growth and improvement."

How to Be an Ally for Yourself and Others

#MeToo accentuated women's experiences in society, illustrating the wrongs that continue to be inflicted on women at all stages of their lives. Fundamentally, #MeToo is about equity: equity of opportunity, equity of behaviors, equity of cause and effect. There is now a greater awareness (especially among men) about egregious, ugly, and sometimes criminal behaviors inflicted on women. However, bias can be subtle, inflicted without intent or malice by well-meaning men; therefore, it requires a more prominent place in the conversation. Progressive men are not immune. It's clear that we have not accomplished equity in the professional setting. Both subtle and egregious behaviors remain, illustrated in this chapter and in other sections of this book. The burden is not one that women should shoulder alone. Men in the wildlife profession have a responsibility to establish equity in a diverse workforce.

First, bad behavior needs to be called out. It cannot be tolerated. Institutional measures to remedy those behaviors are useful, but they are not enough. Men need to correct other men when they see, hear, or sense inappropriate behavior, both subtle and egregious. When institutional remedies are invoked, harm has already occurred. Nevertheless, women should invoke those remedies as the normal course of action as partial redress to that harm.

Second, men need to better understand microaggressions, subtle behaviors that alone may be seemingly inconsequential or innocent but accumulated over time, they become debilitating. Within an organization, open, honest, and safe conversations between men and women need to be enabled so that these microaggressions may be better understood, recognized, and stopped.

Finally, establishing equity is fundamentally a nondiscretionary ethical imperative. An ethical foundation in the workplace requires trust, dialogue, honesty, and corrective actions. Leaders of all genders must take assertive measures to ensure that their organizational culture is grounded in an ethical foundation.

In preparing this chapter, the coauthors reached out to a wide range of women, from all sectors of the profession, at different stages of education and career. They were courageous and honest detailing their experiences suffering the indignity of abuse, embarrassment, discrimination, and harassment. In some cases, the behavior they described was criminal. Telling their stories was not easy, without risk, or without the spiritual and emotional ramifications of reliving their experiences. We are very grateful that they openly and honestly shared their life and career journeys. We hope that their experiences will help shine a light down the long, dark corridor of sexual and professional inequity, not yet fully exposed.

BOX 11.4 **Academic**

- Advisors should never tell a woman she cannot or should not start a family while pursuing a degree. This applies to women in any job.
- Conduct meetings and counseling sessions with graduate students in an open-door or public setting. Meetings behind closed doors can be uncomfortable for the student and should be avoided.
- Participation and leadership in scholarly professional organizations is an expected component of academia. Women must have equal opportunity to assume leadership positions (e.g., president, vice president) in these organizations.
- Similarly, women must be given equal consideration in serving as editors and referees for professional journals.
- Because of the significant power differential between a faculty advisor and a graduate student, each academic institution should ensure that each student has access to a trusted advisor or mentor who can aid and counsel in a safe space on any matter that makes her uncomfortable, especially any instance of inappropriate behavior on the part of any faculty, staff, or other student.
- Faculty should never invite individual students to a social gathering without others present, regardless of the gender of the faculty or student. Social gatherings that occur in association with the academic environment (e.g., following a seminar) should be group gatherings, not a meeting of two individuals.
- When traveling together to meetings and conferences, participants (students and faculty) should be provided with their own rooms or private sleeping arrangements. There should never be shared accommodations between students and faculty members (e.g., under the pretext of saving money on travel costs).

Discussion Questions

1. What specific rules should be in place to govern the behavior of men who serve as advisors to women students? What measures should be put in place to ensure that women thrive in the academic and professional setting?

2. What actions can men take to increase awareness and acceptance by other men of proactive measures designed to establish equity and opportunity for both men and women?

BOX 11.5 **In General**

- Do not engage in sexual or romantic relationships with students, advisees, mentees, or subordinates. Even when consensual, this is not appropriate.
- Persons involved in consensual but appropriate sexual relationships should not express physical or emotional displays of affection in the work setting. This makes others uncomfortable.
- When men witness other men behaving inappropriately, they must speak up to promptly end such behavior. This must occur not only in an office or field setting, but during meetings with members of the public or with volunteers of conservation organizations.
- Conservation requires collaboration and cooperation among governmental, nongovernmental, and academic partners. In this setting, women role models are important, and every effort should be made to achieve balance between men and women.
- Organizations must adopt a culture that enables "awkward conversations" addressing gender and diversity issues. There must be mutual trust and mutual respect. Fostering gender equity and diversity requires establishing a culture that enables our legacy of discrimination to be corrected. Diversifying the demographics of the profession is necessary but not sufficient.
- Alcohol is often a factor preceding inappropriate behavior. If drinks are served, it should never be between only one man and one woman, absent a consensual, appropriate relationship. Men should not ask women out for drinks after work. A faculty advisor should never invite a student for a drink. In social settings, alcohol should be avoided in the advisor/advisee context.
- Among wildlife professionals, an ostensible "social event" is rarely purely social. As with hospitality suites, a social event after hours often includes conversations about work. Alcohol is frequently served at these events. If a woman does not attend because she has concerns about inappropriate behavior, she is cut out of important professional dialogue. Great care must be taken to ensure that all persons can benefit from that dialogue and that it is conducted in an appropriate and professional business climate.
- Similarly, after-hours social events (that mask as professional engagement events) may discriminate against both men and women who have children. Ensure that all employees can participate in the professional dialogue, regardless of social engagements.

3. In academic and professional communities, employers have the authority to establish rules for acceptable behavior. However, wildlife professionals frequently interact with members of the public who do not have to follow those rules. What actions can be taken to help ensure that women are not subjected to abusive behavior in the public setting?

Activity: Planning for the Field Season

Learning Objectives: Participants will read or listen to examples of statements based on real-life experiences and identify microaggressions and other examples of bias. Participants will also discuss how to best respond to these types of comments to better prepare themselves to confront similar situations in the real world.

Time: 1–1.25 hr. If participants complete Part 1 outside of the meeting/class, time is reduced to 30–45 min.

Description of Activity

Materials needed: Paper and writing implement for taking notes.

Process:

Part 1: Personal Response and Reflection (~30 min.)

Participants will read through the example comments below and answer the Reflection Questions at the end of Part 1 individually. This can be completed prior to the meeting or during the meeting with the larger group.

Example Comments:

- "Hi, Christina. This is your new colleague, John. We just hired him last week. I'm sending you two out together on Wednesday for that two-week timber cruise in the backcountry. It'll be a great way to get to know him."
- "Rachel just told me that Jason said something on the drive to the regional meeting that made her 'very uncomfortable.' No one has ever told me anything like that before about him so I'm sure she just misinterpreted what he said."
- "We need to pack in this marine battery to replace the older one at the weather station on Gray's Peak. Are you sure you can do this?"
- "I'm going to run the analysis on the data you collected because you only have a master's degree."
- "Wait, did you say Fred is taking paternity leave? Are you sure? I thought his wife was taking maternity leave."
- "Oh, that's just Mark. He says inappropriate things but he doesn't mean anything by it."

- "Why can't Amy work the extra check station this weekend? She doesn't have kids."
- "How about if we finish discussing the program reorganization over the golf game this evening?"
- "Yeah, you might be one of only a few women at the social in the hotel suite tonight. Why is that a problem?"
- Katie presents an idea during a meeting and is dismissed by the eight men she's meeting with that morning. Five minutes later, Steve presents the same idea, phrased in a slightly different way, and the seven other men readily accept it.
- "Linda's here. She can take notes for the meeting."
- "We're awarding Charles with cufflinks tomorrow for 20 years of service to the department. Good thing women don't last that long because I have no idea what kind of gift we'd buy to honor them!"
- "Is *this* your Discussion? I could write a better Discussion section in my sleep."
- "You completed that hike so quickly for your age!"

Reflection Questions
1. How might the recipient interpret what was said to them?
2. How do you expect it would make the recipient feel, or how might they interpret it?
3. If coworkers, colleagues, students, or other bystanders observed this, how might they interpret this comment?
4. If you were a bystander and observed this interaction, how would you interpret it? What would you say to the speaker and/or to the recipient?

Part 2: Small Group Discussion (15–20 min.)
After everyone has had time to reflect on the statements, form groups of three or four people and share responses and thoughts with the other group members. For this aspect of the activity, emphasize that, as a group, they determine how to be an ally as a bystander in the situation.

Part 3: Large Group Discussion (15–20 min.)
After the large group shares and creates their list, have everyone participate in a discussion.

Debriefing and Discussion Questions
1. How did you feel reading these statements?
2. What did you learn from your small group members?

3. What were some of your group's favorite responses (the strongest or most effective for addressing the bias or the problem presented)?

4. Is there a character trait or skill that would help someone best behave as an ally in situations of bias, discrimination, or inappropriate behavior?

5. How would you use the information you learned or your experience from this activity in your life as a wildlife professional, student, or citizen?

Literature Cited

Blickley, J. L., K. Deiner, K. Garbach, I. Lacher, M. H. Meek, L. M. Porensky, M. L. Wilkerson, E. M. Winford, and M. W. Schwartz. 2013. "Graduate Student's Guide to Necessary Skills for Nonacademic Conservation Careers." *Conservation Biology* 27:24–34.

Clancy, K. B. H., R. G. Nelson, J. N. Rutherford, and K. Hinde. 2014. "Survey of Academic Field Experiences (SAFE): Trainees Report Harassment and Assault." *PLoS ONE* 9(7): e102172. doi: 10.1371/journal.pone.0102172.

DataUSA. N.d. "Wildlife Biology STEM Major." Accessed December 7, 2021. https://datausa .io/profile/cip/wildlife-biology#demographics.

Dezsö, C. L., and D. G. Ross. 2012. "Does Female Representation in Top Management Improve Firm Performance? A Panel Data Investigation." *Strategic Management Journal* 33:1072–89.

Diep, F. 2018. "A Turtle Scientist Loses His Award after He Shows 'Scantily Clad Female Students' in His Acceptance Lecture." *Pacific Standard*, July 16. https://psmag.com /news/a-turtle-scientist-loses-his-award-after-he-shows-scantily-clad-female-students -in-his-acceptance-lecture.

Exec. Order No. 14035, *Fed. Reg.* 86 (June 30, 2021).

Glass, C., A. Cook, and A. R. Ingersoll. 2015. "Do Women Leaders Promote Sustainability? Analyzing the Effect of Corporate Governance Composition on Environmental Performance." *Business Strategy and the Environment* 25:495–511.

Hinsley, A., W. J. Sutherland, and A. Johnston. 2017. "Men Ask More Questions Than Women at a Scientific Conference." *PLoS ONE* 12(10): e0185534. doi: 10.1371/journal.pone.0185534

Jones, M. S., and J. Solomon. 2019. "Challenges and Supports for Women Conservation Leaders." *Conservation Science and Practice* 1:e36. doi: 10.1111/csp2.36.

Levin, J., and R. B. Hussey. 2007. "Improving Advising in the Sciences." Journal of *College Science Teaching* 36:28–35.

Litz, R. A., and C. A. Folker. 2002. "When He and She Sell Seashells: Exploring the Relationship between Management Team Gender Balance and Small Firm Perfor- mance." *Journal of Developmental Entrepreneurship* 7:341.

Mansfield, B., R. Lave, K. McSweeney, A. Bonds, J. Cockburn, M. Domosh, T. Hamilton, R. Hawkins, A. Hessl . . . and C. Radel. 2019. "It's Time to Recognize How Men's Careers Benefit from Sexually Harassing Women in Academia." *Human Geography* 12:82–87.

Melero, E. 2011. "Are Workplaces with Many Women in Management Run Differently?" *Journal of Business Research* 64:385–93.

Meyers, M., T. Boudreax, S. Carmody, V. Dekle, E. Horton, and A. Wright. 2015. "Prelimi- nary Results of the SEAC Sexual Harassment Survey." *Horizon and Tradition: The Newsletter of the Southeastern Archaeological Conference* 57:19–35.

Moylan, C. A., and L. Wood. 2016. "Sexual Harassment in Social Work Field Placements, Prevalence, and Characteristics." *Journal of Women and Social Work* 31:405–17.

National Academies of Sciences, Engineering, and Medicine (NASEM). 2018. *Sexual Harassment of Women: Climate, Culture, and Consequences in Academic Sciences, Engineering, and Medicine*. Washington, DC: National Academies Press. doi: 10.17226/24994.

Nelson, R. G., J. N. Rutherford, K. Hinde, and K. B. H. Clancy. 2017. "Signaling Safety: Characterizing Fieldwork Experiences and Their Implications for Career Trajectories." *American Anthropologist* 119:710–22.

Pew Research Center. 2018. *Sexual Harassment at Work in the Era of #MeToo*. Pew Research Center, Washington, DC.

Rouleau, M., T. L. Sharik, S. Whitens, and A. Wellstead. 2017. "Enrollment Decision-Making in U.S. Forestry and Related Natural Resource Degree Programs." *Natural Sciences Education* 46:170007. doi: 10.4195/nse2017.05.0007.

Saucedo, S., and I. Hentze. 2020. "Raising Awareness about Diversity, Equity and Inclusion." National Conference of State Legislatures 28(38). www.ncsl.org/research/about-state-legislatures/raising-awareness-about-diversity-equity-and-inclusion.aspx.

Sexton, J. M., Newman, H., Bergstrom, C., Pugh, K., Riggs, E., and Phillips, M. 2016. "Mixed Methods Study to Investigate Sexist Experiences Encountered by Undergraduate Geoscience Students." *Geological Society of America Abstracts with Programs* 48(7). https://gsa.confex.com/gsa/2016AM/webprogram/Paper278443.html.

Sharik, T. L., R. J. Lilieholm, W. Lindquist, W. W. and Richardson. 2015. "Undergraduate Enrollment in Natural Resource Programs in the United States: Trends, Drivers, and Implications for the Future of Natural Resource Professions." *Journal of Forestry* 113:538–51.

St. John, K., E. Riggs, and D. Mogk. 2016. "Sexual Harassment in the Sciences: A Call to Geoscience Faculty and Researchers to Respond." *Journal of Geoscience Education* 64:255–57.

Wells, K. M. S., M. R. Ryan, H. Campa III, and K. A. Smith. 2005. "Mentoring Guidelines for Wildlife Professionals." *Wildlife Society Bulletin* 33:565–73.

Westermann, O., J. Ashby, and J. Pretty. 2005. "Gender and Social Capital: The Importance of Gender Differences for the Maturity and Effectiveness of Natural Resource Management Groups." *World Development* 33:1783–99.

Kerry L. Nicholson

12

Melting the Iceberg

For too long, powerful people have expected the people they have mistreated and marginalized to sacrifice themselves to make things whole. The burden of working for racial justice is laid on the very people bearing the brunt of the injustice, and not on the powerful people who maintain it. —Nikole Hannah-Jones

What Lies Beneath

Have you witnessed an event that made you hesitate and think to yourself, "Wait a minute, what happened . . . what am I missing?" For instance, you're stopped at a red light, you look over at the car next to you, and see that the driver is crying. Clearly there is more going on than meets the eye. That the person is crying is likely just the tip of the iceberg. Initially coined by American writer Ernest Hemingway (1960), "iceberg theory," or "theory of omission," has been used to interpret human behavior in the fields of business and marketing, private security, in education, and language studies. In 1976, Ernest T. Hall developed an iceberg analogy of culture in which some aspects are visible, but a larger portion is hidden beneath the water (Hall 1976). The best way to learn the culture of others is to actively participate in them. Once you look a little deeper, you begin to understand cultural foundations and see "below the surface." The same is true of gender inequities.

The message "you don't belong" has crept into our lives indirectly beyond the obvious "MEN ONLY" clubs or "WHITES ONLY" signs that were common just decades ago. The "sexual" forms of harassment like come-ons, assault, unwanted advances, and coercion as discussed in Chapter 11 are less common forms of harassment and gender bias, and yet, as the most notorious, they get most of the publicity. After the #MeToo movement erupted into our social awareness in 2017, many began using the iceberg analogy to describe sexual harassment and gender equity (Dimitry and Murphy 2017). The public consciousness of sexual assault and coercion are the appalling visible part of the iceberg (NASEM 2018).

Contemporary prejudice often manifests in subtler forms that go unseen and may be harder to address. Though not experienced by all, these biases are more common forms of discrimination, which will change through time as social mindsets and expectations change. They range from the well-intended, back-handed compliment to disparities in bathroom safety to products such as clothing made in only certain shapes, sizes, and colors, and safety equipment or tools not designed for a variety of users. Discrimination can be found in the multitudes of media formats (from 10-second commercials to racist and sexist cartoons), in computer software that has built-in biased facial recognition (Bacchini and Lorusso 2019), and technology as simplistic as soap dispensers (Hale 2017) and wristband trackers (Bent et al. 2020) that do not recognize skin colors, or by naming structures and buildings after oppressors. These are just a few.

Sprinkled throughout this book are biases that constitute the iceberg "beneath the surface." In Chapters 9 and 10, the authors described system-wide policies that disadvantage working mothers. In other chapters, authors emphasized the importance of role models and mentors. In this chapter, we dive deep to expose discrimination lurking below the surface unseen by those who don't experience it. Although our examples are not comprehensive, we seek to increase awareness and provide a useful framework for dismantling gender biases and promote allyship to facilitate a more inclusive profession. Using first-hand experiences of women in the wildlife profession, we created an iceberg narrative to shed light on some of the nuances of gender inequities. Examples are not mutually exclusive nor do they fit neatly into one category nor are they exhaustive (Figure 12.1). Each person's iceberg will differ, but perhaps these stories will resonate with you or someone you know and can be a starting point for a conversation.

Lingchi: Death by 1,000 Cuts

Chester M. Pierce coined the term "microaggression" in the 1970s to refer to subtle insults and indignities that can collectively create a hostile atmosphere for minority individuals (Pierce 1970; Lilienfeld 2017). Although there is no

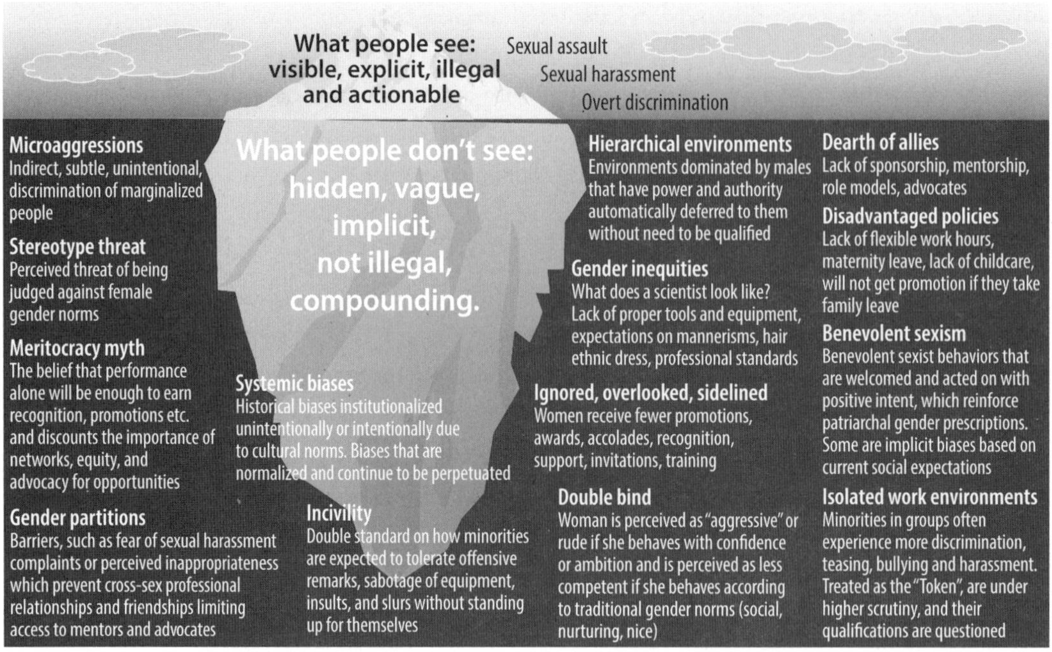

What people see: visible, explicit, illegal and actionable
Sexual assault
Sexual harassment
Overt discrimination

What people don't see: hidden, vague, implicit, not illegal, compounding.

Microaggressions
Indirect, subtle, unintentional, discrimination of marginalized people

Stereotype threat
Perceived threat of being judged against female gender norms

Meritocracy myth
The belief that performance alone will be enough to earn recognition, promotions etc. and discounts the importance of networks, equity, and advocacy for opportunities

Gender partitions
Barriers, such as fear of sexual harassment complaints or perceived inappropriateness which prevent cross-sex professional relationships and friendships limiting access to mentors and advocates

Systemic biases
Historical biases institutionalized unintentionally or intentionally due to cultural norms. Biases that are normalized and continue to be perpetuated

Incivility
Double standard on how minorities are expected to tolerate offensive remarks, sabotage of equipment, insults, and slurs without standing up for themselves

Hierarchical environments
Environments dominated by males that have power and authority automatically deferred to them without need to be qualified

Gender inequities
What does a scientist look like? Lack of proper tools and equipment, expectations on mannerisms, hair ethnic dress, professional standards

Ignored, overlooked, sidelined
Women receive fewer promotions, awards, accolades, recognition, support, invitations, training

Double bind
Woman is perceived as "aggressive" or rude if she behaves with confidence or ambition and is perceived as less competent if she behaves according to traditional gender norms (social, nurturing, nice)

Dearth of allies
Lack of sponsorship, mentorship, role models, advocates

Disadvantaged policies
Lack of flexible work hours, maternity leave, lack of childcare, will not get promotion if they take family leave

Benevolent sexism
Benevolent sexist behaviors that are welcomed and acted on with positive intent, which reinforce patriarchal gender prescriptions. Some are implicit biases based on current social expectations

Isolated work environments
Minorities in groups often experience more discrimination, teasing, bullying and harassment. Treated as the "Token", are under higher scrutiny, and their qualifications are questioned

FIGURE 12.1 An iceberg example showing biases and harassment that women and minorities face daily.

clearly defined rule for what constitutes a microaggression, these acts tend to be defined by impact, not intent. The term "microaggression" may be a misnomer because the act itself may seem small, but the adverse effects are definitely not "micro" and can be long lasting. Microaggressions may take the form of statements, questions, or actions that are more than just insensitive comments because they are painful, demeaning, and directly connected to pervasive, institutionalized biases against people with devalued identities. Microaggressions are short moments that implicitly communicate or engender hostility. They can make someone feel as though they do not belong or are not worthy. Evidence suggests that consistent discrimination and harassment can have a similar impact as a single episode of unwanted sexual attention or coercion (Paula Johnson, president, Wellesley College, in Cheney and Shattuck 2020).

Not all microaggressions are "aggressive" or intentional, but they can still feel like having your toes stepped on inadvertently. When microaggressions occur occasionally, it is often easier to move on, shrug it off, or "forgive and forget." However, if unintentional microaggressions occur daily, their cumulative impact can be like "death by a thousand cuts."

Good Sexism?

Benevolent sexism is a form of discrimination that is protective or generous toward the otherwise underprivileged gender (Mazrui 1993). Typically, it is a subjectively favorable, chivalrous ideology that seems positive in tone but denotes the inferiority of women to men (Glick and Fiske 2001). Relative to men, women are assumed to be less competent, more fragile, or in need of protection. For example, men might be expected to provide security and financial support for women. Benevolent sexist behaviors do not always make us uncomfortable yet ultimately limit the positive effects of other efforts to establish equity and inclusion. Benevolent ideals or social standards often imply women are not capable of psychologically or physically demanding jobs or positions that require intelligence, resulting in an unbalanced gender-based division of labor or duties. As such, benevolent sexism in our field can result in women being mistaken for technicians or assistants or being otherwise overlooked instead of being recognized as an authority or expert.

Here is a real story to illustrate. Evelynn, a professional biologist with a PhD and ten years of field work experience, was in the fifth year working on a black bear project near a mining site in Maine. Each field season she would mentor a junior coworker, teaching them how to capture and safely handle black bears. One year, her coworker James assisted. He had a bachelor's degree and had worked as a professional biologist for two years but never in this remote area. Part of Evelynn's responsibilities included collaborating with safety officers and maintenance personnel to mitigate nuisance wildlife concerns around the remote sites. Evelynn received a report that bear cubs were breaking into buildings and eating garbage, and they had crawled under a building and fallen asleep. It was Evelynn's responsibility to coordinate activities to safely remove the cubs from the site. Evelynn contacted the chief safety officer and together they recruited local volunteers: workers, safety officers, heavy equipment operators, and maintenance personnel. This newly formed team included Evelynn and about 20 men. The chief safety officer gathered everyone around, gave a brief statement that there were cubs under the building and everyone needed to listen to the biologists for instructions and job assignments. All eyes immediately shifted to Evelynn's colleague James, who without hesitation, took this as an invitation to start the briefing. Not more than two sentences in, one of the other officers who had worked with Evelynn stepped forward with a question, interrupting James. He looked specifically at Evelynn and started his question off with "Hey, boss." While the team of volunteers did not intend to snub Evelynn, and although it clearly startled James to have his authority diminished, those new to working with Evelynn just "assumed" James was the lead biologist. All it took was the subtle support from the safety officer to correct the assumption. From that point on, everyone, including James, deferred to Evelynn for guidance as the team worked together to resolve the issue.

Evelynn's story resulted in a positive outcome. However, many women are automatically dismissed because of implicitly held biases. Women get used to being sidelined or grew up with that expectation. Women who are overlooked receive fewer promotions, awards, and accolades (Lincoln et al. 2012) and are viewed negatively by their peers when they self-promote (Exley and Kessler 2019). Contributions by women are so often overlooked or dismissed that it has been labeled the Matilda effect (Rossiter 1993), named for the US suffragist and feminist critic Matilda J. Gage of New York (Figure 12.2), who vocalized her experiences of continually being dismissed.

Moreover, men frequently receive credit for women's work. A famous and egregious example is Rosalind Franklin. In 1952, she took the first photo of DNA, which was used as the basis of the chemical model of the DNA molecule, work that earned James Watson, Francis Crick, and Maurice Wilkins the Nobel Prize in Physiology or Medicine in 1962, while ignoring her contributions. Even within The Wildlife Society, the Aldo Leopold Award has been given to just three women of the 69 recipients since 1947. And one was Leopold's wife, who received the award in his honor (Nicholson 2009).

FIGURE 12.2 Matilda Joslyn Gage (1826–1898), pioneering suffragist, Native American rights activist, abolitionist, and freethinker after whom the Matilda effect is dubbed. Engraving by John Chester Buttre after photo by Napoleon Sarony. Image courtesy of Howcheng, https://en.wikipedia.org/wiki/Matilda_Joslyn _Gage#/media/File:Matilda_Joslyn _Gage_cph.3b20693.jpg.

Birds of a Feather

Women in wildlife often commented that work teams were dominated by men; there were few mixed-gender or woman-dominated teams. Evelynn, in the previous example, was the only woman in a group of 20 men. With few women working in that environment, the men are culturally unaccustomed to seeing a professional woman, especially as a project lead, and likely did not intentionally dismiss her. Recruiting and retaining women and other underrepresented groups is a challenge. But there are other social behaviors that reinforce systemic biases and make them difficult to overcome. For instance, we prefer to connect with others like ourselves (McPherson et al. 2001). When our networks become homogenous, there are powerful implications for attitudes, interactions, and styles of communication (McPherson et al. 2001), which can limit who is perceived as qualified, deserving, and capable. This may be why we see men nominating only male peers and trusting their knowledge and qualifications over those of people unlike themselves (Grunspan et al. 2016; Bradshaw and Courchamp 2018). When women and minorities are not in these homogenous or male-only social circles, explicitly or implicitly they may not immediately be seen as competent, and their knowledge may be dismissed. This lack of diversity leads to isolation, which can be detrimental to work performance and interpersonal interactions.

Gloria, a wetlands biologist II, gets along well with her supervisor. She believes she can trust him, and he supports her actions in the job. However, her supervisor has "go-to guys." When he is asked, "I need someone to help with this problem—who do you recommend?" his immediate suggestion is one of three men. The men he relies on are excellent professionals, dependable and knowledgeable. And because they've performed so well on jobs they know how to do, he suggests them for jobs for which they have no experience, over others with better qualifications. His automatic behavior removes opportunities for others to learn, improve, or develop leadership opportunities. Gloria has broached the subject with her supervisor to no avail. Gloria's coworker, in a lower-ranked biologist I position, not only gained networking exposure as one of her supervisor's "go-to guys" but also gained experiences performing tasks well above his position. Consequently, when Gloria and her coworker applied for the biologist III position, he was hired. Gloria never had the opportunity her coworker had to gain the experience needed for the promotion. A good supervisor grooms more than one or two select people and encourages diverse representation, because a well-rounded team works more effectively.

Ability + Effort = Merit? Leveling the Playing Field

The educational system in the United States is broadly framed as meritocratic: if you work hard and achieve in your studies, you will get ahead in school and in life (Liu 2011). Education is considered an equalizer, enabling those who are financially stressed to achieve equal footing with those who are wealthier, or those from underrepresented groups to catch up with whites. However, the system does not account for opportunity and access. Chapter 5 illustrated how recruiting minorities into the wildlife field requires thinking outside of the current approaches and instead shifting the focus to core beliefs and family roots. Although we think that jobs should go to those best qualified, regardless of their background, the applicant pool may be limited in diversity *because* of their backgrounds. For example, wildlife studies are often underfunded, and it is common to offer unpaid internships and recruit unpaid volunteers (Fournier and Bond 2015). These positions are available only to those who can afford to work for free, which often excludes minorities and people with family obligations (for example, caring for young children or elders; Fournier and Bond 2015). When denied opportunities because of social, gender, economic, or family disparities, the wildlife profession is not able to recruit from the best qualified *regardless* of their backgrounds. In addition, advancement in the wildlife field is slower and more difficult for those who identify with more than one minority characterization (Shiffman 2019). Therefore, obtaining opportunities for professional development is limited to those with greater privilege in our society (Shiffman 2019). When organizations do not open their doors to those needing an alternative path to learn the wildlife profession, we perpetuate a system where biased becomes systemic and normalized.

The (In)visible Spotlight

A group can claim that it is diverse, but this does not mean it is inclusive or equitable in the treatment of each member. The word "token" has come to mean the only person of their category in a homogeneous group. The person has the training and skill to succeed on the job, but because of perceived differences, is never fully accepted as part of the team (Zimmer 1988). To make someone a token is to treat them as a as representative of their category, as symbols rather than as individuals (Kanter 1977). For Black, Indigenous, and people of color (BIPOC), as well as those who identify as LGBTQ+, being considered the token can be dangerous because it makes them hypervisible. Being positioned as a token may seem beneficial but can also make a person feel exposed or needing to prove their worth. This sustained pressure can lead to health problems and impair job performance. For example, coworkers may question their qualifications, which can increase scrutiny and affect performance because of pressure and fear of making mistakes (Kanter 1977). In addition, being the minority is isolating and can lower work performance and lead to reclusive behavior. To be made a token adds pressure to assume the stereotypical behavior expected of them (Kanter 1977). It often amplifies discrimination and sexual harassment, teasing, and bullying (Watkins et al. 2018). Alternatively, a woman in this position can be seen as a competitive threat if she is highly qualified (Duguid 2011) or be judged only by her mistakes (Duguid 2011). Being a supportive ally results in greater team success (Farh et al. 2020), but conversely, being unique (i.e., the hummingbird among ravens) may have strategic benefits. For instance, compared to token women, when men are tokens, they are more likely to advance than women (Watkins et al. 2018).

I Can't Be Friends with Them, I'm Married

Friends in the workplace provide information, networking, and support that are invaluable for both job performance and satisfaction (Kram and Isabella 1985). Yet men, women, and genderqueer people all have difficulties developing cross-gender professional relationships and friendships (Elsesser and Peplau 2006). This hesitation limits access to mentors, advocates, recognition, and coworker bonds. With the recent #MeToo movement, many men have taken a "better safe than sorry" attitude or have felt "damned if they do, damned if they don't": they want to help but are worried that if they do, someone may take offense (Murphy 2019). And both men and women may discriminate against people who are LGBTQ+ because they don't know how to be supportive. Therefore, they pull back from allyship for fear of doing something that could be offensive. When diversity is missing in peer groups, mentors, or professional relationships, it takes active consideration and awareness to recognize others' contributions, and sometimes it takes specific efforts to be inclusive.

After a recent promotion, Simon expressed his frustrations over how his dynamics changed with his immediate coworkers when he shifted from being a peer to their supervisor. Once his position changed, he noticed his own hesitation to meet socially with the same people he had only a month prior, though the hesitation was not as strong with the "guys." He noted that it was now even more uncomfortable being alone with the women under his supervision. He realized he was treating them differently and was struggling with this change in dynamics. When asked what his wife thought, he said he'd never had the conversation with her, though she was uncomfortable with some of the isolated fieldwork situations. So he never had any of the women over to meet his family, although he never hesitated to invite male employees. He recognized he was preventing the women on his team from forging strong coworker bonds that are critical when working in the field. Also, as a supervisor he set the standards of how staff are expected to interact and work together. He noted his actions were being mirrored by the other men, thus isolating the women even more. If his behavior did not change, the women he employed would lose out on his mentorship, and he would lose out on their contributions. After the recent #MeToo movement, managers were surveyed about mentoring. Startlingly, most men (60%) were uncomfortable participating in a common work activity with a woman such as mentoring, working alone, or simply socializing in the breakroom (Sandberg and Pritchard 2019). Senior-level men are now far more hesitant to spend time with junior women even in basic activities like one-on-one meetings or working dinners. Women already do not have the same level of mentorship and sponsorship. It doesn't matter if this inequity is unconscious because men gravitate toward other men or if it is conscious sexism. The effect is detrimental to women rising to leadership.

Shanice has a master's degree and 18 years of experience working for her agency. They are single and have no children. Early in their career, Shanice recognized the most successful collaborative relationship they made in the workplace with Rich (a married man) was after they befriended his partner at their third office holiday party. Shortly after, he started including Shanice in hallway conversations and eventually asked them to collaborate. Shanice and Rich have now published 10 papers together. Sometimes it's not the coworker, rather it is the domestic partner or family of the coworker who can cause barriers. Rich later admitted that if Shanice's hadn't become friends with his partner, he would not have approached Shanice because of how it could possibly look, especially to his wife. As you read in Chapter 2, historically, some women, married or not, were denied field work because the wives of the male coworkers worried their husbands could not control themselves around other women in the workplace or vice versa. Though this may not happen as blatantly today, a person may hesitate to develop friendships or collaborations with coworkers because of their family situations or concerns about appropriate workplace relationships, particularly when the coworker is single. Unfortunately, this overcaution isolates rather than protects, and it prevents a working relationship from even forming. This is a no-win situation for everyone. One simple step at home is to be transparent and take the initiative to introduce your partner to your colleagues. In the workplace, an option for supervisors is to help facilitate relationships among employees.

Too Soft, Too Hard, and Never Just Right

She swims like a guy. Her stroke, her mentality: she's so strong in the water. I've never seen a female swimmer like that. She gets faster every time she gets in, and her times are becoming good for a guy . . . —Ryan Lochte

Ever feel like whatever you do, you will be criticized for it? Caught between impossible choices? Gender stereotypes can be powerful yet invisible threats because of the contradictory expectations of women in leadership positions (Jamieson 1997). Any sort of competitive nature in women can be perceived as dysfunctional, whereas competition among men is dismissed as "boys being boys" (Sheppard and Aquino 2017). Women are expected to be agreeable, helpful, sympathetic, and kind. They should be pleasant, nonthreatening, supportive, and modest. As these are not considered strong, assertive qualities, women are often perceived as less capable in leadership roles.

When women display competence or success, they become a target for derogatory terms such as "ice queen," "bitch," "pushy," and "opinionated," and might be punished for altering the *favorable* image of a woman (Kubu 2018). Not aligning with what is considered gender-appropriate behavior can lead to isolation and alienation; thus, women leaders are rarely perceived as both competent and likable. As a society, we still equate stereotypical masculine behaviors with effective leadership: "think manager—think male" (Schein 1975). Feminine characteristics such as supporting and encouraging others are not seen as essential components of leadership (Tabassum and Nayak 2021). Conversely, problem-solving skills or influential superiors are conventionally considered masculine. So, when a woman seeks advancement, she is viewed as out of place and may feel she must work harder and invest considerably more effort into proving she belongs.

These perceptions can influence how women act, because they can be targets of derogatory and degrading comments or other forms of discrimination or harassment in person or through emails, tweets, or other digital forms. How women handle this varies because they have to censor themselves for fear of being perceived as the "angry Black woman," "someone who makes waves," "a bitch," or "hysterical." These concerns cause women to spend more time and energy formulating communications, for example, using passive ("maybe we should consider") rather than assertive language ("we will"). This author edits her emails to remove equivocating words (e.g., "just," "maybe") and finds that it transfers over time to her speech.

Because of this double bind, women dedicate more time and energy than men to carefully navigating situations with coworkers to maintain an amiable atmosphere. When women are trying to traverse the "damned if you do, doomed if you don't" system, their real work suffers, and they are viewed as less productive. Those not in marginalized communities may not even recognize this necessity.

Queen Bee Syndrome

In some situations, women discriminate against other women. Because of implicit and explicit gender biases and unequal opportunities, some women fail to support others and may, in fact, actively hinder their progress (Derks et al. 2016). Women who have achieved success in traditionally male-dominated fields may distance themselves from and create barriers for other women to succeed. Often this is an unintentional consequence of gender discrimination because to achieve career success, the only path available is to present oneself as masculine (Faniko et al. 2021). "*Without realizing it, women internalize disparaging cultural attitudes and then echo them back*" (Sandberg 2013). In other situations, it occurs because the senior woman believes that her younger counterparts must experience the pain and trauma that she experienced as she advanced in the profession. Alternatively, the possible advantages of being the token woman can increase competitive behavior. Regardless, women should remain empathetic and consider their reactions and interactions with peers, subordinates, and the next generations. They should question how it felt when they experienced discrimination and remember who helped them. Please note Nicholson is not trying to perpetuate any myth that all women can't get along. But the phenomenon exists, nonetheless.

Pretty as a Picture

> *Female Olympians are sexing it up more than ever by wearing makeup during their competitions. . . . Do women who are elite athletes need to wear makeup to feel stronger or is it simply a fashion statement?*
>
> —*Tamara Holder*

Physical appearance plays a role in how others perceive us and can even influence our job performance (Leskinen et al. 2015; Kahalon et al. 2021). How we groom and dress affects our physical, psychological, and social comfort that lets us "fit in" or can make us stand out. As a graduate student, Divya felt pressured to look as masculine as possible so others would take her seriously. When she wore makeup or dresses to school, people would ask her if it was laundry day.

The wildlife field has been dominated by men, so expectations for what makes a woman look like a wildlife biologist are reflected by those expectations and clothing that is practical for the field. Women will often end up with a reputation for being "manly" simply because to outfit themselves properly and safely, often there is no choice except men's clothing. Alternatively, if you can find outdoor gear designed for women, it is often more expensive, form fitting, short, low cut, or designed in only one option. So, when women try to find something that is comfortable and helps get the job done, they risk being seen as unprofessional or inappropriate, which is frequently used as justification for overt sexual harassment. Sometimes outdoor gear does not accommodate those who desire more coverage out of personal values or religious practices such as those of the Islamic, Orthodox, Judaic, Coptic Christian, or other faiths. When what should be a simple trip to the store becomes an arduous task because you have a physical disability, a religious or cultural belief, or a body shape that does not fit the standard available styles, it is just one more of those 1,000 cuts.

Sometimes, the challenge goes deeper than going to the store to select from limited options, because the system was not designed to accommodate diversity in gender, body shapes and sizes, or color of skin. Fabric texture, type, weight, bulkiness, stretch, garment construction, and its fit are all considerations that provide the physical safety and comfort needed to do the work. In 2019, NASA's first ever *all-female* spacewalk was derailed because they did not have enough correctly sized suits to fit both women! The suits were designed in the 1970s when women were not in the space program (Kluger 2019; Cofield 2018). This situation was caused because of a historical perpetuation of biases from a different era. This lack of proper gear is found in the military (Britzky 2020), police (Hussein and Diaz 2019), wildland and rescue operations (FEMA 2019), deep sea diving, and surgical rooms (to name a few). Much of the equipment still used today was built for men, how they perspire, urinate, and support their weight. For example, fireproof clothing used by crews in wildland fires tends to not have small sizes so women must buy larger sizes and alter them (OH! And don't forget to find fireproof thread!). If there are versions made that aren't meant for men, then they tend to be less durable or functional and likely more expensive. For instance, camouflage clothing *designed* to fit a woman's form (i.e., not unisex) is often pink. Is this limited color option trying to say women should not be taken seriously as hunters or women don't take hunting seriously? Working outside in extreme climates makes clothing *designed* to fit and function of paramount importance. Finding clothing that allows a woman to urinate without having to completely undress is a challenge and an expense. Transgender people can have even more difficulty finding appropriate field gear. The lack of proper fit is distracting and potentially deadly for those who do not fit conventional sizing. Constantly having to modify and jerry-rig something to "make it work" is a safety risk that can leave an employee feeling like they don't belong and lead them to leave the job. These lasting legacies of gender bias extend beyond clothing and personal gear to equipment, vehicles, tools, and are now permeating new technologies like watches and facial recognition programs.

Kailani was asked by a fellow woman scientist whether she was planning to cut her hair to interview for faculty positions because long hair was "unprofessional" and "too feminine" for a wildlife biologist. At a conference, a woman received an award for 20 years of work in the wildlife field, but afterward, the first person to speak to her commented only on her appearance and shoes. When Dakota was transitioning, the comments and jokes they navigated for wearing clothing that did not match their assumed gender were devastating. Stories like these are common for women and transgender people, but men rarely hear comments about their looks. There are appropriate times and situations where compliments regarding a person's physical appearance are genuinely appreciated (Kurtzleben 2018). When comments about appearance undermine or imply that looks are more important than a person's scientific contribution and abilities, they are degrading and detrimental.

Why Are Biological Functions Taboo?

Wildlife work often requires physical activity and living or working in remote locations for extended periods. Physically, women are constructed differently than men, and some transgender people undergo gender reassignment surgeries and therefore have different requirements to take care of their bodies. We are all biologists and should understand the basic processes of bodily functions and reproduction. For some, it is easier to discuss biological functions in animals but embarrassing or taboo when discussing with coworkers. Men need to be aware of these things.

In some societies these bodily functions and the changes that happen through maturity are seen as a weakness or held against women. Before we get any further, let's start with simply going to the bathroom. Consider this situation: Casey is a fisheries biologist who was the only woman on a crew of six sampling offshore via a Boston whaler type boat working at least 8 hours before returning to land. Casey is going to have to urinate at some point. For some it may seem relatively easy to directly speak about this, but many people find it uncomfortable. Casey will likely have to ask everyone to turn around, maybe she will need to ask for assistance to make sure she does not fall off as she swings her rear end over the side of the rocking boat, or she must use some sort of container to squat over. Now reconsider this situation knowing that Casey is a transgender woman with an all-male crew.

During one field season, Uki's crew leader addressed this issue before anyone even got onto the boat by providing a bathroom protocol that everyone had to follow. Her crew leader found that this also relieved pressure for the men who felt uncomfortable for their own reasons but were too embarrassed to speak up.

Another possible awkward scenario comes when doing aerial surveys in a two-seater super cub. Sometimes, if the pilot is gracious and the terrain is acceptable, landing for a bathroom break is an option, however there are times when you cannot. A two-seater cockpit is tight, and smells are potent in such a confined space. Wildlife biologist Mei brings a wide mouth water bottle and a pee funnel (yes, this exists!) when flying. Some pilots do not take kindly to stopping for bathroom breaks. Although the portable funnel urinal is intended to allow a woman to urinate while standing up; with some awkward maneuvering and strategically wedged feet, it is possible to do it without making a mess. These situations are even more delicate for those who are transgender or those with certain cultural and religious beliefs. The simple act of relieving your bladder in these situations is so mortifying for some that they will refuse to drink water and risk bladder infections, kidney stones, or other medical issues. Bathroom necessities should not be a barrier to recruiting and retaining capable biologists.

Red Alert! There Will Be Blood

Most female mammals have an estrous cycle, yet only 78 higher-order primates, 4 bat species, elephant shrews (*Macroscelididae* spp.), and one spiny mouse (*Acomys cahirinus*) have a menstrual cycle (Bellofiore et al. 2017). A typical female human has a 36-day menstrual period from the age of ~12, once a month for about 38 years (436 times). You would think that this regular occurrence in half of the world's human population (King 2020) would be normalized.

Menstruation affects every woman differently, but for most of us it can affect us physically and mentally. For some unknown reason, likely motivated by the myth of the irrational or hysterical female, emphasis has focused on the psychological symptoms of menstruation more than the physiological implications (King 2020). Anyone who menstruates, is in menopause, or takes hormone replacement therapy can experience swings in their emotional and mental states. Some experience painful periods with a multitude of physically debilitating symptoms like cramping, which feels like being repeatedly stabbed in the groin; migraines that knock you out flat; or breast sensitivities that turn wearing certain field gear into torture.

Ironically, "science" was used to justify gender discrimination with medical claims that the female body rendered women pathologically emotional and therefore less capable of reason (Ussher 2005). The term "hysteria" shifted from a gynecological condition (suffocation of the womb) to a nervous condition prevalent among women, due to their weaker nervous constitution and subsequent emotional instability (Gilman et al. 1993). In the United States still today, premenstrual symptoms are used as a sexist slander that undermine anyone's legitimacy. Statements like "she's a little moody today, it's probably her time of the month" perpetuate this narrative. Period shaming is a thing. In the United States, we use code names ("Aunt Flo is coming for a visit"), whisper about it, and downplay symptoms. At home and with friends, Mei openly talks about periods with everyone, from her dad to neighbors and friends, yet in a professional setting there is hesitation to even say the word.

We fear making colleagues uncomfortable around us, fear being stigmatized, and get angry at ourselves for brushing aside the pain to avoid appearing weak, sometimes to the detriment of our own health. When it does come up at work it is typically with women colleagues. Unfortunately, this places the burden on those who menstruate, rather than those who don't—those who often occupy leadership positions in professional settings—for learning, discussing, or being aware of menstruation. Women, transmen, and nonbinary individuals all must think about their access to toilets, if they will have water or a sink to clean up, and how they will dispose of any product they used. It's not acceptable to wash out a Mooncup (www.mooncup.co.uk/) in the sink of a communal men's or women's bathroom. Women find themselves trying to bury used pads beneath other garbage in the communal field camp garbage, if there is a garbage bin. Women in public life have long faced questions based on their hormonal cycle. It is demeaning when people are slandered for sticking up for themselves, laughed at, told to toughen up or lighten up, forced to prove their pain is real, or told their symptoms are psychosomatic. It's pretty maddening to live in a society that recoils at any mention of a period, yet no one blinks at poop emojis, stuffed poop toys, or penis and fart jokes. Field season training should include these topics and open conversations about working around menstruation and other bodily functions.

The Change of Life

There are five known species that go through menopause: the narwal (*Monodon monoceros*), beluga (*Delphinapterus leucas*), killer whale (*Orcinus orca*), short-finned pilot whale (*Globicephala macrorhynchus*), and, the only primates, humans (Ellis et al. 2018). Broadly speaking, menopause is the programmed end of female fertility. Menopause is defined as having started once a woman has not had a period for 12 months.

How we perceive menopause is shaped by dominant societal attitudes (Lazar et al. 2019). In Victorian times it was associated with sin and decay, considered a neurosis in the early 20th century, a disease with intervention approaches in the 1960s, and it remains simply taboo in many societies (Utian 1997; Grandey et al. 2019; Krajewski 2019). Over the years, more women have joined the workforce and are staying in the workforce longer. Given that the average age of retirement is 68 years old, this guarantees that working women will go through this change during their work life. Frustratingly, there are no hard and fast rules as to what age women will enter menopause, what kinds of changes a woman will experience or their intensity. Nor is there a definitive time limit to how many years this bodily change will take. What we need to be aware of is menopausal symptoms that might affect a woman's professional productivity. Physically, the ovaries stop producing eggs and estrogen stores plummet, which has both short- and long-term effects. Many women experience hot flashes, but there can be bone loss, reduced stature, and diminished strength, heart palpitations, muscle or joint pain, irritability, slowing metabolism and subsequent weight gain, lowering libido, loss of confidence, urinary incontinence, increased urination, and more. Many of these symptoms lead to sleep disruption, mental fogginess, fatigue, forgetfulness, difficulty focusing or concentrating (Atkinson et al. 2021; Jack et al. 2016). Emotional triggers during this time make many women feel out of control, depressed, insecure, anxious, or irritable. Symptoms associated with menopause may cause difficulties for working women (Jack et al. 2016; Grandey et al. 2019), yet the topic is typically avoided in the workplace. Women who are going through it may downplay its severity for fear of marginalization (Hardy et al. 2017). Employers may be uncomfortable talking about it or may not know enough to know how they can help. An initial key step for employers is to provide information to other employees on menopause including the difficulties at work, coping strategies, and acknowledging that this may be uncomfortable (Hardy 2020).

How Can We Melt the Iceberg?

Each generation thinks it is different, that when the old men retire, they will change the system. However, biases and inequities start early in life. Systemic societal undercurrents that begin when we are young children are reinforced as we grow. In some instances, prejudice and biases are reinforced so much that discrimination in a myriad of forms becomes normalized and difficult to change (Buck et al. 2002). Some biases are embedded in culture, and we end up being our own unreliable narrators because we "grew thick skin" or more often knew no other way. Ignorance is bliss, right? So how do we fix a systemic problem when we inherited these social norms and are often unaware of them? Obviously, there isn't an easy solution.

If You See It, You Can Be It

Girls aspiring to careers in wildlife had few role models among wildlife scientists, and they were often limited middle- or upper-class white men (Diekman et al. 2010; Carli et al. 2016). In 2007, author Kerry Nicholson was working on her PhD, and her advisor asked her to teach some of his classes. She asked what topics she should cover, and he offhandedly suggested women in the wildlife profession. In preparation for the lecture, she began digging through the literature and quickly realized she would have very little to talk about if she confined her subject to women "scientists" and nearly zero if she focused her discussion on women in the wildlife profession (Nicholson et al. 2008; Nicholson 2009, 2014). At the time she found no BIPOC, LGBTQ+, or disabled minority role models. It didn't occur to her until that point that she could count on one hand how many women minority faculty that were in the wildlife departments of the three universities she attended (it was three, by the way). A colleague of hers, who currently teaches a principles of ecology, conservation biology, and wildlife course, noted that, in the textbook she uses for fundamental theories and foundations of ecology and wildlife, *all* the authors were male, and predominantly white. Because of this, she deliberately created a segment that highlighted scientists who are not cis white men just to provide aspiring students with diverse role models.

A person's perceptions of scientists are influenced by what they see, particularly in the media (Steinke 2005; Pietri et al. 2021). Early childhood exposure influences gender roles, occupations, and personality traits as well as aspirations for our future selves (Lemish and Götz 2017). As early as the second grade, children form an image of the stereotypical scientist (Chambers 1983). Minorities are more likely to succeed when provided evidence that success is attainable and when there is a strong representation of a possible future self (Drury et al. 2011; Riegle-Crumb et al. 2011; Jones and Solomon 2019; Midgley et al. 2021). Knowing the power of media, colleges and universities portray diversity in their recruitment materials, although these images may be aspirational (Pippert et al. 2013). Having a visual example of someone you can relate to may not seem like a big deal, but it is a subtly powerful motivator. Over time, when the people in leadership positions are diverse, the positive effects trickle down: to who is assumed competent at the job, to alternative approaches and solutions, to increased productivity, who we hire, job satisfaction, performance, and thus increased retention in the profession.

If You See Something, Say Something

If you are a bystander of a situation that is making you uncomfortable, then likely it is also making some of the people involved uncomfortable. Oppressed people are often conditioned to not show weakness, to appear stoic and build a protective shell, so they will not complain and make waves. You may think they are okay or they seem to be handling it okay; maybe they may even have laughed it off. However, no one should have to laugh harassment off. For example, some faculty and supervisors have a reputation for acting inappropriately around students or employees and a whisper network arises to warn the next group of subordinates. Instead of stopping the problem and holding the senior member accountable at a higher level, all the responsibility is left in the subordinates' hands. The catch-22 is that a subordinate has to put their job and reputation the line to confront the aggressor. If they indulge the "harmless" flirtation or try to make light of the situation, they may find their reputation tarnished as having encouraged it for personal advancement. Damned if they do, damned if they don't. We cannot wait on oppressed people to make all the changes.

After a committee meeting on improving hunting and trapping opportunities in the Cascades, a male coworker asked Rosalind why the other women in the group did not speak up when disparaging and derogatory statements about women's hunting abilities were made by the cochair. Her immediate reply was, "Why don't you? If you heard it, recognized it for what it was, why didn't you?" Because he wasn't the target of the assault, he believed he didn't have the right to be offended and didn't want to appropriate the indignation on behalf of women—as if by stepping in to defend the women he would somehow undermine their ability to defend themselves. Simply put, Rosalind's coworker was aware of the sexism but didn't know how to take action. In this situation, Rosalind told him it is okay to speak with her, not *for* her. There are tons of ways to be an effective ally, including seeking out bystander intervention training for educated responses. It may not be easy, but it is important not to stay silent. Silence gives permission to the bullies to continue their behavior. Rosalind's coworker has more influence than he realizes. If he vocalized that the behavior is not acceptable, it may have encouraged others to also speak up.

We Are Not Perfect: We Can Do Better and We Will Be Better Together

At some point in our lives, we have all been guilty of discriminatory behaviors from the grossly obvious to the unintentional microaggression. However, once we are aware of how our words, actions, or inaction can negatively affect others, we can reflect, learn, and move forward in our efforts to create a more inclusive and supportive working environment and society. Societal change is inevitable and necessary as globalization continues to diversify the world. In the United States, many women and minorities are hearing that they don't belong in the wildlife profession. We need to take action to change this narrative.

If we are passive when observing discrimination, we reinforce a culture that tolerates such behavior. But if we are empowered to engage in solutions that prevent biases and are given the tools and resources to do so, we can become engaged bystanders, which will make a significant difference in changing the culture of the wildlife profession. We need to stop looking at harassment as a problem of targets, harassers, and legal compliance. Coworkers, supervisors, students, faculty, clients, stakeholders, collaborators, field technicians, and customers all have roles to play in stopping discrimination.

We all make mistakes. None of us knows everything. We still have much to learn. We cannot improve if we are not told when we are making a mistake. We are not mind readers, and sometimes we struggle with taking off our own blinders. Allies should pledge to be open to being corrected and accept that someone will have to help teach them. Those who have been marginalized should pledge to give an ally the opportunity to change, to help them improve; because unlearning is as important as learning. We all should be committed to learning and sharing. We all should pledge to listen, to understand, to be mindful, and to try to do better the next time.

How to Be an Ally for Yourself and Others

- Take responsibility for your own actions.
- Take responsibility to educate yourself on how to be a better ally.
- Acknowledge the presence of gender discrimination in society and work to change it.
- Stop chasing the myth that science is apolitical and free from bias or that only the best rise to the top. Science is a human endeavor which means it is subject to our brilliance and our bias.
- It is natural to see injustice that is done towards ourselves. Reflect on situations or another person's behavior with mercy.
- Move from a culture of compliance to a culture of change.
- Recognize that you might be the voice needed by marginalized people. Use your voice.
- Teach about implicit bias and how it disadvantages women and minority groups.
- Change power dynamics by encouraging advisor networks so it is not all dictated by one voice.
- Form strategic alliances for yourself by supporting the leaders above you who will listen, support them, and it will allow you to thrive.
- Advocate for the support you need. Most people are not mind readers, nor should you expect them to be, even seasoned supervisors.
- Seek opportunities to mentor women and girls, from serving as a reference to being a role model.
- Spread the wealth of experience among employees instead of repeatedly using the same few people.
- Facilitate bonding activities at work such as professional development workshops. Organize something fun like welcome lunches or scavenger hunts to help your team find common ground.
- Incorporate gender-identify-specific nondiscrimination policies and practices to increase employees' understanding and acceptance of LGBTQ+ people.
- Take the extra step to ensure properly tailored clothing is available for all body shapes to do the work safely and effectively.
- Listen. You don't have to fix everything for someone.
- Invite your coworkers to dinner with your family.
- Raise awareness and provide training to help managers and colleagues understand how to support menopausal women.

- Consider reasonable work adjustments for menopausal women such as a desk fan, extra uniforms, access to water and toilet facilities, flexible working schedules, temporary reallocation of duties, and open dialogue.

Discussion Questions

1. Imagine yourself as Casey, a transgender person, on a boat with your coworkers. You must go to the bathroom. Whose responsibility is it to start a discussion about restroom use? How should the subject be broached? As Casey's coworker, how would you support them? As Casey's supervisor, how would you assist them and how might you address any negative behaviors by Casey's coworkers?

2. Apply the metaphor of the iceberg to your life and write in examples of the obvious to subtle or invisible experiences of harassment you have had.

3. This chapter has predominantly focused on discrimination in binary terms (male/female). Modify your iceberg to what you think a person from an(other) underrepresented group (women of color, indigenous people, people in the LGBTQ+ community, people with disabilities, or those with multiple intersecting social categories) might experience.

4. Imagine you are preparing for a field season, working in a remote area. What would you include in your training program for technicians to ensure a safe and healthy work environment?

Activity: Melting the Iceberg: Solutions for Inclusion

Learning Objectives: Participants will review the list of actions to increase inclusivity of the wildlife field provided in this chapter (or any other chapter's "How to Be an Ally" list) and develop possible solutions for addressing inclusivity directly in their office, classroom, or work setting.
Time: 60–75 min.
Description of Activity
Working or teaching in the field of wildlife biology and management often involves physically demanding work, living and working in remote or rural field settings and in close quarters with other people, power dynamics between team members, a unique professional social culture, attending meetings and conferences that involve large numbers of people and travel, and early career positions or internships that pay minimum wage or are volunteer positions and do not include benefits. These experiences have the

potential to exclude wildlife students or professionals from pursuing job opportunities or may negatively impact their success or advancement in the field. However, there are inclusive solutions to many of these exclusionary experiences. In this activity, participants will develop solutions.

Materials needed: Paper and a writing utensil for taking notes.

Process:

Part 1: Initial meeting (20 min.)

Organizers identify items from the "we can do better list" above and assign one item each to small groups of four to six. Small groups will identify at least three ways (more are encouraged) they can address this issue. Solutions can involve conversations, training, workshops, incentives, changes, or modifications to current practices. When considering possible solutions, participants may reflect on the SMART goal they developed in Chapter 1 for ideas. All small group members should take notes because they will need to describe these solutions to a new group in Part 2 of the activity.

Part 2: Mixing for solutions (20 min.)

For this next activity, rearrange into new small groups. No two group members should be from the same original group from Part 1, if possible. Members will each take a turn sharing the list item they discussed in their first small group and the potential solutions they developed. New group members are encouraged to suggest additional solutions.

Part 3: Return to the original group (5–10 min.)

After the discussion in Part 2 ends, re-form the original groups, and discuss any additional ideas that were developed during Part 2.

Part 4: Sharing solutions (10 min.)

Each group presents the issue they discussed and shares their favorite solution with the larger group. Organizers should encourage everyone to revisit their SMART goal in Chapter 1 and evaluate whether they want to modify or add to the solutions they developed as part of that earlier activity.

Part 5: Debriefing and Discussion Questions (15 min.)

1. What did you learn?
2. What was most surprising about this exercise? Had you previously considered how the work associated with the wildlife field might exclude some people from participating? Why or why not?
3. What are some potential challenges to implementing your solutions? What are possible ways to overcoming them?
4. How will you use the information you learned or your experience from this activity in your life as a wildlife professional, student, or citizen?

Literature Cited

Atkinson, C., V. Beck, J. Brewis, A. Davies, and J. Duberley. 2021. "Menopause and the Workplace: New Directions in HRM Research and HR Practice." *Human Resource Management Journal* 31(1): 49–64. doi: 10.1111/1748-8583.12294.

Bacchini, F., and L. Lorusso. 2019. "Race, Again: How Face Recognition Technology Reinforces Racial Discrimination." *Journal of Information, Communication and Ethics in Society* 17(3): 321–35. doi: 10.1108/JICES-05-2018-0050.

Bellofiore, N., S. J. Ellery, J. Mamrot, D. W. Walker, P. Temple-Smith, and H. Dickinson. 2017. "First Evidence of a Menstruating Rodent: The Spiny Mouse (*Acomys cahirinus*)." *American Journal of Obstetrics & Gynecology* 216(1): 40.e1–40.e11. doi: 10.1016/j.ajog.2016.07.041.

Bent, B., B. A. Goldstein, W. A. Kibbe, and J. P. Dunn. 2020. "Investigating Sources of Inaccuracy in Wearable Optical Heart Rate Sensors." *NPJ Digital Medicine* 3:18. doi: 10.1038/s41746-020-0226-6.

Bradshaw, C. J., and F. Courchamp. 2018. "Gender Bias when Assessing Recommended Ecology Articles." *Rethinking Ecology* 3:1.

Britzky, H. 2020. "The Military Says It's Finally Designing Body Armor for Women for the Umpteenth Time This Decade." Analysis. *Task and Purpose*. https://taskandpurpose.com/analysis/women-combat-gear-problem/.

Buck, G. A., D. Leslie-Pelecky, and S. K. Kirby. 2002. "Bringing Female Scientists into the Elementary Classroom: Confronting the Strength of Elementary Students' Stereotypical Images of Scientists." *Journal of Elementary Science Education* 14(2): 1. doi: 10.1007/BF03173844.

Carli, L. L., L. Alawa, Y. Lee, B. Zhao, and E. Kim. 2016. "Stereotypes about Gender and Science: Women ≠ Scientists." *Psychology of Women Quarterly* 40(2): 244–60. doi: 10.1177/0361684315622645.

Chambers, D. W. 1983. "Stereotypic Images of the Scientist: The Draw-a-Scientist Test." *Science Education* 67(2): 255–65. doi: 10.1002/sce.3730670213.

Cheney, I., and S. Shattuck., dirs. 2020. *Picture a Scientist*. Uprising Production.

Cofield, C. 2018. "NASA's New Spacesuit Has a Built-In Toilet." Space.com. www.space.com/39710-orion-spacesuit-waste-disposal-system.html.

Derks, B., C. Van Laar, and N. Ellemers. 2016. "The Queen Bee Phenomenon: Why Women Leaders Distance Themselves from Junior Women." *Leadership Quarterly* 27(3): 456–69. doi: 10.1016/j.leaqua.2015.12.007.

Diekman, A. B., E. R. Brown, A. M. Johnston, and E. K. Clark. 2010. "Seeking Congruity between Goals and Roles: A New Look at Why Women Opt Out of Science, Technology, Engineering, and Mathematics Careers." *Psychological Science* 21(8): 1051–57. doi: 10.1177/0956797610377342.

Dimitry, S., and A. Murphy. 2017. *Beyond #MeToo: The Gender Equity Iceberg*. www.academia.edu/35921937/Beyond_MeToo_The_Gender_Equity_Iceberg.

Drury, B. J., J. O. Siy, and S. Cheryan. 2011. "When Do Female Role Models Benefit Women? The Importance of Differentiating Recruitment from Retention in STEM." *Psychological Inquiry* 22(4): 265–69. doi: 10.1080/1047840X.2011.620935.

Duguid, M. 2011. "Female Tokens in High-Prestige Work Groups: Catalysts or Inhibitors of Group Diversification?" *Organizational Behavior and Human Decision Processes* 116(1): 104–15. doi: 10.1016/j.obhdp.2011.05.009.

Ellis, S., D. W. Franks, S. Nattrass, T. E. Currie, M. A. Cant, D. Giles, K. C. Balcomb, and D. P. Croft. 2018. "Analyses of Ovarian Activity Reveal Repeated Evolution of Post-reproductive Lifespans in Toothed Whales." *Scientific Reports* 8(1): 12833. doi: 10.1038/s41598-018-31047-8.

Elsesser, K., and L. A. Peplau. 2006. "The Glass Partition: Obstacles to Cross-Sex Friendships at Work." *Human Relations* 59(8): 1077–100. doi: 10.1177/0018726706068783.

Exley, C. L., and J. B. Kessler. 2019. "The Gender Gap in Self-Promotion." National Bureau of Economic Research Working Paper Series, no. 26345. doi: 10.3386/ w26345.

Faniko, K., N. Ellemers, and B. Derks. 2021. "The Queen Bee Phenomenon in Academia 15 Years After: Does It Still Exist, and If So, Why?" *British Journal of Social Psychology* 60(2): e12408. doi: 10.1111/bjso.12408.

Farh, C. I. C., J. K. Oh, J. R. Hollenbeck, A. Yu, S. M. Lee, and D. D. King. 2020. "Token Female Voice Neactment in Traditionally Male-Dominated Teams: Facilitating Conditions and Consequences for Performance." *Academy of Management Journal* 63(3): 832–56. doi: 10.5465/amj.2017.0778.

Federal Emergency Management Agency (FEMA). 2019. *Emerging Health and Safety Issues among Women in the Fire Service.* Emmitsburg, MD: US Fire Administration and Women in Fire.

Fournier, A. M. V., and A. L. Bond. 2015. "Volunteer Field Technicians Are Bad for Wildlife Ecology." *Wildlife Society Bulletin* 39(4): 819–21. doi: 10.1002/wsb.603.

Gilman, S. L., S. L. Gilman, H. King, R. Porter, G. Rousseau, and E. Showalter. 1993. *Hysteria beyond Freud.* Berkeley, CA, University of California Press. http://ark.cdlib.org /ark:/13030/ft0p3003d3/.

Glick, P., and S. T. Fiske. 2001. "An Ambivalent Alliance: Hostile and Benevolent Sexism as Complementary Justifications for Gender Inequality." *American Psychologist* 56(2): 109.

Grandey, A. A., A. S. Gabriel, and E. B. King. 2019. "Tackling Taboo Topics: A Review of the Three Ms in Working Women's Lives." *Journal of Management* 46(1): 7–35. doi: 10.1177/0149206319857144.

Grunspan, D. Z., S. L. Eddy, S. E. Brownell, B. L. Wiggins, A. J. Crowe, and S. M. Goodreau. 2016. "Males Under-estimate Academic Performance of Their Female Peers in Undergraduate Biology Classrooms." *PLoS ONE* 11(2): e0148405. doi: 10.1371/ journal.pone.0148405.

Hale, T. 2017. "This Viral Video of a Racist Soap Dispenser Reveals a Much Bigger Problem in Tech." Technology. *IFLS.* https://www.iflscience.com/technology/this-racist-soap -dispenser-reveals-why-diversity-in-tech-is-muchneeded/.

Hall, E. T. 1976. *Beyond Culture.* Garden City, NY: Anchor Books.

Hardy, C. 2020. "Menopause and the Workplace Guidance: What to Consider." *Post Reproductive Health* 26(1): 43–45. doi: 10.1177/2053369119873257.

Hardy, C., A. Griffiths, and M. S. Hunter. 2017. "What Do Working Menopausal Women Want? A Qualitative Investigation into Women's Perspectives on Employer and Line Manager Support." *Maturitas* 101:37–41. doi: 10.1016/j.maturitas.2017.04.011.

Hemingway, E. 1960. *Death in the Afternoon.* New York: Charles Scribner's Sons.

Hussein, F., and J. Diaz. 2019. "Employers Exposed when Women's Safety Equipment Doesn't Fit." Occupational Safety. *Bloomberg Law.* https://news.bloomberglaw.com /safety/employers-exposed-when-womens-safety-equipment-doesnt-fit.

Jack, G., K. Riach, E. Bariola, M. Pitts, J. Schapper, and P. Sarrel. 2016. "Menopause in the Workplace: What Employers Should Be Doing." *Maturitas* 85: 88–95. doi: 10.1016/j. maturitas.2015.12.006.

Jamieson, K. H. 1997. *Beyond the Double Bind: Women and Leadership.* New York: Oxford University Press.

Jones, M. S., and J. Solomon. 2019. "Challenges and Supports for Women Conservation Leaders." *Conservation Science and Practice* 1(6): e36. doi: 10.1111/csp2.36.

Kahalon, R., J. C. Becker, and N. Shnabel. 2021. "Appearance Comments Presented as Compliments at Work: How Are They Perceived by Targets and Observers in and outside of Workplace Settings?" *Journal of Applied Social Psychology* 00: 1–12. doi: 10.1111/jasp.12732.

Kanter, R. M. 1977. *Men and Women of the Corporation.* New York: Basic Books.

King, S. 2020. "Premenstrual Syndrome (PMS) and the Myth of the Irrational Female." In *The Palgrave Handbook of Critical Menstruation Studies*, edited by Chris Bobel, Inga T. Winkler, Breanne Fahs, Katie Ann Hasson, Elizabeth Arveda Kissling, and Tomi-Ann Roberts, 287–302. Singapore: Springer.

Kluger, J. 2019. "Why NASA's Planned All-Female Spacewalk Really Fell Apart." Science. *Time*, March 27. https://time.com/5558827/nasa-female-spacewalk-suits/.

Krajewski, S. 2019. "Killer Whales and Killer Women: Exploring Menopause as a 'Satellite Taboo' That Orbits Madness and Old Age." *Sexuality & Culture* 23(2): 605–20. doi: 10.1007/s12119-018-9578-3.

Kram, K. E., and L. A. Isabella. 1985. "Mentoring Alternatives: The Role of Peer Relationships in Career Development." *Academy of Management Journal* 28(1): 110–32.

Kubu, C. S. 2018. "Who Does She Think She Is? Women, Leadership and the 'B'(ias) Word." *Clinical Neuropsychologist* 32(2): 235–51. doi: 10.1080/13854046.2017.1418022.

Kurtzleben, D. 2018. "Poll: The 'Inappropriate' Office Behaviors Most Pervasive in Workplaces." Politics. NPR. February 23. www.npr.org/2018/02/23/586174752.

Lazar, A., N. Makoto Su, J. Bardzell, and S. Bardzell. 2019. "Parting the Red Sea: Sociotechnical Systems and Lived Experiences of Menopause." *Proceedings of CHI Conference on Human Factors in Computing Systems (CHI '19)*, Glasgow, Scotland, May 4–9.

Lemish, D., and M. Götz, eds. 2017. *Beyond the Stereotypes? Images of Boys and Girls, and Their Consequences*. The International Clearinghouse on Children, Youth and Media. University of Gothenburg: Nordicom.

Leskinen, E. A., V. C. Rabelo, and L. M. Cortina. 2015. "Gender Stereotyping and Harassment: A 'Catch-22' for Women in the Workplace." *Psychology, Public Policy, and Law* 21(2): 192–204. doi: 10.1037/law0000040.

Lilienfeld, S. O. 2017. "Microaggressions: Strong Claims, Inadequate Evidence." *Perspectives on Psychological Science* 12(1): 138–69. doi: 10.1177/1745691616659391.

Lincoln, A. E., S. Pincus, J. B. Koster, and P. S. Leboy. 2012. "The Matilda Effect in Science: Awards and Prizes in the US, 1990s and 2000s." *Social Studies of Science* 42(2): 307–20. doi: 10.1177/0306312711435830.

Liu, A. 2011. "Unraveling the Myth of Meritocracy within the Context of US Higher Education." *Higher Education* 62(4): 383–97.

Mazrui, A. A. 1993. "The Black Woman and the Problem of Gender: An African Perspective." *Research in African Literatures* 24(1): 87–104.

McPherson, M., L. Smith-Lovin, and J. M. Cook. 2001. "Birds of a Feather: Homophily in Social Networks." *Annual Review of Sociology* 27(1): 415–44. doi: 10.1146/annurev.soc.27.1.415.

Midgley, C., G. Debues-Stafford, P. Lockwood, and S. Thai. 2021. "She Needs to See It to Be It: The Importance of Same-Gender Athletic Role Models." *Sex Roles*. doi: 10.1007/s11199-020-01209-y.

Murphy, W. 2019. "Advice for Men Who Are Nervous about Mentoring Women." *Harvard Business Review*. March 15. https://hbr.org/2019/03/advice-for-men-who-are-nervous-about-mentoring-women.

National Academies of Sciences Engineering and Medicine (NASEM). 2018. *Sexual Harassment of Women: Climate, Culture, and Consequences in Academic Sciences, Engineering, and Medicine*. Washington, DC: National Academies Press.

Nicholson, K. L. 2009. "Wanted: Female Role Models." *Wildlife Professional* 3(1): 40–42.

Nicholson, K. L. 2014. "Exploring the Gender Gap in Leadership: A Survey of The Wildlife Society's Leadership Institute Alumni." *Wildlife Professional* 8(4): 46–49.

Nicholson, K. L., P. R. Krausman, and J. A. Merkle. 2008. "Hypatia and the Leopold Standard: Women in the Wildlife Profession 1937–2006." *Wildlife Biology in Practice* 4:57–72.

Pierce, C. M. 1970. "Black Psychiatry One Year after Miami." *Journal of the National Medical Association* 62:471–73.

Pietri, E. S., I. R. Johnson, S. Majid, and C. Chu. 2021. "Seeing What's Possible: Videos Are More Effective Than Written Portrayals for Enhancing the Relatability of Scientists and Promoting Black Female Students' Interest in STEM." *Sex Roles* 84(1): 14–33. doi: 10.1007/s11199-020-01153-x.

Pippert, T. D., L. J. Essenburg, and E. J. Matchett. 2013. "We've Got Minorities, Yes We Do: Visual Representations of Racial and Ethnic Diversity in College Recruitment Materials." *Journal of Marketing for Higher Education* 23(2): 258–82. doi: 10.1080/08841241.2013.867920.

Riegle-Crumb, C., C. Moore, and A. Ramos-Wada. 2011. "Who Wants to Have a Career in Science or Math? Exploring Adolescents' Future Aspirations by Gender and Race/Ethnicity." *Science Education* 95(3): 458–76. doi: 10.1002/sce.20431.

Rossiter, M. W. 1993. "The Matthew Matilda Effect in Science." *Social Studies of Science* 23(2): 325–41. doi: 10.1177/030631293023002004.

Sandberg, S. 2013. *Lean In: Women, Work, and the Will to Lead.* New York: Alfred A. Knopf.

Sandberg, S., and M. Pritchard. 2019. "The Number of Men Who Are Uncomfortable Mentoring Women Is Growing." *Fortune*, May 17. https://fortune.com/2019/05/17/sheryl-sandberg-lean-in-me-too/.

Schein, V. E. 1975. "Relationships between Sex Role Stereotypes and Requisite Management Characteristics among Female Managers." *Journal of Applied Psychology* 60(3): 340.

Sheppard, L. D., and K. Aquino. 2017. "Sisters at Arms: A Theory of Female Same-Sex Conflict and Its Problematization in Organizations." *Journal of Management* 43(3): 691–715. doi: 10.1177/0149206314539348.

Shiffman, D. S. 2019. "A Bad Deal for Early-Career Researchers." *American Scientist* 107: 325. doi: 10.1511/2019.107.6.325.

Steinke, J. 2005. "Cultural Representations of Gender and Science: Portrayals of Female Scientists and Engineers in Popular Films." *Science Communication* 27(1): 27–63. doi: 10.1177/1075547005278610.

Tabassum, N., and B. S. Nayak. 2021. "Gender Stereotypes and Their Impact on Women's Career Progressions from a Managerial Perspective." *IIM Kozhikode Society & Management Review* 10(2): 192–208. doi: 10.1177/2277975220975513.

Ussher, J. M. 2005. *Managing the Monstrous Feminine: Regulating the Reproductive Body.* London: Routledge. doi: 10.4324/9780203328422.

Utian, W. H. 1997. "Menopause—a Modern Perspective from a Controversial History." *Maturitas* 26(2): 73–82. doi: 10.1016/S0378-5122(96)01092-4.

Watkins, M. B., A. Simmons, and E. Umphress. 2018. "It's Not Black and White: Toward a Contingency Perspective on the Consequences of Being a Token." *Academy of Management Perspectives* 33(3): 334–65. doi: 10.5465/amp.2015.0154.

Zimmer, L. 1988. "Tokenism and Women in the Workplace: The Limits of Gender-Neutral Theory." *Social Problems* 35(1): 64–77.

Carol L. Chambers and Kerry L. Nicholson

13

Looking toward the Future

Leadership is about empathy. It is about having the ability to relate to and connect with people for the purpose of inspiring and empowering their lives.

—*Oprah Winfrey*

Change Happens

Why would a scientific journal, with more than 100 years of history, change its name? First issued in 1913, the journal of the American Society of Ichthyologists and Herpetologists (ASIH) was titled *Copeia* to honor Edward Drinker Cope. Yet in 2021, ASIH members voted to change the name of the journal, referencing the need to "remove the stigma associated with Cope's views" (Smith 2020; Parker 2021). Tamekia Wakefield, then a student in biology, and one of the few Black students in a predominantly white university, was aware of Cope's reputation as a racist and misogynist (Cope 1890; Davidson 1997). She was shocked when a professor delivered a lecture venerating him. She brought her knowledge and proof of Cope's racism to a trusted white male professor, Alan Jaslow, who listened, empathized, and acted by contacting ASIH and asking for a journal name change (Smith 2020; Parker 2021). The journal is now titled *Ichthyology & Herpetology*. This is a powerful example of a brave woman acting to identify a problem and empathy from a man stepping up to help change it. Action can support women and people of color achieve equity and inclusion (Figure 13.1).

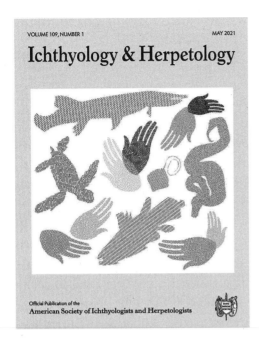

FIGURE 13.1 The American Society of Ichthyologists and Herpetologists voted to change its journal's name to *Ichthyology & Herpetology* in 2020 to better reflect values of the organization, becoming more inclusive and equitable.
Wikimedia Commons, courtesy of the American Society of Ichthyologists and Herpetologists.

Where Are We Going from Here?

After reading this book, you are more aware of the institutional barriers that some wildlife biologists face. Women and underrepresented groups are often held to different standards and their work or leadership styles may be questioned. Gender norms create conflicts and difficulties in managing both a career and family (e.g., Hall 2010). Throughout this book we have described situations faced by those in marginalized communities. We have touched on barriers, suppression, and struggles experienced by marginalized people. Our goal has not been to focus on problems but to depict realities in our profession and inspire *all* of us in the wildlife community to solve our inequities. We have highlighted the importance of those in positions of power—men, whites, administrators, supervisors, graduate advisors—to emphasize their ability to address these issues, provide support, change the cultural atmosphere, and actively be part of the solution. Discrimination, exclusion, biases, harassment—these behaviors will *not* stop on their own.

We also hope this book has reached those who have felt overlooked, ignored, or abandoned; know that you are not alone. We have tried to provide words of support and ideas for change. We have tried to encourage empathy, openness, and empowerment to spark positive changes in the wildlife profession. Ultimately, words are not enough. All of us must take responsibility for building diversity, equity, and inclusion (DEI). We cannot be complacent bystanders and expect our professional culture to change on its own.

We are making progress. We are seeing relationships change and inequalities being addressed in the workplace, at home, and in society. Yet gender gaps and inequities persist, even in the face of startling social and economic transformations and concerted movements to challenge discrimination and the forced subordination of certain groups. There are still flaws in the system. We must look deeper to ensure we are not inadvertently perpetuating inherent bias. We need to identify and address fundamental flaws that result in unintentional discrimination.

A privileged scientist takes for granted that science is a meritocracy and all work has equal value. But for hundreds of years, institutions of Western science have excluded or precluded women and minorities from full participation (Malcom et al. 1976; Noordenbos 2002; Watts 2007). Our contemporary conceptions of femininity and science have been historically constructed in opposition to each other. If science has come to mean objectivity, reason, analysis, and power, femininity has come to mean everything that science is not: subjectivity, feeling, passion, and weakness (Archer et al. 2012; Francis et al. 2017). Women do indeed see things differently, as do people from diverse cultures, ethnicities, and genders (e.g., Handelsman et al. 2005; Medin and Bang 2014). Their perspectives bring a variety of opinions, ideas, and values to the table, providing an enriching perspective to science (Valantine and Collins 2015; Nielsen et al. 2017). We have struggled to recruit and retain women in the scientific pipeline, but somehow the problem persists. When women feel left out or purposefully excluded, they withdraw (Blickenstaff 2005; Handelsman et al. 2005).

Can we approach science without forcing conformity? By excluding half of all potential scientists because of perceived differences, we weaken our science. If we achieve equal and balanced representation in the sciences and positions of leadership, our institutions will improve, as will our research priorities and methods and codes of professional conduct. As we seek to achieve equity, diversity, and inclusion, we enrich our science and make better decisions.

The journey you have traveled through the chapters of this book has likely caused a range of emotions, from outrage to recognition of injustice to sadness, acknowledgment, commiseration, clarity, hope, and determination. For some it may have triggered confusion, disbelief, anger, and denial. You may have experienced some of the themes explored in the book. There are almost certainly aspects of this book that deserve further reflection, context, and insight from those who have lived experiences. Regardless of your conclusions, the data are clear: women and minorities are underrepresented in science. It is now your job to consider the research and results presented in this book. The question is, now what?

Are the Times Changing? Yes and No

Between 1966 and 1977, nearly 5,000 elementary school students from the United States and Canada were asked to draw a scientist (Chambers 1983). Only 28 girls (0.6%) drew a woman. A meta-analysis of data from the 1980s onward, which included >20,000 students, found a positive increase (to 28%) in drawing scientists as women (Miller et al. 2018). Children develop perceptions of which genders should be in a discipline early in life, in educational settings from elementary through high school. Stereotyping which disciplines are "for women" can negatively affect entry of and attitudes toward women in typically male-dominated professions (Ganley et al. 2018). Although improving, the increase indicated by Miller et al. (2018) is still far from adequate. We should provide support and education at early ages to make more dramatic shifts toward equity.

Most would agree that we have seen increases in representation of women in the wildlife profession. This is evident in reflections by mid- and late-career women wildlife professionals and statistics (see Chapter 3, NAUFWP 2021), who observed an increase in the number of women working in the wildlife field during their careers. Most stated that they were no longer the only woman in their organization and saw more women in the pipeline. Their colleagues were more likely to assume that they could perform as capably on the job as men. However, the increase in women employees did not always correlate with number of women in leadership positions. Advancing to leadership roles varied by geographic location, agency, or organization, with some doing more to foster equity and others undermining women's retention and promotion (e.g., Matlock 2016; Burton 2018; Jenkins 2018; Abernethy et al. 2020). Having supportive, often male, leadership in these organizations played an important role in increasing women at all career levels and moving towards gender equity. Professional organizations have often been even more disproportionately dominated by men than academic and agency settings. More groups are aware of these inequities and of the importance of DEI in professional societies and making changes in outreach and support (Abernethy et al. 2020; Madzima and MacIntosh 2021; TWS n.d.) (Box 13.1).

Statistics echo our perceptions of increased gender equity, especially at the student level in university fisheries and wildlife programs in the United States.

BOX 13.1 **Picture a Scientist**

More agencies, universities, and workplaces are dedicating time and resources to DEI training but may be struggling with the right approach to reach employees. Formation of a DEI committee is a good start, but concern over administrative issues or the unknowns of how to approach and discuss sensitive subjects can slow or stop progress. There is no rule book, directive, or protocol on how to handle some of these topics that are guaranteed to work. As an example, if a department wants to host a screening of the 2020 documentary *Picture a Scientist* for their employees and encourage interaction and discussion after, how do they go about it? Should it be mandatory, restricted to known colleagues, or open to people outside the organization? How do they avoid triggering memories of traumatic events? As biologists, we are trained to work with wildlife, not emotional psychological situations or cases of inappropriate behavior. Concerns such as these can push an inexperienced DEI committee into inaction.

More agencies are developing DEI support by establishing committees or hiring full-time staff dedicated to building equity and inclusion. The Washington Department of Fish and Wildlife, Texas Parks and Wildlife Department, and Virginia Department of Wildlife Resources are just a few examples. Wildlife-focused nongovernmental organizations such as the Association of Fish and Wildlife Agencies and The Wildlife Society are also strengthening their DEI efforts through strategic plans and actions.

As support builds, some sensitive issues could be solved by facilitators trained to moderate discussion sessions. In addition, agency leaders and supervisors can show strong support for DEI training, even without making it mandatory. However, DEI efforts will fail with agency inaction or glacially slow progress. Recognize that those taking their first steps in DEI work may fear failure or negative repercussions. Don't expect perfection. If you make room for people to make mistakes by demonstrating empathy and understanding, people are more willing to try.

Women now make up more than half of the undergraduate and graduate student population among 25 programs. However, women are still underrepresented in faculty positions, with 71% of full-time, tenure-track positions held by men. We are seeing equity in the classroom but not in retention and advancement. Women students thus outnumber women faculty by almost twice (NAUFWP 2021), placing a higher burden on women faculty as mentors.

At all levels, Black, Indigenous, and people of color (BIPOC) are dramatically underrepresented the wildlife profession (see Chapter 3, NAUFWP 2021).

In universities, BIPOC students constitute only 21% of undergraduate and ≤16% of graduate enrollment (NAUFWP 2021). Mentors and role models are even scarcer: 0%–12% (NAUFWP 2021). The paucity of BIPOC individuals in faculty positions suggests barriers to entering or remaining in academia at the tenure-track level (NAUFWP 2021). Many of the women we talked with, when asked about the biggest change they had seen during their careers, immediately described the increase in white women in wildlife and then expressed concern that this was not the case for women of other races and ethnicities. Efforts to recruit minorities have garnered some success, but retention is an important aspect of increasing diversity. If you are made to feel like a "diversity hire" or if you are harassed on the job, leaving is a rational, though unfortunate, option. We have a way to go to successfully establish BIPOC individuals in wildlife careers.

Surprisingly, almost every woman we talked with described experiences of harassment in the past and recently. They reported a constant level of harassment since they started their careers, which for some was decades earlier. A young woman, new to the profession, described trying to shut down advances from a man at a conference. She had to ask a friend to walk her to her hotel room because the man was so aggressive she was concerned for her safety. A graduate advisor propositioned a young graduate student when his wife was not around. They were in a remote field setting that made the situation awkward and uncomfortable. Condoms are still carried in first aid kits. A supervisor refused to promote a young woman biologist because she had children. A female supervisor, feeling threatened by a young professional, gave her menial tasks, telling her, "You have to pay your dues because I had to pay mine." Shockingly, supervisors of two state agencies were charged with sexual harassment and assault in the past 10 years (Matlock 2016; Burton 2018); one was convicted and sentenced to 10 years to life for raping an employee (Burton 2018; Jenkins 2018). These are examples of situations that continue to exclude, intimidate, or otherwise negatively affect women.

There are positive stories. One woman described the support she received from her all-male crew. *"Anytime me and my guys (I supervised a group of fabulous guys) would be in a fast-food joint or some such, in uniform, random men would come up to them, ignoring me, and ask questions about elk, the agency, hunting laws. They would sometimes respond, 'I don't know, you should ask my boss, she's standing right here.' They genuinely respected me and got a kick out of upending the assumptions of those random men on the street. Yeah, they were awesome! And give me hope that things really are changing for the better."* A young Black woman has support from her supervisors and colleagues to submit complaints when she is prevented from working the same duties as other employees. The support of supervisors built her trust in them and her ability to handle situations appropriately.

What has changed? Not changed? The presence of women has changed. Women have increased in number in the wildlife profession, and sometimes there have also been increases in women in leadership roles. Harassment has not changed (see Chapter 11). Organizations like the Societies Consortium on Sexual Harassment in STEMM can provide recommendations to reduce harassment in our profession (Societies Consortium on Sexual Harassment in STEMM 2021). The number of BIPOC individuals has not changed. We are still a dominantly white profession. Why have some factors changed and others not? What can be done?

Pandemic Pressure

The COVID-19 pandemic changed us. The disease affected people of color and women disproportionally. People of color contracted the virus in higher numbers, with Black and Indigenous people hospitalized at rates five times those of non-Hispanic whites (Morales and Ali 2021; Tai et al. 2021; Wilder 2021), because of biological, social (i.e., effects of racism), and structural mechanisms (Wilder 2021) (Figure 13.2). Many women exited the workforce (Karageorge 2020; Madgavkar et al. 2020) because of family or job-related issues. Women were more likely to be exposed to the virus through their occupations because they tend to occupy jobs more likely to be considered "essential" (e.g., health care workers; Carli 2020; Madgavkar et al. 2020).

How we work has changed. How we communicate and operate in our personal and professional lives has altered as we focused on human safety and health. Many of us learned, for the first time, how to use Kaltura, WebEx, Zoom, or other video platforms supporting virtual communication. We shifted our workplaces from office to home. On the positive side, with everyone working from home, those with children saw a reduction in gender gap in childcare and other housework, with men performing more tasks. On the negative side, women continued to do more than men and found their work time, performance, and productivity was reduced (Carli 2020).

COVID-19 will have long-term effects on the wildlife profession. Although we do not yet know how the virus affects wildlife, it has been reported in free-roaming bats, white-tailed deer (*Odocoileus virginianus*), and captive mink (*Mustela lutreola, Neovison vison*) (Mallapaty 2021; USDA APHIS 2021). The virus has disproportionately affected those working in the wildlife profession because

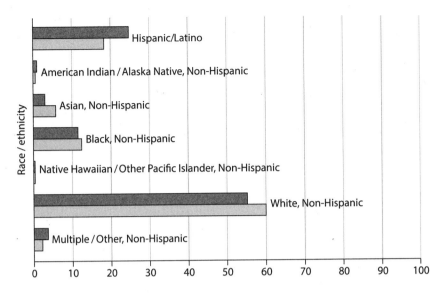

● Percentage of cases, all age groups
○ Percentage of the US population, all age groups

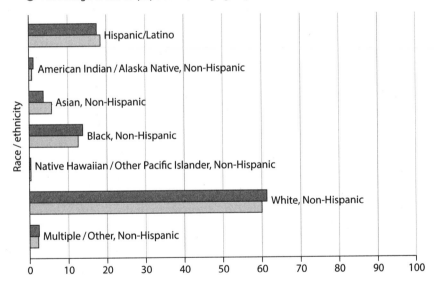

● Percentage of deaths, all age groups
○ Percentage of the US population, all age groups

FIGURE 13.2 Deaths and cases of COVID-19 by race/ethnicity in the United States, disproportionately affecting people of color.
https://covid.cdc.gov/covid-data-tracker/#demographics.

we don't have to protect just ourselves from the virus but also the wildlife we work with. DEI issues will be hampered as women, especially women of color, leave the workforce. The time that women scientists (particularly those with young children) devote to research has been reduced (Myers et al. 2021). Since there is a well-known gender gap in science, the effects of the pandemic may be additive, with greatest effects on female scientists with young children (Knobloch-Westerwick et al. 2013; Cardel et al. 2020; King and Frederickson 2021; Myers et al. 2021). These factors can reverse the progress we have been making.

What's Our Future?

DEI is growing as a field of interest in workforce recruitment and training. Although the work has gone on for decades, focusing on disability, race, religion, or nationality, we are broadening our scope to encompass gender, cognitive abilities, age, the intersectionality of these traits, and virtually any other attribute that differentiates us. As biologists we understand ecological evolution, so we should be comfortable with the need for DEI strategies to evolve.

Women and LGBTQ+ and BIPOC individuals continue to face challenges in the wildlife profession. A seemingly simple request is to be accepted for who we are, without needing to prove we are twice as good as a man to be considered equal. The energy required to continually show you can do the job and, in addition, educate others about your qualities and abilities can be overwhelming (Amon 2017). Assertive women do not want to be considered "bitches." BIPOC individuals do not want to have to "code switch" (Amon 2017; Blanchard 2021). LGBTQ+ individuals do not want to have to hide their personal lives. We want to be taken seriously and be known for the wildlife work we do, rather than have assumptions placed on where our careers should go. We may choose to have a family, but that does not mean our profession is unimportant or optional.

Another challenge is being unique to an organization, being in the minority. Wildlife is dominated by white men. Individuals from other backgrounds frequently feel isolated. Being the only one (woman, person of color) in the room can be exhausting and it is harder to stand your ground when feeling alone. Finding role models and mentors, especially for BIPOC individuals, can reduce the energy drain. We sometimes hear people talk about feeling like the "unicorn" in the room. In the business world, a unicorn is someone who provides enormous benefits to the workplace (Ragland 2019). In the case of a woman or BIPOC individual, they can also offer tremendous unique perspectives that lead to better decision making (Phillips et al. 2009; Aminpour et al. 2021).

We know that mentors and role models can pay significant roles in helping women succeed. What other actions will help increase diversity, equity, and inclusion in the wildlife profession?

How to Be an Ally

Education
- Educate yourself about your social identity (e.g., gender, race, ability) and how it influences your career.
- Learn more about people with identities different from yours.
- Let children know at early ages (grades K–12) about the field of wildlife, and showcase women in wildlife.
- Develop course requirements for a university wildlife or natural resources department that integrate DEI throughout the program.
- Increase diversity in students, faculty, and department heads in academia.

Mentor
- Actively recruit women and BIPOC individuals to the wildlife profession.
- Reduce isolation for women and BIPOC individuals.
- Regardless of your gender, spend time as a mentor to women and BIPOC individuals. You will learn more about their perspectives while providing support.
- Support individuals through difficult work situations by actively stepping in to assist when needed.
- Encourage dialogue in your workplace, school, organization, or professional society.
- Help build a critical mass of BIPOC individuals and women in the workplace.

Role model
- Show women of all colors as role models early in school (K–12).
- Use social media to highlight the contributions of BIPOC individuals and women in positions of leadership.

Financial support
- Help build funding for scholarships, programs to recruit and retain students, postdoctoral scholars, and faculty.
- Avoid creating unpaid internships that favor those with greater privilege and resources.

- Support those responsible for childcare and elder care by allowing adjustable or flexible course schedules and progress toward a degree or within a job.

Codes of conduct
- Develop codes of conduct that support DEI initiatives and explicitly share them with students, faculty, and employees.

Take Action

In the first chapter, we described a SMART goal approach to help you not just think about what you can do but to encourage you to act. We conclude this book with an example that we developed for ourselves. We start by identifying a barrier in the wildlife field that might limit DEI, setting a goal to overcome it, then developing an approach to achieve that goal. For our example, we encouraged dialogue among individuals in our professional society about challenges and barriers for BIPOC individuals in the wildlife field. We want to develop a specific, measurable, achievable, relevant, and timely (SMART) goal.

> *Specific*: Survey authors including women and BIPOC wildlife professionals about challenges they face in publishing and actions to support overcoming them
>
> *Measurable*: Publication of a paper describing results of the survey
>
> *Achievable*: Institutional Research Board approval has been obtained, survey questions developed, so next step is to develop the survey instrument
>
> *Relevant*: Discussion and identification of barriers will help reduce challenges faced by BIPOC individuals
>
> *Timely*: Within 1.5 years

We now ask you to use this approach to set your own SMART goals for the coming year and keep an eye out for the results of our survey.

In this book we have identified barriers and challenges experienced by women and BIPOC individuals. Chapter authors provided ideas both for seeking support from allies and actively engaging as allies. We hope you find this book useful in your classroom, workplace, or for public discussion. In closing, we ask you to commit to taking action. We invite you to be the change.

Discussion Questions

1. What do you remember about allyship from previous chapters? Discuss allyship from the perspective of the mentor, role model, and member of an underrepresented group. What actions do you think will be most meaningful to promote diversity and inclusion?
2. List pronouns that you are aware of for describing gender identity. How would you explain the importance of using correct pronouns to someone who is unfamiliar with their use? What should you do if you make a mistake when using pronouns?
3. Women describe their experiences in deflecting harassment by ignoring it, using humor to diffuse it, challenging the harasser, leaving the situation, or relying on someone else to help. How do you think variation in age, time in their career, position in an organization, or other factors influences their approach to harassers?
4. Select a movie from the 1980s or 1990s that has men and women characters. How are women or BIPOC people presented—do you see examples of inequity and exclusion or diversity, equity, and inclusion? How does this compare with a movie you've seen recently (in the previous two years)? Do you think we making progress? Why or why not?
5. Describe an action that you can take to increase diversity, equity, and inclusion in the wildlife profession in the next month, year, and five years.

Activity: SMART Goals

Return to Chapter 1 and revisit the SMART Goals activity. Knowing what you know now, how have your SMART goals changed or remained the same?

Literature Cited

Abernethy, E. F., I. Arismendi, A. G. Boegehold, C. Colón-Gaud, M. R. Cover, E. I. Larson, E. K. Moody, B. E. Penaluna, A. J. Shogren, A. J. Webster, and M. M. Woller-Skar. 2020. "Diverse, Equitable, and Inclusive Scientific Societies: Progress and Opportunities in the Society for Freshwater Science." *Freshwater Science* 39(3): 363–76. doi: 10.1086/709129.

Aminpour, P., S. A. Gray, A. Singer, S. B. Scyphers, A. J. Jetter, R. Jordan, R. Murphy Jr., and J. H. Grabowski. 2021. "The Diversity Bonus in Pooling Local Knowledge about Complex Problems." *PNAS* 118(5): e2016887118. doi: 10.1073/pnas.2016887118.

Amon, M. J. 2017. "Looking through the Glass Ceiling: A Qualitative Study of STEM Women's Career Narratives." *Frontiers in Psychology* 8:236. doi: 10.3389/fpsyg.2017.00236.

Archer, L., J. DeWitt, J. Osborne, J. Dillon, B. Willis, and B. Wong. 2012. "'Balancing Acts': Elementary School Girls' Negotiations of Femininity, Achievement, and Science." *Science Education* 96(6): 967–89.

Blanchard, A. K. 2021. "Code Switch." *New England Journal of Medicine* 384:e87. doi: 10.1056/NEJMpv2107029.

Blickenstaff, J. C. 2005. "Women and Science Careers: Leaky Pipeline or Gender Filter?" *Gender and Education* 17(4): 369–86.

Burton, L. 2018. "Woman Raped by Fish & Wildlife Boss: 'I Used to Be So Happy.'" *Seattle PI.* March 16. www.seattlepi.com/local/crime/article/Judge-Former-Fish-and-Wildlife -official-a-12752972.php.

Cardel, M. I., N. Dean, and D. Montoya-Williams. 2020. "Preventing a Secondary Epidemic of Lost Early Career Scientists: Effects of COVID-19 Pandemic on Women and Children." *Annals of the American Thoracic Society* 17(11): 1366–70.

Carli, L. L. 2020. "Women, Gender Equality and COVID-19." *Gender in Management: An International Journal* 35(7/8): 647–55. doi: 10.1108/GM-07-2020-0236.

Chambers, D. W. 1983. "Stereotypic Images of the Scientist: The Draw-a-Scientist Test." *Science Education* 67(2): 255–65. doi: 10.1002/sce.3730670213.

Cope, E. D. 1890. "Two Perils of the Indo-European." *Open Court* 3:2052–54. https://bit.ly /TwoPerils.

Davidson, J. P. 1997. *The Bone Sharp: The Life of Edward Drinker Cope.* Philadelphia: Academy of Natural Sciences.

Francis, B., L. Archer, J. Moote, J. de Witt, and L. Yeomans. 2017. "Femininity, Science, and the Denigration of the Girly Girl." *British Journal of Sociology of Education* 38(8):1097–110.

Ganley, C., C. E. George, J. R. Cimpian, and M. B Makowski. 2018. "Gender Equity in College Majors: Looking beyond the STEM/Non-STEM Dichotomy for Answers Regarding Female Participation." *American Educational Research Journal* 55(3): 453–87. doi: 10.3102/0002831217740221.

Hall, L. 2010. "'The Problem That Won't Go Away': Femininity, Motherhood and Science." *Women's Studies Journal* 24(1): 14–30.

Handelsman, J., N. Cantor, M. Carnes, D. Denton, E. Fine, B. Grosz, V. Hinshaw, C. Marrett, S. Rosser, D. Shalala, and J. Sheridan. 2005. "More Women in Science." *Science* 309(5738): 1191. doi: 10.1126/science.1113252.

Jenkins, A. 2018. "Former Fish and Wildlife Deputy Director Convicted of Rape, Burglary." *Northwest News Network.* January 24. www.nwnewsnetwork.org/crime -law-and-justice/2018-01-24/former-fish-and-wildlife-deputy-director-convicted-of -rape-burglary.

Karageorge, E. X. 2020. "COVID-19 Recession Is Tougher on Women." US Bureau of Labor Statistics. September. www.bls.gov/opub/mlr/2020/beyond-bls/covid-19-recession-is -tougher-on-women.htm.

King, M. M. and M. E. Frederickson. 2021. "The Pandemic Penalty: The Gendered Effects of COVID-19 on Scientific Productivity." *Socius: Sociological Research for a Dynamic World* 7:1–24.

Knobloch-Westerwick, S., C. J. Glynn, and M. Huge. 2013. "The Matilda Effect in Science Communication: An Experiment on Gender Bias in Publication Quality Perceptions and Collaboration Interest." *Science Communication* 35(5): 603–25.

Madgavkar, A., O. White, M. Krishnan, D. Mahajan, and X. Azcue. 2020. *COVID-19 and Gender Equality: Countering the Regressive Effects.* McKinsey & Company. July. www .apucis.com/frontend-assets/porto/initial-reports/COVID-19-and-gender-equality -Countering-the-regressive-effects-vF.pdf.pagespeed.ce.DDg-UdcoaA.pdf.

Madzima, T. F., and G. C. MacIntosh. 2021. "Equity, Diversity, and Inclusion Efforts in Professional Societies: Intention Versus Reaction." *Plant Cell* 33:3189–93.

Malcom, S. M., P. Q. Hall, and J. W. Brown. 1976. *The Double Bind: The Price of Being a Minority Woman in Science.* Report of a Conference of Minority Women Scientists,

Arlie House, Warrenton, Virginia. AAAS-Misc-Pub-76-R-3. https://files.eric.ed.gov
/fulltext/ED130851.pdf.

Mallapaty, S. 2021. "The Hunt for Coronavirus Carriers." *Nature* 591:26–28.

Matlock, S. 2016. "Game and Fish Workers Detail Harassment under Former Director."
Santa Fe New Mexican. July 23. www.santafenewmexican.com/news/local_news/game
-and-fish-workers-detail-harassment-under-former-director/article_6bcddc55-2283
-55c1-bea1-058b56d42e0e.html.

Medin, D. L., and M. Bang. 2014. *Who's Asking? Native Science, Western Science, and Science
Education.* Cambridge, MA: MIT Press.

Miller, D. I., K. M. Nolla, A. H. Eagly, and D. H. Uttal. 2018. "The Development of
Children's Gender-Science Stereotypes: A Meta-analysis of 5 Decades of U.S. Draw-a-
Scientist Studies." *Child Development* 89(6): 1943–55. doi: 10.1111/cdev.13039.

Morales, D. R., and S. N. Ali. 2021. "COVID-19 and Disparities Affecting Ethnic Minorities."
Lancet 397(10286): 1684–85.

Myers, K. R., W. Y. Tham, Y. Yin, N. Cohodes, J. G. Thursby, M. C. Thursby, P. Schiffer, J. T.
Walsh, K. R. Lakhani, and D. Wang. 2021. "Unequal Effects of the COVID-19 Pandemic
on Scientists." *Nature Human Behaviour* 4:880–83.

National Association of University Fisheries and Wildlife Programs (NAUFWP). 2021. *2021
Diversity & Equity Survey Report.* www.naufwp.org/diversity.html.

Nielsen, M. W., S. Alegria, L. Börjeson, H. Etzkowitz, H. J. Falk-Krzesinski, A. Joshi, E.
Leahey, L. Smith-Doerr, A. W. Woolley, and L. Schiebinger. 2017. "Opinion: Gender
Diversity Leads to Better Science." *PNAS* 114(8): 1740–42. doi: 10.1073/
pnas.1700616114.

Noordenbos, G. 2002. "Women in Academies of Sciences: From Exclusion to Exception."
Women's Studies International Forum 25(1): 127–37. doi: 10.1177/0146167208328062.

Parker, M. R. 2021. "Refusing to Cope with a Name." *Icthyology & Herpetology* 109(1): 3–4.

Phillips, K. W., K. A. Liljenquist, and M. A. Neale. 2009. "Is the Pain Worth the Gain? The
Advantages and Liabilities of Agreeing with Socially Distinct Newcomers." *Personality
and Social Psychology Bulletin* 35(3): 336–50.

Ragland, C. 2019. "How to Attract and Motivate Unicorn Employees." *Forbes.* April 18. www
.forbes.com/sites/forbescommunicationscouncil/2019/04/18/how-to-attract-and
-motivate-unicorn-employees/?sh=678245ee123e.

Smith, W. L. 2020. "Motion to Change the Name of the ASIH Journal." *Ichthyology &
Herpetology*, June 29. www.ichthyologyandherpetology.org/blog/2020/6/29/motion-to
-change-the-name-of-the-asih-journal.

Societies Consortium on Sexual Harassment in STEMM. N.d. Accessed December 4, 2021.
https://societiesconsortium.com/.

Tai, D. B. G., A. Shah, C. A. Doubeni, I. G. Sia, and M. L. Wieland. 2021. "The Disproportionate Impact of COVID-19 on Racial and Ethnic Minorities in the United States."
Clinical Infectious Diseases 72(4): 703–6. doi: 10.1093/cid/ciaa815.

USDA Animal and Plant Health Inspection Service (APHIS). 2021. *Questions and Answers:
Results of Study on SARS-CoV-2 in White-Tailed Deer.* APHIS 11-55-014. August. www.aphis
.usda.gov/animal_health/one_health/downloads/qa-covid-white-tailed-deer-study.pdf.

Valantine, H. A., and F. S. Collins 2015. "National Institutes of Health Addresses the
Science of Diversity." *PNAS* 112(40): 12240–42.

Watts, R. 2007. *Women in Science: A Social and Cultural History.* New York: Routledge.

Wilder, J. M. 2021. "The Disproportionate Impact of COVID-19 on Racial and Ethnic
Minorities in the United States." *Clinical Infectious Diseases* 72(4): 707–9.

The Wildlife Society (TWS). N.d. "Diversity, Equity, and Inclusion." Accessed December 4,
2021. https://wildlife.org/dei/.

INDEX

Page numbers followed by *a b* refer to boxes; by an *f*, figures; and by *a t*, tables.

barriers to wildlife professions (*cont.*) 190*f*; recruitment and retention, 125, 128, 130–33; resources to counter, 103–4; sexual harassment and, 10, 315–16, 367; for students, 11, 86–88

Barton, Anne Christine Kristiansson, 62

Bassi, Laura, 24

Batcheller, Gordon R., 308

bathroom necessities, 327*b*, 348–51

Beatty, Mollie, 44

Becoming an Outdoors-Woman (BOW), 67*b*

Ben-David, Merav, 68*b*

Benga, Ota, 118, 119*b*

Bennett, John, 26

Bernard, Riese, 211

Bernardi, Gene C., 43

BIA (Bureau of Indian Affairs), 160

Bian, L., 84

bias: affinity bias, 9, 293, 341; awareness of, 102, 218, 283, 295; barriers to success and, 84; benevolent sexism and, 339–40; cultural, 119*b*, 352; federal employment and, 98; gender-diverse teams and, 5; in hiring and promotions, 70, 90, 98–99, 198–99, 283–84, 288–89; iceberg analogy and, 336–37; inclusion and, 283–84; in job announcements, 288; lack of diversity resulting from, 9; racial, 122; in research and publications, 294; stereotypes and, 8, 284; toward LGBTQ+ people, 218, 225, 228; toward Native Americans, 154, 159; women's bias toward other women, 346. *See also* gender discrimination and bias

Birdcraft (Wright), 32

Bird Ways (Miller), 31

Black, Indigenous, and people of color (BIPOC), 118, 197, 310, 342, 366–68. *See also* Black people; Native Americans and Alaska Natives

Black people: anti-transgender violence against, 218; COVID-19 pandemic and, 368; faculty demographics, 61; mental health issues and, 258; recruitment and retention of, 130–31; stereotypes of, 10, 12, 193, 198; student demographics, 57, 367; success assessments of, 123–24; wildlife profession demographics, 310, 366–67

Blake, Jamila, 113, 131, 243, 245–46, 249, 253, 256–59

bodily functions, 85, 168, 262–63, 265, 320–21, 327*b*, 348–52

Bolnick, D. I., 29

Bonta, Marcia, 29

Booms, Travis L., 14, 211, 230–32

Borkhataria, Rena, 180

Bradley, Nina Leopold, 83

breastfeeding, 263, 265, 320–21, 327*b*

Bronstein, J. L., 29

Brooks, David, 248

Brown-Iannuzzi, J. L., 199

Bryant, Jenny, 147

Bryant Ditsuh Taii, Joanne, 147, 167–68

Bureau of Indian Affairs (BIA), 160

Cajas-Cano, Lubia N., 52

Cameron, E. Z., 70

Canada, 8–9*b*, 71, 90, 92, 94, 156

Cardin, Ben, 8*b*

Carr, P. L., 93

Carroll, Stephanie, 113

Carson, Rachel, 34, 38*b*

Causae et Curae (Saint Hildegard), 23

Caveny, Dale, 308

Ceci, S. J., 90

Celestin, Modeline, 180, 203*b*

Census Bureau, 54, 64, 121, 125

Chambers, Carol L., 3, 45, 244, 362

Changing the Narrative about Native Americans: A Guide for Allies (Reclaiming Native Truths), 173

Charter for Gender Equality (NRC, 2019), 71

childcare, 263–64

children's education, 84, 125–27, 169, 181–82, 352–53, 365

Christiansen, Vicki, 44

Citizen Science Association, 293

Civil Rights Act (1964), 41, 223–24, 313

Clancy, K. B. H., 316

Clark, Edward, 25

Clark, Emily K., 22

Clark, William, 113

climate change, 196*b*, 197

clothing decisions, 250–51*b*, 347–48

codes of conduct, 69, 293–94, 372

Colden, Jane, 30

colonialism, 117, 155–58

conferences, 66, 153–54, 231–32, 265, 292–93, 325*b*, 330*b*

Considine, Erin, 243, 244, 249, 253–54, 256, 263

Cooper, Susan Fenimore, 30

Cope, Edward Drinker, 362

Cordova, Ruth, 113

Courageous Conversations model, 203

COVID-19 pandemic, 11, 368–70, 369*t*

Cowan, Joyce McTaggart, 35

Cox, William T. L., 284

Crow, Claire, 211

culture: barriers to wildlife professions and, 129–30; biases based on, 119*b*, 352; children's education and, 125–27; cross-cultural dialogue activity, 175–77; cultural appropriation, 154; cultural capital, 5; diversity and teamwork, 5; iceberg analogy of, 336–37, 338*f*; of Native Americans

Americans and Alaska Natives, 149, 168; outdoor work and, 10; women in workforce and, 249

generational differences, 214, 269–75, 272–74t. *See also* millennials

Generation X, 246–47t, 248, 252, 254–55, 258, 264, 266, 269, 272–74t

Generation Z, 5, 246–47t, 254, 256, 269, 272–74t

Gewin, V., 198

The Giraffe (Dagg), 38

Glasscock, Selma N., 22, 45

Glimcher, Laurie, 70

Goodall, Jane, 40

Gorsuch, Neil, 225

Granillo, Kathy A., 82, 99f

Graves, Tabitha A., 279, 283

Gumpertz, M., 94

Haldane, J. B. S., 211

Hall, Ernest T., 336

Hall, Minna, 33

Hallett, Diana, 44–45

Hamerstrom, Frances "Fran" Carnes Flint, 35, 36–37b, 67b

Hannah-Jones, Nikole, 336

Harris, C., 93

Hate Crimes Prevention Act (2009), 224

Haynes, N. A., 188

Hemenway, Harriet, 33

Hemingway, Ernest, 336

Hentze, I., 310

Hester, Princess, 180, 194b

Hewlett, S. A., 88

Hildegard (saint), 23

hiring practices: in academia, 89, 90–92, 131; allies and, 198–99; bias and, 70, 90, 98–99, 198–99, 283–84, 288–89; in federal government, 96–101, 97b, 99f; gender bias and, 70, 283; inclusion and, 285, 288–90; job ads and announcements, 198, 226–27, 288–90; job search and application activity, 104–7; leaky pipeline and barriers to wildlife professions, 88; LGBTQ+ people and, 228–29; race and ethnicity and, 198–99. *See also* recruitment and retention

historical trauma, 148, 155, 156–58

history of women in science, 15, 22–51; activity on, 46–48; allies of, 45–46; discussion questions on, 46; Equal Employment Opportunity Act and, 41–44; exclusion and, 364; gender diversity among faculty, 62, 63t; golden age of nature study and, 26–28; in leadership positions, 44–45; scarcity of women in science, 23–26; in United States, 29–34; wildlife professionals and, 34–41, 36–37b, 39–40b

Hoagland, Serra J., 147

Hodges, Clare Marie, 41

Holder, Tamara, 346

Holmes, M. A., 83

Holvino, Evangelina, 291

Home Studies in Nature (Treat), 30

Homyack, Jessica A., 279

Huff, Sheila Minor, 16, 180, 191, 192–93b

Human Rights Campaign, 217–18

hunting and fishing, 67b, 162–65, 168, 243–44

Hypatia, 23

identity, 65, 193, 204–6, 214–18, 216f, 219–22b, 223f, 238–40

Implicit Association Test (IAT), 218, 283

impostor syndrome, 131, 195, 198, 259, 296

inclusion, 16, 279–307; activities on, 300–303, 356–57; ADVANCE program for, 280–81b; allies for, 133–36, 295–97, 299; changes for, 286–88, 287t; culture of, 285–86; defined, 6; DEI and, 298; discussion questions on, 299–300; hiring and promotion practices, 288–90; individual actions for, 294–95; men in support of women and, 311; mental health awareness and, 297–98; mentors and sponsors, 295–96; organizational culture and, 281–82; representation and, 297; research and publishing, 294; society and, 291–94; tokenism and, 10; unconscious bias and, 283–84; work-life balance and, 290–91. *See also* diversity, equity, and inclusion initiatives

Inclusion, Diversity, Equity, and Awareness Working Group (TWS), 69

Indian Self-Determination Act (1975), 174

International Union for the Conservation of Nature, 55

internships, 6, 99f, 128, 185–86, 189b, 260, 289, 342

intersectionality, 13, 121, 125, 180–81, 197, 257

Introduction to Botany (Wakefield), 27

isolation, 10–12, 128, 187b, 288, 317, 341, 343–45, 370

Jacobson, S., 188

James, Laurel, 171b

Jewell, Sally, 100

job ads and announcements, 96, 104–7, 198, 226–27, 288–90

Joecks, J., 93

Johnson, Larry, 245

Johnson, Maria, 180, 186

Johnson, Megan, 245

Jones, M. S., 324

Kalev, Alexandra, 286

Kay, Aaron C., 289

Myers, K. R., 11
Myers, Vernā, 6

NAFWS (Native American Fish and Wildlife
  Society), 69, 170, 172*b*
National Academies of Sciences, Engineering,
  and Medicine (NASEM), 87, 316
National Aeronautics and Space Administration
  (NASA), 285–86, 347
National Association of University Fish and
  Wildlife Programs (NAUFWP), 55–57, 56*t*, 60,
  64
National Defense Authorization Act (2019), 265
National Park Service, 188, 250*b*
National Science Foundation (NSF), 64, 71, 90,
  92, 170, 279, 280–81*b*, 285
Native American Fish and Wildlife Society
  (NAFWS), 69, 170, 172*b*
Native Americans and Alaska Natives, 16,
  147–79; activity on, 175–77; allies of, 172–74;
  colonialism and, 117; COVID-19 pandemic
  and, 368; cultural overview of, 149–59, 152*t*,
  153*f*; discussion questions on, 174–75;
  environmental ethics and, 159; faculty
  demographics, 61; indigenous knowledge of,
  114–15, 118–20; recruitment and retention of,
  130–31; student demographics, 367; student
  retention and, 128; traditional ecological
  knowledge and, 160; tribal wildlife departments and, 160–65, 161–65*b*; underrepresentation of, 169–71, 171–72*b*; wildlife as
  relatives, 165–67; wildlife profession
  demographics, 310, 366–67; women in Native
  societies, 167–68
Native Peoples' Wildlife Management Working
  Group, 69, 151–53
The Natural History of Birds (Lewis), 31
Natural Resource Canada, 71
Natural Sciences and Engineering Council
  (NSERC, Canada), 90, 92
The Nature Conservancy, 232
NAUFWP (National Association of University
  Fish and Wildlife Programs), 55–57, 56*t*, 60,
  64
Navajo Nation v. Urban Outfitters (2016), 154
Nelson, R. G., 199, 316
networking, 66, 72, 92, 103, 232, 291
Nice, Margaret Morse, 41, 42–43*b*
Nicholson, Kerry L., 3, 52, 336, 362
Nielsen, M. W., 4–5
Noah, Joakim, 159
Novarro, Alexander, 211
NSERC (Natural Sciences and Engineering
  Council, Canada), 90, 92
NSF. *See* National Science Foundation

Obergefell v. Hodges (2015), 229
O'Donnell, Katherine M., 211
Office of Personnel Management (OPM), 127–28
organizational culture. *See* inclusion
Orwell, George, 248

Pacific Flyway Council, 217
paid family leave, 94, 263, 265
parenting, 10, 93–94, 261–65, 279, 290–91
Parker, Mamie, 180, 182, 183–85*b*
patriarchy, 99, 130
Patton, David, 244
Paul, Alice, 8–9*b*
people with disabilities, 124
Peterman, Audrey, 180, 196–97*b*
Phillips, P., 28
Physica (Saint Hildegard), 23
physical appearance, 10–11, 250–51*b*, 346–48
physical strength, 85, 186, 311, 311*f*, 327*b*
Pierce, Chester M., 337
Pipkin, M. A., 199
Pitka, Rhonda, 161–62*b*
Plenty Sweetgrass-She Kills, Ruth, 113, 121–22,
  128
power differentials, 10, 215–18, 216*f*, 313–18, 330*b*
Pratt, Anne, 27
pregnancy, 85, 262–63
privilege: activity on, 300–303; allies and, 102;
  defined, 8; diversity and, 12–13; LGBTQ+
  people and, 215–18, 216*f*; race and ethnicity,
  121–22; self-testing on, 12; training on, 292;
  unpaid internships and, 6, 185, 342
professional associations, 66–69, 92, 129, 135–36,
  291–94. *See also* conferences
promotions: in academia, 90–92; barriers to, 195;
  in federal government, 96–101, 97*b*, 99*f*;
  gender discrimination and bias, 43, 70, 94,
  294, 341, 345; inclusion and, 288–90; leaky
  pipeline and barriers to wildlife professions,
  88; millennials and, 265–66; professional
  relationships and, 343–44; sponsors and,
  296–97
pronouns activity, 238–40

race and ethnicity, 16, 180–210; activity on,
  204–6; allies and, 198–202, 200–201*b*;
  barriers to careers based on, 188; COVID-19
  pandemic and, 368; discussion questions on,
  204; entering wildlife professions and, 181–86,
  183–85*b*, 186*f*, 187–89*b*; of faculty, 52, 54,
  60–61, 61*f*, 131, 367; intersectionality and, 13,
  121, 125, 180–81, 197, 257; of students, 4, 57,
  58–59*f*, 72, 182–85, 186*f*, 366–67; of wildlife
  professionals, 4, 15, 54, 121, 169, 366–67. *See
  also* exclusion; *specific groups*

227–28; on diversity and inclusion, 291–92; first woman member of, 35; first women in leadership positions, 44–45; Inclusion, Diversity, Equity and Awareness Working Group, 69, 256, 292; land acknowledgment and, 151–53, 153*f*; Leadership Institute, 260; LGBTQ+ people in, 218, 230–32, 292; Native Student Professional Development program, 170; networking groups of, 66, 92; women contributors to journals of, 91; Women of Wildlife, 66

Thomas, Christine, 67*b*

Thompson, Ed, 231

Title IX (1972), 71, 223–24, 313

tokenism, 6, 10, 131, 288, 297, 342–43

traditionalist generation, 246–47*t*, 269, 272–74*t*

transgender people, 57, 215, 218, 222–23, 229–30, 348–49. *See also* LGBTQ+ people

Trautman, Milton B., 42*b*

Treat, Mary, 30–31

tribal wildlife departments, 160–65, 161–65*b*

TWS. *See* The Wildlife Society

United Kingdom, 71

US Fish and Wildlife Service (FWS), 35, 44, 99*f*, 127, 173–74, 182, 188, 244, 249

US Forest Service, 99–100, 191, 282

US Geological Survey (USGS), 66, 98–101, 188, 291

Vania, Tom, 319–20

Vogt, Richard, 315

von Bingen, Hildegard (saint), 23

Waits, Lisette P., 52, 60, 65–66

Wakefield, Priscilla, 27

Wakefield, Tamekia, 362

WEBS (Women Evolving Biological Sciences), 72–73

Weeks-McLean Act (1913), 33–34

Werner, Patricia, 62

Whipple, Margaret, 62

White, Laura Lyon, 33

white people: biases of, 122; colonialism and, 117, 155–58; discipline identity and, 87; feminist movement and, 121; historical dominance of science by, 117–20, 187*b*; stereotypes of, 198; student demographics, 57–59; success assessments of, 123–24; wildlife profession demographics, 4, 15. *See also* privilege

The Wild Mammals of Missouri (Schwartz & Schwartz), 37

Williams, W. M., 90

Wilson, Marcia, 62

Winfrey, Oprah, 362

Winkelmann, Maria, 24

Women Evolving Biological Sciences (WEBS), 72–73

Women in the Field (Bonta), 29

women's rights, 8–9*b*, 121

work-life balance, 10, 69–70, 90–94, 101, 255–56, 261–66, 279, 290–91

Wright, Mabel Osgood, 32

Zimmermann, Anne, 44

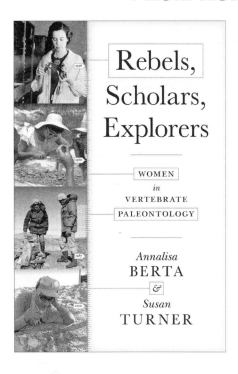
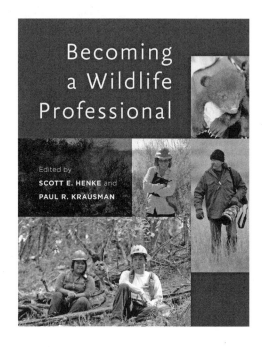